C O M A P ' S

Mathematics: Modeling Our World

DEVELOPED BY

COMAP, Inc.

57 Bedford Street, Suite 210
Lexington, Massachusetts 02173

PROJECT LEADERSHIP

Solomon Garfunkel
COMAP, INC., LEXINGTON, MA

Landy Godbold
THE WESTMINSTER SCHOOLS, ATLANTA, GA

Henry Pollak
TEACHERS COLLEGE, COLUMBIA UNIVERSITY, NY, NY

JOIN US ON THE INTERNET
WWW: http://www.thomson.com
EMAIL: findit@kiosk.thomson.com   A service of I⬀T⬀P®

South-Western Educational Publishing
an International Thomson Publishing company I⬀T⬀P®

Cincinnati • Albany, NY • Belmont, CA • Bonn • Boston • Detroit • Johannesburg • London • Madrid
Melbourne • Mexico City • New York • Paris • Singapore • Tokyo • Toronto • Washington

Published and distributed by

South-Western Educational Publishing
Cincinnati, OH 45227

This book was prepared with the support of NSF Grant ESI-9255252. However, any opinions,
findings, conclusions, and/or recommendations herein are those of the authors
and do not necessarily reflect the views of the NSF.

ISBN 0-538-68218-3

Printed in the United States of America.

1 2 3 4 5 6 7 8 C5 02 01 00 99 98

IP°

International Thomson Publishing

South-Western Educational Publishing is an ITP Company.
The logo is a registered trademark used herein under License by
South-Western Educational Publishing.

EDITOR: Landy Godbold
AUTHORS: Allan Bellman, WATKINS MILL HIGH SCHOOL, GAITHERSBURG, MD; John Burnette, KINKAID SCHOOL,
HOUSTON, TX; Horace Butler, GREENVILLE HIGH SCHOOL, GREENVILLE, SC; Claudia Carter, MISSISSIPPI SCHOOL FOR MATH
AND SCIENCE, COLUMBUS, MS; Nancy Crisler, PATTONVILLE SCHOOL DISTRICT, ST. ANN, MO; Marsha Davis, EASTERN
CONNECTICUT STATE UNIVERSITY, WILLIMANTIC, CT; Gary Froelich, COMAP, INC., LEXINGTON, MA; Landy Godbold,
THE WESTMINSTER SCHOOLS, ATLANTA, GA; Bruce Grip, ETIWANDA HIGH SCHOOL, ETIWANDA, CA; Rick Jennings,
EISENHOWER HIGH SCHOOL, YAKIMA, WA; Paul Kehle, INDIANA UNIVERSITY, BLOOMINGTON, IN; Darien Lauten, OYSTER
RIVER HIGH SCHOOL, DURHAM, NH; Sheila McGrail, CHARLOTTE COUNTRY DAY SCHOOL, CHARLOTTE, NC; Geraldine
Oliveto, THOMAS JEFFERSON HIGH SCHOOL FOR SCIENCE AND TECHNOLOGY, ALEXANDRIA, VA; Henry Pollak, TEACHERS
COLLEGE, COLUMBIA UNIVERSITY, NY, NY, J.J. Price, PURDUE UNIVERSITY, WEST LAFAYETTE, IN; Joan Reinthaler, SIDWELL
FRIENDS SCHOOL, WASHINGTON, D.C.; James Swift, ALBERNI SCHOOL DISTRICT, BRITISH COLUMBIA, CANADA;
Brandon Thacker, BOUNTIFUL HIGH SCHOOL, BOUNTIFUL, UT; Paul Thomas, MINDQ, FORMERLY OF THOMAS JEFFERSON
HIGH SCHOOL FOR SCIENCE AND TECHNOLOGY, ALEXANDRIA, VA

Dear Student,

Mathematics: Modeling Our World is a different kind of math book than you may have used, for a different kind of math course than you may have taken. In addition to presenting mathematics for you to learn, we have tried to present mathematics for you to use. We have attempted in this text to demonstrate mathematical concepts in the context of how they are actually used day to day. The word "modeling" is the key. Real problems do not come at the end of chapters in a math book. Real problems don't look like math problems. Real problems ask questions such as: How do we create computer animations? Where should we locate a fire station? How do we effectively control an animal population? Real problems are messy.

Mathematical modeling is the process of looking at a problem, finding a mathematical core, working within that core, and coming back to see what mathematics tells you about the problem with which you started. You will not know in advance what mathematics to apply. The mathematics you settle on may be a mix of several ideas in geometry, algebra, and data analysis. You may need to use computers or graphing calculators. Because we bring to bear many different mathematical ideas as well as technologies, we call our approach "integrated."

Another very important and very real feature of this course is that frequently you will be working in groups. Many problems will be solved more efficiently by people working in teams. In today's world, this is very much what work looks like. You will also see that the units in this book are arranged by context and application rather than by math topic. We have done this to reemphasize our primary goal: presenting you with mathematical ideas the way you will see them as you go on in school and out into the work force. There is hardly a career that you can think of in which mathematics will not play an important part and understanding mathematics will not matter to you.

Most of all, we hope you have fun. Mathematics is important. Mathematics may be the most useful subject you will learn. Using mathematics to solve truly interesting problems about how our world works can and should be an enjoyable and rewarding experience.

Solomon Garfunkel
CO-PRINCIPAL INVESTIGATOR

Landy Godbold
CO-PRINCIPAL INVESTIGATOR

Henry Pollak
CO-PRINCIPAL INVESTIGATOR

CONTENTS

Mathematics: Modeling Our World

2

UNIT

Gridville

Maximize your gain. Minimize your losses. These two short phrases represent the goals of many people. Important business and safety decisions are made with the purpose of minimizing or maximizing quantities, time, and costs.

The primary focus of this unit is on the mathematics of measuring and representing distance in order to minimize time and cost. You investigate the geometry of distance on a line and in a plane using the absolute value function. Through tables and graphs you observe patterns that result from adding functions together. You discover the usefulness and the limitations of extending concepts and strategies from one dimension to two.

WHAT IS THE BEST LOCATION?

The alarm sounds. Quickly the fire station comes to life. Workers committed to protecting the community move swiftly into action. Sirens blare as the trucks speed off to meet the challenge of defeating another fire. Time, of course, is critical. If a firetruck has to travel a long distance to reach a fire, precious time is lost. Thus, the location of a fire station becomes an important decision for all members of a community.

Communities must also decide where to locate other basic services. What is the best location for the post office, police station, hospital, or school? Any location that is chosen will benefit some people more than others. In order to make these decisions, community members must decide which factors are most important. In the case of the fire station, how would you decide on the best location? What makes one location better than another, and for whom is it better?

In this unit you will be asked to make decisions about finding the best location and to give reasons for your choices. When your reasons or criteria can be measured, then mathematics will help you determine a best location. There may be more than one best location depending on the criteria you use and how you determine what is fair.

Mathematical modeling is used to simplify the problem and discover methods for finding locations that satisfy the criteria. Gridville is a nice place to blend the mathematical tools of measurement, multiple representations, and the modeling process together with community values.

May the best location win!

LESSON ONE
In Case of Fire

KEY CONCEPTS

Distance in firetruck geometry

Circles in firetruck geometry

Total and average distance in two dimensions

The Image Bank

PREPARATION READING

A Model Village

Where is a restroom when you need one? Where is a gas station when you need to fill up your tank? Where is a restaurant when you crave a hamburger or an ice cream cone? Where are the firetrucks in an emergency?

A business must decide the best location for a store. Choosing the best location is critical to the survival of the business. In the same way, every community must decide the best location for a fire station. Choosing the best location for emergency services is critical to the members of the community.

Paul Klaver Collection

The focus of this first lesson is on deciding what is meant by "best" as you search for a location to build a fire station. Decide which criteria are most important; then use mathematics to find a location that meets the criteria.

Figure 1.1.
Aerial view of a town.

It is not easy to determine the best location for anything in a city with meandering streets, a variety of traffic patterns, and five thousand houses or apartments, such as the city pictured in the drawing in **Figure 1.1**. Also, all cities are different. You may develop a method for finding the best location in your town, or the town shown in Figure 1.1, but can you use your method with any town or city?

Start by designing a simple model of a city. Create a model that contains the essential elements of a typical town or city. Use your model city to develop strategies to determine the best location for a fire station.

At the end of this unit you will return to the city in Figure 1.1, or maybe your own city, and choose the best location for a fire station based on community values and the mathematics you learn in Gridville.

CONSIDER:

1. What are important criteria to be considered when determining the location for a fire station?

2. Where is the fire station nearest to your school? Is it in a good location? Give reasons for your answer.

3. Fire stations and hospitals provide emergency services. Many people find it desirable to live near a fire station and a hospital. Name other buildings or services in your community that people might want located near their homes.

4. Suppose you want to create a simple model for a city. Describe the essential components of the typical city and how you arrange the essential components to create the simplest model for a city. In particular, what features of the city in Figure 1.1 would you simplify, how would you simplify them, and what features would you retain?

5. Describe how you would determine the best location for the fire station in a model city.

ACTIVITY

1

THE NEW FIRE STATION

Welcome to Gridville! This small village has grown in the past year. The people of Gridville have agreed they now need to build a fire station. What is the best location for the fire station?

Gridville

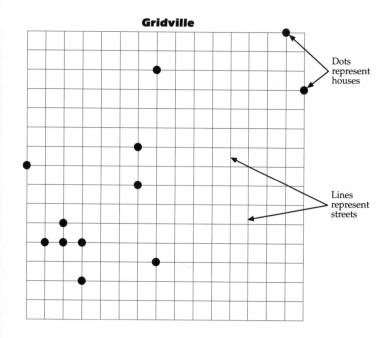

Dots represent houses

Lines represent streets

Figure 1.2.
Map of Gridville.

All towns and cities are different. Gridville is a simplified version of a small village with streets running parallel and perpendicular to one another. It represents the start of the modeling process. What you learn in Gridville may be useful in other settings that require finding a best location. The streets of Gridville are two-way streets represented by the lines of the grid. Houses are represented by points and are located at the intersections of grid lines in order to identify their locations easily. Firetrucks may only travel along grid lines.

Diagonal movement is not allowed. It is simpler than a real town in its size and in its layout (geometry), but it still has houses and roads, and it needs a fire station.

THE TASK:

The map of Gridville (**Figure 1.2**) shows the location of all houses. Determine the best location for Gridville's fire station, and write a persuasive argument defending your choice. The city leaders will follow the advice of the group that delivers the most convincing argument.

ACTIVITY

THE NEW FIRE STATION

1

GUIDELINES:

1. Begin your argument by answering the questions, "What are important factors to consider in deciding the best location?" and "What does 'best' mean?"

2. You are encouraged to use charts, diagrams, tables, graphs, equations, calculations, and logical reasoning in making your decision.

3. Clearly state your choice of best location. Your written summary should include the arguments and mathematics that support your decision. The summary should also explain how your charts, diagrams, tables, graphs, equations, calculations, and logical reasoning relate to the factors you considered and led your group to your choice.

4. Your written presentation may be posted in a display area in the classroom.

5. In addition to the written presentation, your group will give an oral presentation of approximately two minutes. The oral presentation should summarize your arguments and explain the reasons for your decision.

CONSIDER:

Answer the following questions based on the classroom presentations for Activity 1, *The New Fire Station*.

1. Which criteria were used most often to determine the location of the fire station?

2. Which factors invite mathematical investigation?

3. Even a model such as Gridville can be simplified further in order to study the essential elements of the location problem in detail. What are some ways you can simplify the Gridville model to investigate distance relationships?

Where do you build a new high school when the only high school in town becomes overcrowded? A class of students at Redlands High School in Redlands, California, answered this question and won $10,000. In 1993, the 25 students of teacher Donna St. George and teacher Judy Kanjo won second place in the American Express Geography Competition sponsored by American Express. The students conducted surveys of local residents and gathered information provided by local companies, local organizations, and Redlands Unified School District officials. They analyzed several sites proposed by the school district. Their 80-page report addressed major factors including traffic patterns, busing patterns, soil conditions, air traffic patterns, availability of public utilities, anticipated population growth, and potential for flood damage.

INDIVIDUAL WORK 1

Keeping Your Distance

1. Gridville is a model for a city. In some ways it is similar to a real town or village. In many ways it is different.

   a) How is Gridville different from your city?

   b) How is Gridville similar to your city?

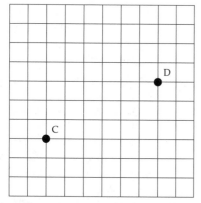

Figure 1.3.
Map to determine shortest path.

2. **Firetruck distance** is the shortest path along grid lines between two points.

   a) What is the length of the shortest path from point C to point D in **Figure 1.3**? Explain how you can be sure it is the shortest.

   b) In Figure 1.3, is there more than one shortest path? If your answer is no, explain why there can be only one shortest path. If your answer is yes, find one or more paths that are the same length as the shortest path.

3. Suppose you use a coordinate system to identify locations in another Gridville, one with only three houses. The houses are located at (1, 4), (2, 7) and (8, 5) and are labelled A, B, and C respectively.

   a) Two locations are proposed for the fire station: (3, 5) and (5, 4). How do you determine which location is better?

   b) Find a different location you think is better. Explain why.

4. The community may decide that the best location is one that minimizes the total distance. The total distance associated with a fire station location is the sum of all the lengths of the shortest paths from the fire station to all of the houses.

   Suppose the fire station F is located at (5, 5), house A is located at (3, 8), and house B is located at (10, 4).

   a) What is the distance from F to A? F to B? What is the total distance?

   b) Suppose the fire station is moved from (5, 5) to (8, 2). Determine the change in the total firetruck distance.

c) Miles claimed he found a location for the fire station that would make the total distance 10. Do you agree with Miles? Explain your answer.

5. Carla delivers pizza for J.J.'s Pizzeria. J.J.'s is located at (6, 1) on a grid. The houses in Carla's delivery area are located at (1, 1), (2, 7), (5, 1), (3, 5) and (6, 8) (see **Figure 1.4**).

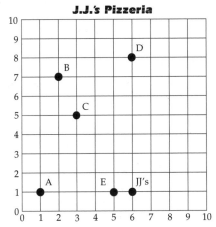

J.J.'s Pizzeria

Figure 1.4.
J. J.'s Pizzeria map for Item 5.

a) For which house is the firetruck distance from J.J.'s the greatest? Explain how you determine your answer. (Measure distance using units on the grid.)

b) Determine the total distance from the fire station to the five houses.

c) The **average distance** is the total distance divided by the number of houses or trips. Determine the average distance from the fire station to the five houses.

d) **Round-trip distance** is the length of the shortest path from the fire station to a house combined with the length of the shortest path from the house back to the fire station. What is the round-trip distance between F and A?

e) If Carla makes separate trips from the Pizzeria to each house, what is the total round-trip distance she would travel?

f) Do you think total round-trip distance is always twice as much as total distance? Explain your answer.

g) Suppose Carla gets pizza orders from all five houses at the same time. She decides she will make one trip and deliver pizza to all five houses on the same trip. What is the shortest total distance she will travel? Explain how your answer represents the shortest total distance.

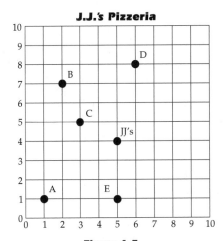

J.J.'s Pizzeria

Figure 1.5.
New map for J. J.'s Pizzeria.

6. Suppose J.J.'s Pizzeria moves from location (6, 1) to location (5, 4) (see **Figure 1.5**).

a) If Carla makes separate round-trips from the Pizzeria's new location to each house, what is the total round-trip distance she travels? How much does the total round-trip distance change?

b) Calculate the change in total distance and average distance.

c) Explain how the total distance changes if Carla moves the Pizzeria from location (5, 4) to location (5, 3).

d) You found the total distance when the Pizzeria is located at (5, 4). Find a new location for the Pizzeria that makes the total distance shorter. Justify your answer.

e) Explain to Carla how to find a location for the Pizzeria that makes the total distance as short as possible.

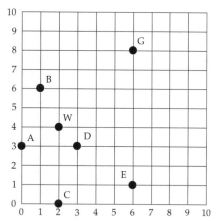

Figure 1.6.
Map of stores and warehouse.

7. **Figure 1.6** shows the location of several stores and a warehouse. The stores are identified using the letters A, B, C, D, E, and G. The warehouse is located at W.

   a) Calculate the total distance, average distance, and total round-trip distance if a truck makes a separate delivery trip from the warehouse to each of the six stores.

   b) What is the maximum distance the delivery truck must travel to get from the warehouse to the farthest store?

   c) Suppose the warehouse were moved one unit to the right. For which stores does the distance to the warehouse increase? For which stores does the distance to the warehouse decrease? For which stores does the distance remain the same?

   d) Describe how the total distance changes when the warehouse is moved one unit to the right.

   e) Describe how the distance to the farthest store changes when the warehouse is moved one unit to the right.

   f) Describe how the total distance changes when the warehouse is moved up one unit.

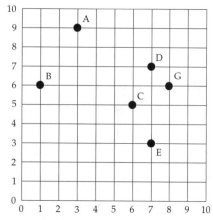

Figure 1.7.
Boxville.

8. Boxville (**Figure 1.7**) is building a recycling station. The planning department is considering two locations for the recycling station: (7, 5) and (4, 6).

a) Which of these two locations is the best location for the recycling center? Give reasons for your answer.

b) What location is better than (7, 5) and (4, 6) for the recycling center? Give reasons for your answer.

c) Is it easier to find the best location in Boxville or Gridville? Explain your answer.

9. Rowville (**Figure 1.8**) is building a bus station. The planning department is considering two locations for the bus station: (5, 6) and (6, 6).

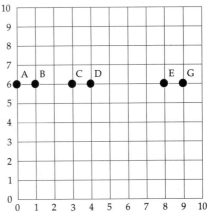

a) Which of the two locations is the best location for the bus station? Give reasons for your answer.

b) What location is better than (5, 6) and (6, 6) for the bus station? Give reasons for your answer.

c) Is it easier to find the best location in Rowville or Gridville? Explain your answer.

Figure 1.8.
Rowville.

10. The grid in **Figure 1.9** represents a community of houses and a school. Houses are identified using the letters A, B, C, D, E, and G. The school is located at point F.

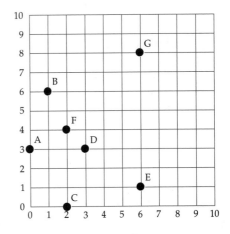

a) In your opinion, which houses seem to benefit the most when the school is located at F? Does the placement of the school seem fair to you? Explain your reasoning.

b) Suppose you live in house G. You want the school to be built at the location that is best for you. Where do you build it? Explain your reasoning.

c) Suppose you live in house G and a member of your family lives in house C. You want to build the school at a location that is best for both houses. Where do you build the school? Explain your reasoning.

Figure 1.9.
Map of houses and school.

d) Suppose you want to build the school at a location that is best for the three houses A, C, and D. Where do you build the school? Explain your reasoning.

INDIVIDUAL WORK 1

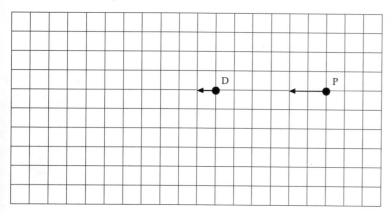

Figure 1.10.
Pursuit map for Item 11(a).

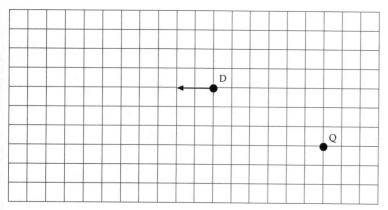

Figure 1.11.
Pursuit map for Item 11(b).

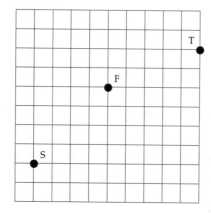

Figure 1.12.
Map for Item 12.

11. a) In **Figure 1.10**, a police car at point P is chasing a driver at point D who is driving west. If the police car goes twice as fast as the driver, how far will the police car travel before it catches up? Why?

   b) Suppose the police car described in part (a) begins the chase at point Q (**Figure 1.11**). The police car can travel twice as fast as the driver. The driver is located at D when the chase begins and is traveling west. How far will the police car need to travel to catch up? Describe at least two paths the police car could follow during the pursuit.

12. The police dispatcher receives an emergency call. An individual at location F (see **Figure 1.12**) needs the immediate assistance of a police officer. Cruiser S and cruiser T are in the vicinity.

   a) If shortest distance is the criterion, which cruiser should the dispatcher send?

   b) Suppose it takes twice as long to travel the east-west streets (horizontal) than it takes to travel the north-south streets (vertical). Which cruiser could arrive at location F in the least amount of time? Explain your answer.

13. Distance is a factor in other decisions and guidelines.

a) Some businesses restrict delivery service to a certain number of blocks or miles. Draw your own grid and label location P (see **Figure 1.13**). Mark all points at a firetruck distance of 2 from P. Remember that houses are located only at the corners of intersections, so do not connect the points in the graph with segments. Use another color to mark all points at a firetruck distance of 3 from P. Use a third color to mark all points at a firetruck distance of 5 from P. Describe the visual patterns created by the colored marks on your grid.

b) **Figure 1.14** shows a town with two schools, S1 and S2. Students attend the school closest to their homes, in firetruck distance. Draw S1 and S2 in the same locations on your own grid paper and mark on the grid all the points that are the same distance from both schools. (This is the boundary or dividing line between the student attendance areas for the two schools.)

c) Mr. and Mrs. Walker work at locations A and B (see **Figure 1.15**). They want to rent an apartment or house close enough to their jobs so that the total firetruck distance they walk to work is no more than 10 units. Draw A and B on your own grid paper and mark on the grid all possible locations for a suitable apartment.

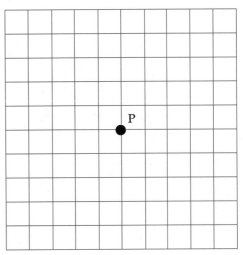

Figure 1.13.
Grid for Item 13(a).

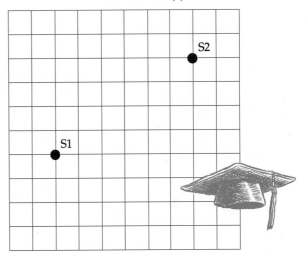

Figure 1.14.
Map of two schools.

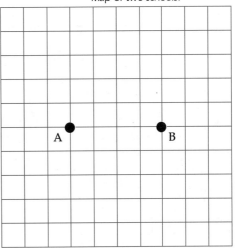

Figure 1.15.
Walker house locations.

INDIVIDUAL WORK 1

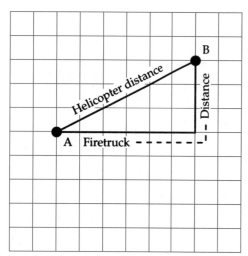

Figure 1.16.
Comparing firetruck distance to helicopter distance.

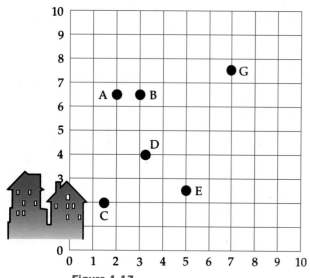

Figure 1.17.
Gridville with houses between the grid lines.

14. a) In **Figure 1.16**, find the firetruck distance between points A and B and find the **helicopter distance**, or direct distance, between points A and B.

  b) Is helicopter distance always less than firetruck distance? Explain your answer.

15. Gridville is only a model for a real city. It is simplified. One simplification involves placing houses only at the intersections of streets. For this item that restriction has been removed. **See Figure 1.17.**

  a) Find the firetruck distance between points A and B.

  b) Find the firetruck distance between points C and D.

  c) Find the firetruck distance between points A and E.

  d) Describe problems you may encounter when you try to find the location that minimizes total distance.

16. Suppose you build more streets in Gridville. In **Figure 1.18**, Map A shows Gridville with streets 1/4 mile apart and Map B shows Gridville with streets 1 mile apart.

  a) What is the distance between points C and D? Which map of Gridville did you use to determine your answer?

  b) What are the advantages of placing streets closer together?

MAP A   **MAP B**

Figure 1.18.
Map A streets are 1/4 mile apart; Map B streets are 1 mile apart.

17. a) Interview a city planner, city manager, planning consultant, or fire chief in your community and ask what factors they usually consider when determining the location for a new fire station.

   b) What are some services or businesses people typically don't want near their homes? Explain how the process of finding a location for a service people don't want near their homes is different from finding a location for a fire station.

18. In Item 13, you located the attendance boundary between two schools by identifying all locations that were the same distance from each of the two schools.

   a) Determine similar boundaries for Gridville cities with 3, 4, or 5 schools (see **Figure 1.19**). Copy each figure on your own paper. Draw or mark the boundary lines.

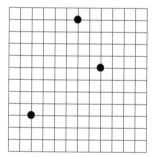 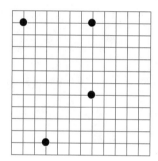 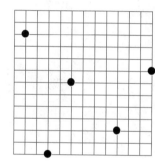

Figure 1.19.
School locations are represented by dots.

   b) Draw your own map with six schools and draw the boundaries.

19. The diagonal streets (**Figure 1.20**) are one level or unit above the horizontal and vertical streets. One house is located at A and a second house is located at B. The fire station is located at F.

Horizontal and vertical streets are on the lower level and are represented by wider lines. Diagonal streets are thinner lines and represent the streets on the upper level. Point A is on the lower level. Points B and F are on the upper level.

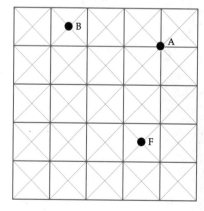

Figure 1.20.
Map of a two-level city.

   a) Determine the total distance.

   b) Find the location(s) for the fire station that would minimize total distance in this two-house, two-level village. Explain your answer.

LESSON TWO

Linear Village

KEY CONCEPTS

**Total and average
distance in
one dimension**

**Absolute-value
equations
and inequalities**

**Absolute-value
graphs**

**Piecewise-defined
functions**

**Addition of
functions**

**Slope and rate
of change**

Median

David Barber

PREPARATION READING

Main Street

D istance is one important factor to consider as you look for the best location. Time is another, since time is critical in an emergency. If you can minimize the distance a firetruck must travel to reach an emergency then you reduce the time needed to respond.

By now, your group may have decided the best location for the fire station minimizes the total distance a firetruck would travel if it had to make a separate visit to each house. Or, maybe your group has decided the best location makes the longest distance as short as possible. Perhaps your group decided another factor was more important. In any case, mathematics is useful when the criteria involve quantities. Distance is one quantity that can be measured and compared.

This unit focuses on developing methods for finding locations that minimize the total distance or the longest distance regardless of the number or arrangement of houses in Gridville. The issue of fairness will be argued throughout the unit because of its importance to the context.

Remember that Gridville is a model, a simplified version of a real-life city. Even in a village of only twelve houses, numerous calculations must be performed to find a location that makes total distance or maximum distance as small as possible. Imagine performing those same calculations for a city of five thousand! Sometimes it helps to simplify the model.

In Boxville (see Individual Work 1, Item 8) you realized that the calculations take less time in a village with only six houses. What is the smallest number of houses for which this problem (locating a fire station) makes any sense?

In Rowville (see Individual Work 1, Item 9) you realized that the calculations are simpler when there are fewer streets. What is the fewest number of streets for which this problem makes any sense?

Suppose you simplify Gridville by placing all of the houses along one street to form a linear village. Take an important step in the modeling process and create Linear Village. Use mathematical tools, search for patterns, and gain insights in Linear Village that you can apply in Gridville. Develop the mathematics of distance along a straight line. Linear Village has only one street. The location that minimizes the total distance a firetruck would travel should be easy to find in a village with only one street!

Your goal is to find the best location for the fire station in Gridville. This process of creating a simpler context (Linear Village) in which to develop methods for solving problems in a more complicated context (Gridville) is a familiar step in the process of mathematical modeling. You may be surprised by how much you learn in Linear Village! Get ready to add some absolutely valuable mathematical tools as you continue your quest for the best station location.

CONSIDER:

1. Which criteria for best fire station location would be appropriate to investigate in Linear Village? Which criteria do not make sense to investigate in Linear Village?

2. Already you have made several decisions in the modeling process. The creation of Gridville, a model for a city, is a simplification. The decision to focus on a few criteria instead of all possible criteria is another simplification. The decision to simplify Gridville from a two-dimensional town to a one-dimensional Linear Village is one more simplification. How will you organize your investigation of more than one criterion as you move between Linear Village and Gridville and the real world? Discuss possible modeling paths you might follow in your search for the best fire station location in Gridville.

> In modeling the diagonal motion of multiple points, you might simplify by decreasing the number of points to one and by restricting the motion to only one dimension—horizontal, for example. However, you could also continue your investigation of a single point by modeling diagonal movement. There is more than one way to proceed through the individual steps to reach your ultimate goal.

3. You know one way to find the location that minimizes total distance in a linear village—try different locations for the fire station and calculate the total distance until you find the location that yields the smallest total. This method of "guess and check" can be time consuming and is subject to errors due to the impossibility of checking all possible solutions. How will you develop a faster, more reliable procedure to determine the location that minimizes total distance for any size linear village? What is your plan?

4. In Course 1, Unit 5, *Animation*, the starting location and velocity of an object were represented by constants in an equation; they described the situation. The variables were time and location; they described possible scenarios within that situation. In the context of Gridville and with the task of finding the best location for the fire station, what are the constants and what are the variables?

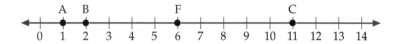

DEVELOPING LINEAR VILLAGE

Suppose all the houses in your town are on one street and the criterion is "minimum total distance." Where do you build the fire station in this situation? The purpose of this activity is to develop an easy method for answering this question.

Total distance is the sum of the distances measured between the fire station and each house. For example, consider a linear village with three houses and a fire station (see **Figure 1.21**). The letters A, B, and C identify the locations for three different houses. The letter F identifies a possible location for the fire station.

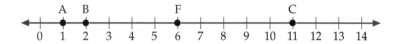

Figure 1.21.
Linear village with three houses and a fire station.

The distance between F and A is 5 units.

The distance between F and B is 4 units.

The distance between F and C is 5 units.

Thus, the total distance is $5 + 4 + 5 = 14$ units.

Suppose you build the fire station at location 4 instead of location 6 (see **Figure 1.22**).

Figure 1.22.
Linear village with the fire station in a different location.

The distance between F and A is 3 units.

The distance between F and B is 2 units.

The distance between F and C is 7 units.

The total distance is $3 + 2 + 7 = 12$ units.

In this three-house village, the total distance is less if you build the fire station at location 4 than if you build the fire station at location 6. Where do you build the fire station if you want the

ACTIVITY

2

DEVELOPING LINEAR VILLAGE

total distance to be as small as possible? What if the village contains 10 houses? Fifty houses? Two-thousand houses?

SUMMARY OF DEFINITIONS FOR LINEAR VILLAGE:

Points on a number line represent houses in a linear village.

The **distance** between a house and the fire station in a linear village is a non-negative number representing the difference in their coordinates on a number line.

The **total distance** is the sum of the one-way distances between the fire station and each of the houses.

The **average distance** is the total distance divided by the number of houses.

The **round-trip** distance is twice the distance between a house and the fire station.

The **total round-trip** distance is twice the total distance.

THE TASK:

Develop a rule or strategy for finding the location that minimizes the total distance in any size linear village.

THE PROCEDURE:

Recall from earlier units that searching for patterns is easier when you keep models simple and when you vary only one key feature at a time. Here is a suggested procedure for examining fire station locations. Start with villages that have two houses. Move the fire station to different locations and determine the total distance. Change the locations of the houses and repeat the process of finding the total distance for each location of the fire station. Add more houses. Look for patterns as you try to find the location for the fire station that

ACTIVITY

DEVELOPING LINEAR VILLAGE

2

makes the total distance smallest regardless of the number of houses and the placement of the houses in Linear Village. Tables and graphs may help your search.

THE REPORT:

Present your specific procedure and conclusions to the class. Be prepared to defend your rule and prove that it works for any size linear village.

Apply your rule to the linear village in **Figure 1.23** by finding the location that minimizes total distance.

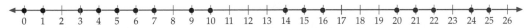

Figure 1.23.
Linear village with 17 houses.

THE CHALLENGE:

Design a linear village to give to another group to see if they can find quickly the location that minimizes total distance.

INDIVIDUAL WORK 2

Odd or Even

1. Suppose Linear Village has houses located at 3, 6, 7, 10, and 12 (see **Figure 1.24**).

Figure 1.24.
Linear Village for Item 1.

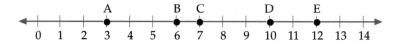

a) Find the total distance and average distance when you place the fire station at location 8.

b) Find the total distance and average distance when you place the fire station at location 10.

c) Which location, 8 or 10, is the better location for building the fire station?

d) Is there a location that yields a smaller total distance? Explain your answer.

2. The Linear Village in Item 1 is growing! Add a new house at location 15 (see **Figure 1.25**).

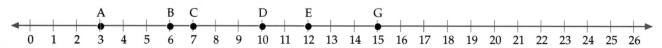

Figure 1.25.
Linear Village for Item 2.

a) Calculate the change in the total distance for the fire station at location 8 when the new house is added at location 15.

b) Calculate the change in the total distance for the fire station at location 10 when the new house is added at location 15.

c) Which location is better for the fire station now that a new house is added at location 15?

d) What do you predict will happen to the total distance for locations 8 and 10 when you add another house at location 1?

Note: The next three items involve an in-depth study of distances for linear villages with 2 or 3 houses. Look for patterns that you can extend to a linear village of any size.

3. The houses in Linear Village are located at 3 and 7 (see **Figure 1.26**).

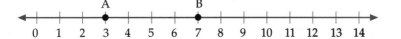

Figure 1.26.
Linear Village with two houses.

For any particular Linear Village, your search for a fire station location assumes that the houses do *not* move, but trial locations for the station *do* change. Thus, station location acts as an explanatory variable and total distance is a response variable.

a) Predict the location that minimizes the total distance.

b) Prepare a table that you can use to look for patterns while you search for the location that minimizes the total distance. An example table layout is shown in **Figure 1.27**.

Fire Station Location	Distance From F to A	Distance From F to B	Total Distance	Average Distance
0				
1				
2				
3				
4				
5				
6				
7				
8				
9				
10				
11				
12				
13				
14				

Figure 1.27.
Table for Item 3(b)

INDIVIDUAL WORK 2

c) Describe any patterns you notice in the table.

d) The table in Figure 1.27 uses only integer values for the house locations and integer values for the fire station locations. You used integer pairs to locate houses at the intersection of streets in Gridville. Can houses and fire stations be located between integer locations in Linear Village? Explain your answer.

e) Use the values from the table to prepare a graph relating the fire station location and total distance. Be sure to label the axes appropriately; see **Figure 1.28**. Explain whether it is appropriate to connect the points to represent locations between those listed in the table when you complete your graph for total distance versus station location.

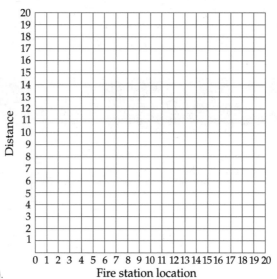

Figure 1.28.
Example of graph set-up for Item 3(e).

f) Describe the appearance of the graph representing total distance versus station location. Write your description as if you are describing the appearance of the graph to someone on the telephone.

g) Describe the location or locations where you should build the fire station if you want the total distance to be as small as possible. Explain how you determined this answer.

h) Suppose you locate the fire station F somewhere between A and B. Explain what happens to the total distance as you move the point F closer to B.

i) Suppose you could locate the fire station F anywhere between A and B. What would be the total distance if the point F is located at 5.8?

j) Suppose you locate the fire station F to the right of point B, which means the coordinate for F is greater than the coordinate for B. Describe what happens to the total distance as you move F farther and farther to the right of B.

k) Suppose you locate the fire station F to the left of point A, which means the coordinate for F is less than the coordinate for A. Describe what happens to the total distance as you move F farther and farther to the left of A.

l) Predict how the appearance of the graph for total distance would change if B were located at coordinate 5 instead of 7.

4. a) Use your table and graph from Item 3 and add a graph representing the average distance for different fire station locations. Using the same axes as for Item 3 makes comparisons easier.

  b) Compare the graph of average distance to the graph of total distance. Describe similarities between the graphs. Describe differences between the two graphs.

5. Consider the three-house Linear Village shown in **Figure 1.29**. House A is located at 2, house B at 4 and house C at 9.

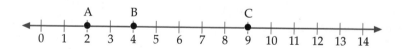

Figure 1.29.
Linear Village with 3 houses.

a) Predict the location that minimizes the total distance.

b) Construct a table similar to your table for Item 3. Include an additional column for the distance from the fire station to house C (see **Figure 1.30**).

Fire Station Location F	Distance From F to A	Distance From F to B	Distance From F to C	Total Distance	Average Distance
0					
1					
2					
3					
4					
5					
6					
7					
8					
9					
10					
11					
12					
13					
14					

Figure 1.30.
Table with additional column.

c) Use the values from the table to plot (on a single set of axes) graphs representing total distance and average distance between various fire station locations and the three houses.

d) Describe how to use the graph to determine the location that minimizes total distance.

e) The graph of total distance is made up of linear segments or "pieces." Identify the slope of each linear piece, and state the interval on which that slope applies.

f) Slope represents a rate of change of one quantity with respect to another quantity. What rate does the slope represent for the graph of total distance in part (d)?

g) Why is the slope of the graph for average distance equal to +1 for all fire station locations to the right of 9?

6. Suppose you add 1 or 2 or 4,997 more houses to Linear Village. Preparing new tables and graphs for villages with 4 and 5 houses is time consuming. Preparing a table and a graph for a village of 5000 houses is unthinkable. How can you use the technology of the graphing calculator or computer to investigate patterns for linear villages of different sizes?

7. Four groups were trying to determine a location for the fire station that minimizes total distance in the linear village in **Figure 1.31**.

Figure 1.31.
Linear Village for Item 7.

Group 1 wants to build the fire station across from house D. Group 2 wants to build the fire station across from house E. Group 3 claims they can minimize total distance by building the fire station anywhere between houses D and E. Group 4 says to build the fire station anywhere between houses D and G.

Which group is right? Explain your answer.

8. As its name implies, total distance represents several distances added together. In this item you examine those distances separately and together, using the linear village from Item 3.

a) Use your table from Item 3 to add a graph of distance between F and A versus fire station location to your earlier graph. Label this new portion of the graph as Graph A.

b) Now add a plot of distance between F and B versus fire station location to the same graph. Label the new addition to the graph as Graph B.

c) Label the graph of total distance as Graph C, and the graph of average distance as Graph D.

d) Determine the slopes of the linear "pieces" for each of Graph A, Graph B, Graph C and Graph D. Describe any patterns and relationships that you notice among the slopes or graphs.

   For example, for Graph A, the slope is –1 for locations to the left of 3 and +1 for locations to the right of 3.

e) Explain why the graphs of total distance and average distance are horizontal for locations between house A and house B.

9. Erik started with a Linear Village of two houses. He graphed total distance between the two houses and various locations for the fire station. Then he added a new house to Linear Village and graphed total distance. He repeated this process several times. Each time he added a new house to Linear Village he prepared a new graph for total distance.

   a) Erik noticed the graph representing total distance became steeper each time he added a house. He wondered if there was a relationship between the number of houses and the steepness of the graph. Are the slopes of the linear pieces of the graph for total distance related to the number of houses? Explain your answer.

   b) Erik wondered if the graphs for average distance followed a pattern. Investigate graphs for average distance in different size linear villages. Describe any patterns you discover.

10. Finding distances between two points in Gridville (two-dimensional) is difficult if the two points do not lie on the intersections of grid lines. That is the reason for restricting locations to integer values in your initial investigation of Gridville. Consider the same restriction in Linear Village. Suppose there are 5 houses in a linear village. House A is located at 3.75, house B at 4.12, house C at 6.83, house D at 10.09, and house E at 11.35.

    a) What is the total distance if you place the fire station at location 7.40?

    b) How do you determine the location that minimizes the total distance? (When you worked with integers only, you could construct a table of values and try all the possible integer locations for the fire station.)

11. The following items refer to a linear village shown in **Figure 1.32**, with a fire station F and two houses, A and B. Distances between locations are measured in meters (m).

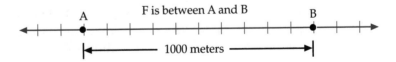

Figure 1.32.
Linear village for Item 11.

    a) Suppose A and B are 1000 m apart and F is between them. A firetruck leaves the station at F, drives to A, then to B, then back to F. Is this enough information to figure out the total distance the truck goes?

Is it enough to figure out the starting location of F? Give reasons for your answers.

b) Houses A and B are 1000 m apart. If F is between them, what is the average distance from F to the houses?

c) Suppose that A and B are 800 m apart (see **Figure 1.33**) and F is between them, but more than 100 m from each of them. If F is moved 100 m to the right, how is the average distance from F to the houses affected? Explain.

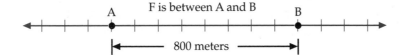

Figure 1.33.
Linear village for Item 11(c).

d) Suppose F is to the right of B. Now, if F is moved 100 m to the right, exactly how does the average distance change? What if F is moved 200 m to the right, how does the average distance change? What if F is moved 500 m to the right of B? What is the pattern here? If you made a graph of average distance versus the location of F, for F to the right of B, what would be the slope of the graph? Explain the meaning of the slope of the graph to the right of B.

e) Suppose F is between A and B and the average distance from F to A and F to B is 750 m. How far apart are A and B?

f) Suppose F is 100 m to the right of B and the average distance from F to the two houses is 500 m. How far apart are A and B? Explain how you found your answer.

g) Suppose Linear Village has five houses, A, B, C, D, and E, and fire station F (see **Figure 1.34**). Each unit on the number line below represents 500 m. (The distance from A to B is 1000 m.)

Figure 1.34.
Linear village for Item 11(g).

The fire station is located at 10. Suppose, on a given day, the firetruck traveled a total of 10,000 m. Assume that on each trip the firetruck traveled from F to a house and back to F. Which house or houses did the firetruck visit on that day?

12. Suppose a linear village has two houses, A and B, located 1000 m apart. The fire station F is located between A and B. It takes 45 seconds to get the firetruck out into the road once an alarm is received. After that, the truck drives at a speed of 20 m/sec. What is the average time it will take the truck to get to a fire?

13. In the linear village shown in **Figure 1.35**, the owner of house A thinks it is unfair that the fire station is closer to B than to her house. The owner of house B says it shouldn't matter because he can prove by math that the average distance to their houses is the same no matter where F is located between A and B. Do you think that owner A is justified? If so, where should F be located to be fair to both parties? What criteria are you using to determine a "fair" location?

Figure 1.35.
Linear village for Item 13.

In several items, you have been asked to describe a graph. Many of the graphs representing distance appear to consist of pieces joined together to form one graph. Functions having such graphs are called **piecewise-defined functions**. In the cases you have seen, the pieces have all been lines, so you could call these "piecewise-linear functions." You used words and phrases such as "to the right of," "to the left of," and "between" when you described these graphs. Mathematics provides a concise language for describing regions or intervals. It uses the inequality symbols: $<$, $\leq$, $>$, and $\geq$.

The notation $x > A$ means that the number x is greater than A. Thus, x is to the right of A on the number line and x is not A. See **Figure 1.36**. The dark section represents all possible locations of x. Note that A is "open" and not filled in.

Figure 1.36.
Number line representing $x > A$.

In Linear Village, the expression $x > 3$ would represent all fire station locations x with addresses greater than 3 and to the right of 3 on the number line.

The notation $x \geq A$ means that x is greater than or equal to A on the number line. Thus, x is to the right of A and may even be A, as illustrated in **Figure 1.37**. Note that A is "closed" and is now filled in.

$x \geq A$

Figure 1.37.
Number line representing $x \geq A$.

In Linear Village, the expression $x \geq 5$ represents all fire station locations x with address greater than or equal to 5; that is, the location 5 and all locations to the right of 5 on the number line.

The symbol $<$ represents "less than" and the symbol $\leq$ represents "less than or equal to."

In Linear Village, the expression $x < 4$ represents all fire station locations x with address less than 4 or to the left of 4 on the number line. The expression $x \leq 4$ includes the address 4 and all locations to the left of 4.

The notation $A < x < B$ means that x is both greater than A and less than B, as illustrated in the diagram (see **Figure 1.38**). This notation also can be read, "x is between A and B, not including points A and B." The diagram represents all possible locations, x, between locations A and B. Note that it must be true that $A < B$.

$A < x < B$

Figure 1.38.
Number line representing $A < x < B$.

In Linear Village, the expression $2 < x < 6$ represents all fire station locations x with addresses between 2 and 6 on the number line.

The notation $A \leq x \leq B$ is similar to $A < x < B$; x is between A and B including the points A and B.

The expression $2 \leq x \leq 6$ represents locations 2 and 6 and all addresses between 2 and 6.

14. Inequality symbols may be used to represent locations that minimize total distance, average distance, and round-trip distance. For example, the expression $x > 7$ represents all fire station addresses or locations to the right of 7. The expression $x \leq 3$ represents all locations to the left of 3 and including 3. The expression $2 \leq x \leq 5$ represents all locations between 2 and 5, including 2 and 5.

Write an expression using inequalities to represent all fire station locations (or addresses) that satisfy the given condition.

INDIVIDUAL WORK 2

a) to the right of 5 on the number line.

b) to the left of 10, including 10.

c) between 3 and 9.

d) to the right of 2.5, including 2.5.

e) that minimize the average distance to houses located at A = 4 and B = 9.

15. Describe in words the locations represented by the following inequalities:

a) $x < 10$.

b) $5 \leq x \leq 9$.

c) $x \geq 6.5$.

16. The one street in Rectangleville forms a rectangle (**Figure 1.39**). All buildings are designed for one family except Building A, which is a 5-unit apartment building, Building B, which is a 3-unit triplex and Building C, which is a 10-unit apartment building.

Figure 1.39.
Rectangleville with three multi-unit buildings, A, B, and C.

Where should the fire station be built in Rectangleville to minimize the average response time per family unit?

VILLAGE GRAPHS

In Activity 2 you looked for a method for solving the minimum-total-distance problem for linear villages. But how do you know that such a rule is valid for *all* linear villages? Analytic justification can help.

Solutions thus far have relied mainly on tables and graphs or trial and error. Building tables and graphs by hand is a time-consuming process. You now understand the one-point-at-a-time process used to create a table or graph of distance. This guided activity develops algebraic equations representing distances. A second purpose is to develop an analytic verification of your rule for finding the location that minimizes total distance in a linear village of n houses. The equations and their representations are the key to that justification.

Distance along the number line is so important that it has its own notation and name. The distance between A and B is written $|A - B|$ and is read "the **absolute value** of A subtract B." Some calculators and computers use different symbols. For example, abs(A – B).

Because you will use your calculator, "X" will represent the location of the fire station. "Y" will represent distance. (Lower-case letters will be used in pencil-and-paper work.)

Usually the distance between two points is determined by taking the larger coordinate and subtracting the smaller. The distance between X and A is A – X when A > X (that is, when X is to the left of A) and is X – A when X > A (namely, X is to the right of A). Absolute-value notation assures a positive result from the subtraction of the two different coordinates without regard to which number is larger or which location is to the right of the other;

$|X - A|$ and $|A - X|$ both mean the distance between X and A.

For example, suppose A is located at 3 and X is located at 5, which is 2 units from 3.

ACTIVITY

3

VILLAGE GRAPHS

Therefore $|5 - 3| = |3 - 5| = 2$. Use your calculator to verify.

If the locations are reversed with A at 5 and X at 3, the distance between them is the same.

1. a) Suppose house A is at 6 and the fire station is at x. Write a symbolic equation to represent the distance y between the house and the fire station.

   b) If house A is at 6 and house B is at 10, write a symbolic equation to represent the total distance y from the fire station at x to both houses.

   c) Write an equation representing the average distance to both houses from the fire station at x.

2. a) Graph: $y = |x - 3| + |x - 7|$ or Y1 = abs(X – 3) + abs(X – 7). Clearly identify the window you use.

   b) Describe what this graph represents. Compare this graph to your total-distance graph from Individual Work 2, Item 3.

   c) Where would you build the fire station if you wanted to minimize the total distance in a linear village with two houses, one located at 3 and the other located at 7? How does the graph help you determine the location?

   d) Do not erase Y1. What function could you enter into the calculator as Y2 to create a graph that represents the *average* distance from the fire station to the two houses? Graph Y2.

   e) Where would you build the fire station if you wanted to minimize the average distance in this linear village with two houses?

   f) Use inequality symbols to describe the fire station locations x where total distance is a minimum.

   g) Use your graph of the function representing average distance from Item 2(d) to find the location(s) where the average distance is equal to 4 units. What equation are you solving?

VILLAGE GRAPHS

ACTIVITY 3

h) One point on the graph of $y = |x - 3| + |x - 7|$ is the point (2, 6). Explain the meaning of the point (2, 6).

3. a) What happens to the graphs from Item 2 if you change the location of one of the houses? Move house B to location 11. What is a location for the fire station that minimizes the average distance when house A is located at 3 and house B is located at 11? Make a prediction. Write a sentence to describe your prediction and sketch a graph to illustrate your prediction.

   b) Write an equation to represent the total distance between houses located at 3 and 11 and a fire station located at x. Write a second equation to represent the average distance for the same situation. Graph both equations on the graphing calculator and compare the locations that produce minimum values with your predictions in part (a).

   c) What locations for the fire station would make the *average* distance 5 units? What equation are you solving?

4. What location(s) will minimize the average distance in a two-house linear village when house A is located at 3 and house B is located at 5? Create an appropriate function and graph it. Identify and sketch a picture of your window.

5. How would the graph of total distance change if there were three houses to consider in Linear Village? Locate house A at 3, house B at 7, and house C at 9. Where should the fire station be located to minimize the total or average distance for three houses? Write and graph an appropriate function. Sketch and label a picture of your window.

6. a) Predict the shape of the total-distance function for a linear village in which the houses are located at 4, 8, 10, and 11.

   b) Describe how to use its graph to determine the fire station location that minimizes total distance.

   c) Find the proper location of the fire station using your method.

The first primitive fire engines on wheels were pulled by the volunteer firefighters themselves; firefighters strongly resisted using the new horse-drawn steam engines. Around 1850, for example, firefighters in Cincinnati, Ohio, attacked their new steam engine with rocks and clubs at its first big fire. The tide was turned, however, when angry citizens at the blaze then attacked the firefighters and successfully manned the steam engine themselves to put out the fire. Thus, Cincinnati became the first city in the United States to use horse-drawn steam engines to replace its volunteer-pulled engines.

from Fire! How Do They Fight It? by Annabel Dean

ACTIVITY

3

VILLAGE GRAPHS

7. For each item, find the fire station location that minimizes *average* distance. Write an appropriate equation to enter in the graphing calculator:

a) when the houses are located at 3, 4, 7, 8, 11, 12, and 16.

b) when the houses are located at 2, 5, 6, 14, 16, 17, 19, 25, 27, and 31.

8. Describe a fast and easy way to find the fire station location that minimizes total distance or average distance in any size linear village without using the graphing calculator. (Look back at patterns you discovered in Activity 2, *Developing Linear Village*, and in Items 1 through 7 of this activity. The rule must work for any size linear village.)

9. Apply your rule to the following items. Explain how to find the fire station location that minimizes total distance:

a) when there are 12 houses in Linear Village.

b) when there are 39 houses in Linear Village.

c) when there are 90 houses in Linear Village.

10. a) Describe the shape of the graph of the total-distance function in a linear village of n houses. Be as specific as possible.

b) Describe the shape of the graph of the average-distance function in a linear village of n houses. Be as specific as possible.

c) Suppose there is only one house in Linear Village and the house is located at 3. Predict how the graphs of average and total distance would look. Write a sentence to describe your prediction and sketch a graph to illustrate your prediction. Explain why you think your prediction will be correct. Then write and graph an appropriate function representing the distance from the fire station located at x to the one house located at 3. Compare the actual graph with your prediction.

INDIVIDUAL WORK 3

Hermitville

1. Absolute value is used to find the distance between two points on a line. The expressions $|9 - 5|$ and $|5 - 9|$ represent the distance between 5 and 9 on the number line.

   a) Describe the meaning of the absolute-value expression $|3 - 12|$. Evaluate the expression.

   b) Write two different absolute-value expressions to represent the distance between 17 and 53 on the number line. Evaluate the expressions.

   c) Negative numbers identify locations to the left of zero on the number line. Write two absolute-value expressions to represent the distance between –3 and 4. Evaluate the expressions.

   d) Describe the meaning of the absolute-value expression $|x - 8|$.

   e) Write an equation to represent the distance between –3 and any point on the number line. Use the letter d for distance and the letter x for the location of any point on the number line.

2. Absolute value equations represent distances on a number line. For example, the equation $y = |x - 5|$ says y is the distance between the number 5 and any location x on the number line.

   a) Describe the meaning of the equation $4 = |x - 15|$. Solve the equation.

   b) Write an equation to represent all points 3 units from 21. Solve the equation.

   c) Describe the meaning of the equation $6 = |x + 2|$. Solve the equation.

3. Imagine a linear village with only one house. It is called Hermitville (see **Figure 1.40**).

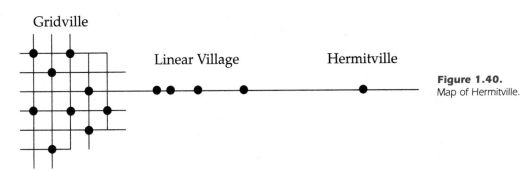

Figure 1.40.
Map of Hermitville.

a) Suppose the one and only house in Hermitville is located at coordinate 2. Write two equivalent equations, using absolute-value notation, for the distance d between the house and possible fire station locations x. Recall that locations in Hermitville (and Linear Village) are not limited just to integer values. Distances are measured along a continuous line.

b) Without calculating individual points to plot or using your calculator, predict how the graphs of your two equations will look. Draw a picture and write a sentence to record your prediction and explain your reasoning. Then enter the functions into your calculator and graph. Be sure to identify the window you use. Compare your graph to your prediction. Remember, since non-integer addresses make sense in this context, points in the plot may be connected.

c) Copy and complete the table in **Figure 1.41** to record sample values of your distance function.

Location of Fire Station	Distance From Station to House
0.0	
0.5	
1.0	
1.8	
2.0	
2.6	
3.1	
4.3	
5.0	

Figure 1.41.
Table for Hermitville.

d) Identify two possible locations for the fire station that are the same distance from the only house. Use your equation, graph, or table to explain your answer.

e) Given a location for the fire station, explain how to find another location exactly the same distance from the single house in Hermitville.

f) The graph indicates that the distance function is piecewise linear. Identify the regions of Hermitville that determine the individual "pieces" of the function, then write separate equations for these pieces.

g) Using the same window as in part (b), graph the equations you wrote in part (f) and compare that graph to the one you made in part (b). Explain your observations.

4. Move the address of the only house in Hermitville to coordinate "0."

   a) State and graph a function, using absolute-value notation, representing total distance for fire stations in this setting. Use the window [-5, 5] x [-2, 10].

   b) Use the TABLE feature on your calculator to investigate distances for positive and for negative fire station addresses. Summarize your observations in words and with a piecewise-defined equation.

   c) The points (–5, 5) and (5, 5) are both points on the graph of $y = |x|$. Explain the meaning of each point in the context of Hermitville.

5. Suppose that the only house in Hermitville is at location –3.

   a) Write a function involving absolute-value notation to represent the distance from all possible fire station locations x to this address.

   b) Graph your function on the window [–7, 3] x [–2, 8].

   c) Write a piecewise-defined equation that is equivalent to your answer to part (a). Identify the slope of each piece and the region of Hermitville for which each piece is to be used.

6. Suppose the function $y = |x + 5.8|$ represents the distance from the only house to possible locations x for the fire station.

   a) What is the new address for the house?

   b) Explain the meaning of the equation $8 = |x + 5.8|$, then solve the equation.

7. How are Situation 1 and Situation 2 related to absolute value?

   Situation 1:
   Herman and Helen are talking to each other on cell phones.

   Helen: "Where are you, Herman?"

   Herman: "I'm five miles outside of Gridville on highway 40. Same place I told you one hour ago."

   Helen: "I went there and I didn't see you."

   Herman: " I haven't moved in the last hour. You could not have come here."

   Helen: "At this very minute I am standing here exactly 5 miles outside of Gridville on highway 40. You are not here! Herman, you are a liar!"

   Situation 2:
   Herman and Helen are three years apart in age. One of them is 51 years old. Who is older? How old?

8. Absolute value expressions may be combined to represent the sum of two or more distances. The expression $|12 - 3| + |12 - 14|$ represents the combined distance from 12 to 3 and 12 to 14. The sum is 11 units.

   a) Describe the meaning of the expression $|7 - 17| + |17 - 13|$. Evaluate the expression.

   b) Describe the meaning of the equation $d = |x - 8| + |x - 20|$.

   c) Describe the meaning of the equation $14 = |x - 3| + |x + 5|$. Solve the equation and describe your method.

   d) Suppose that each of a, b, and c is a number. How must $|a - c|$ be related to $|a - b| + |b - c|$? Explain.

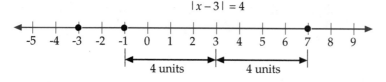

Figure 1.42.
Solutions to $|x - 3| = 4$.

Absolute-value notation is used to represent distance. The notation $|x - 3| = 4$ means the "distance between 3 and x is 4 units." The numbers −1 and 7 are each 4 units from point 3 (see **Figure 1.42**). The numbers −1 and 7 are the values of x that make the equation true.

Therefore, we say $x = -1$ and $x = 7$ are solutions to the equation

$|x - 3| = 4$.

9. Write each expression using words and then practice using a number line such as the one shown in **Figure 1.43** to solve the given equations and inequalities.

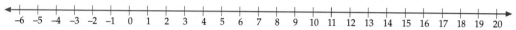

Figure 1.43.
Blank number line.

a) $|x - 8| = 11$

b) $|x| = 7$

c) $|x + 3| = 6$

d) $|15 - x| = 4$

e) $|x + 2| = -3$

f) $|x - 5| < 4$

g) $|x + 3| \geq 2$

10. **Figure 1.44** shows houses, identified using the letters A, B, C, D, E, and G, in a linear village. The letter F represents the location for the fire station.

Figure 1.44.
Linear village for Item 10.

a) Find the total distance and average distance between the fire station and each of the six houses in this linear village.

b) Where do you build the fire station if you want to minimize the total distance? Explain how you determine the answer.

11. The houses in a linear village are located at -4, 2, 5 and 10.

a) Write an equation to represent the total distance y for a post office located at x to the four houses in linear village.

b) Write an equation to represent the average distance y.

c) Describe the similarities between the graphs of total distance and average distance.

12. Esther says, "The location that minimizes total distance is always across the street from the middle house." Esther is referring to a linear village of n houses and the task of finding the location for a fire station that minimizes the total distance. Do you agree with Esther? Explain your answer.

13. Zach says, "The location or locations that minimize the total distance are always the same as the location(s) that minimize the average distance." Zach is referring to a linear village of n houses and the task of finding the location for a fire station that minimizes the total distance or average distance. Do you agree with Zach? Explain your answer.

14. Find the location that minimizes total distance for each of the following linear villages. Explain how you determined the answer for each village.

   a) What is the location for **Figure 1.45**?

Figure 1.45.
Linear village for Item 14(a).

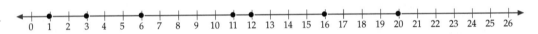

   b) What is the location for **Figure 1.46**?

Figure 1.46.
Linear village for Item 14(b).

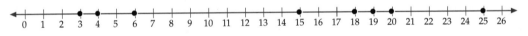

   c) What is the location for **Figure 1.47**?

Figure 1.47.
Linear village for Item 14(c).

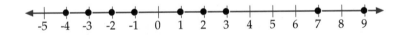

   d) What is the location for **Figure 1.48**?

Figure 1.48.
Linear village for Item 14(d).

15. Explain where to locate a fire station that minimizes average distance in a linear village with:

   a) 3 houses.

   b) 10 houses.

c) 29 houses.

d) n houses.

16. The following graphs represent average distance between possible fire station locations and the houses in a linear village. The graph in **Figure 1.49** represents "before." The graph in **Figure 1.50** represents "after." Explain what happened in linear village between the before and the after views.

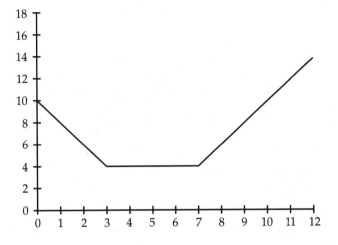

Figure 1.49.
Graph of Linear Village before.

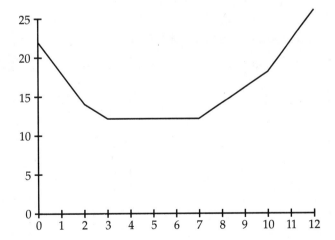

Figure 1.50.
Graph of Linear Village after.

INDIVIDUAL WORK 3

17. The context is a linear village and the graphs below (see **Figure 1.51**) represent information about distance between locations x in linear village and two Automated Teller Machines (ATMs).

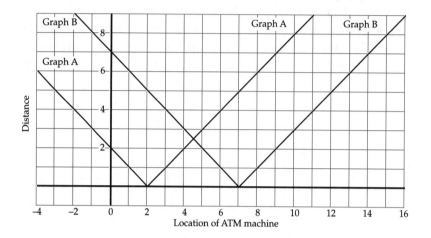

Figure 1.51.
Distance to ATM machines.

a) Explain the meaning of Graph A.

b) Explain the meaning of Graph B.

c) Explain the meaning of the intersection of Graphs A and B.

d) A person who services or monitors the two ATM machines might investigate total distance. Sketch a new graph showing the sum of the distances, along with Graph A and Graph B.

18. The graph of $y = |x + 2| + |x - 1| + |x - 3|$ is shown in **Figure 1.52**.

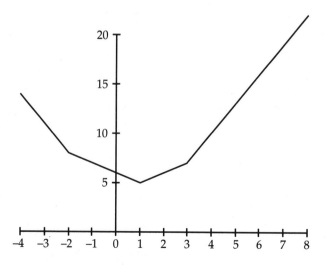

Figure 1.52.
Graph for total distance to three houses.

a) What does this graph represent in the context of building a fire station in a linear village?

b) Explain how the equation and the graph representing total distance from the fire station to all the houses will change when you add more houses to a linear village.

c) The graph of $y = |x + 2| + |x - 1| + |x - 3|$ is piecewise linear. Identify the x-values that separate the various pieces, then use the given equation to determine the corresponding y-values for these "corners."

d) Use the points you found in part (c), and any additional points you need, to determine the slope of each linear piece.

e) Write a piecewise-linear description for $y = |x + 2| + |x - 1| + |x - 3|$ by writing a linear equation to represent each linear segment or piece and inequalities to identify the interval or x-coordinates for which each equation is to be used. For example, the description for the first piece is: $y = -3x + 2$ when $x \leq -2$. Show algebraically how you get each equation.

19. Piecewise descriptions occur in life. Rourke earns $8.00 per hour working in an office. He is paid time-and-a-half for overtime. That is, for each hour above 40 that he works during the week he is paid $12.00 per hour.

a) Prepare a graph to represent Rourke's weekly earnings. Label the horizontal axis "Number of Hours" (up to 60 hours) and the vertical axis "Total Weekly Earnings."

b) Write a piecewise linear description for the two lines that appear in the graph. Use inequality symbols to identify the x-values that apply to each equation.

c) How would the graph and the equations change if Rourke is paid double-time ($16.00 per hour) for hours worked beyond 40 per week?

d) Describe another situation in life that produces a piecewise graph, then graph it.

20. Suppose Nancy made a list of locations she would like to live near. She assigned a point value to each location, giving higher values to the more desirable locations. Here are her valuations of locations.

School (S) = 5 points

Fire Department (F) = 4 points

Park (P) = 3 points

Police Department (D) = 2 points

Market (M) = 1 point.

Figure 1.53 shows the linear village in which she will live. At what address should she live in order to maximize her total point value? Show or explain how you determine your answer, using functions as appropriate.

Figure 1.53.
Linear village for Item 20.

21. a) Develop a spreadsheet or write a calculator program that uses the definition of total distance (not a graph or the median rule) to find the location that minimizes total distance in any size linear village.

b) Use your spreadsheet or calculator program and demonstrate how to find the location that minimizes total distance for the linear village with houses at –3, –2, 4, 6, 7, 11.

c) Compare the results of your spreadsheet or calculator program to those obtained by the median rule. Which method do you prefer? Give at least two reasons for your choice.

LESSON THREE
Absolute Value

KEY CONCEPTS

Absolute-value
equations and
inequalities

Absolute-value
graphs

Piecewise-defined
functions

Transformations
of functions

The Image Bank

PREPARATION READING

Welcome to the Family

Welcome the absolute-value function to your toolkit of functions! In this lesson you investigate the newest member of your family of functions. You graphed the absolute-value function $y = |x|$ (**Figure 1.54**) when you visited Hermitville in Individual Work 3 and represented the distance between the fire station and a house with address 0. Formally, $|x|$ is defined as the distance between x and 0. The alternate characterization from Item 4 of Hermitville, $|x| = x$ for $x \geq 0$ and $|x| = -x$ for $x < 0$, is also useful.

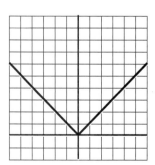

Figure 1.54.
Graph of $y = |x|$ on $[-6, 6]$ x $[-2, 10]$.

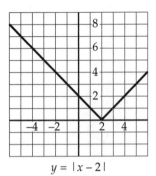

$$y = |x - 2|$$

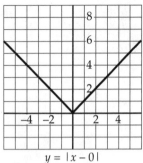

$$y = |x - 0|$$

Figure 1.55.
Three graphs on [-6, 6] × [-3, 9] for three different addresses in Hermitville.

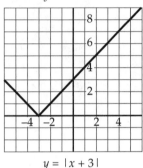

$$y = |x + 3|$$

When the address of the one house in Hermitville changed from 2 to 0 to –3, the function representing distance from the house changed from $y = |x - 2|$ to $y = |x - 0|$ and then to $y = |x + 3|$. Each time the function changed, the graph moved (see **Figure 1.55**). The location changed but the shape did not change.

Based on what you studied about transformations of functions in previous units in Course 1 (linear functions in Unit 4, *Prediction*, exponential functions in Unit 6, *Wildlife*, and quadratic functions in Unit 8, *Testing 1, 2, 3*), can you predict what the graph will look like for each of the following equations?

1. $y = |x - 2|$

2. $y = |x| + 3$

3. $y = |x - 4| + 1$

4. $y = 2|x + 5|$

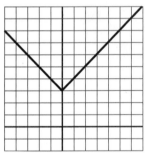

[–5, 7] × [–2, 10]

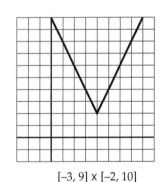

[–3, 9] × [–2, 10]

Figure 1.56.
Three transformed absolute-value graphs.

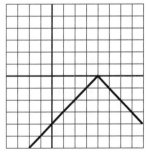

[–4, 8] × [–6, 6]

You should also be able to find the absolute-value equation to match a graph. Can you figure out what function was used to create each of the graphs in **Figure 1.56**?

When you complete this lesson you should be able to (1) sketch a graph given an absolute-value function, and (2) determine the absolute-value function that describes a given graph. You should also be able to provide a piecewise description for any absolute-value function.

Welcome to a mathematical excursion into symbolic and graphical representations.

ACTIVITY

THE ABSOLUTE-VALUE GAME

4

OBJECT OF THE GAME:

First, stump your group! Then, stump the class! Create an absolute-value graph. Challenge other students to find the absolute-value function that was used to create the graph.

GAME SUMMARY:

Create a graph from an absolute-value function. Show the graph to other members of your group. Ask the other group members to guess the function you used to create the graph. They might use the TRACE feature or "Grid On." If your group guesses your function easily, then create another absolute-value graph that is more challenging. Create several. Record the function that is the most challenging so you can present it to the entire class.

RULES:

(1) The function must use absolute-value expressions of the form $y = a|x - h| + k$.

(2) The distinguishing corners and segments of the graph must appear in the calculator window.

(3) All graphs must be displayed in the window $[-9.4, 9.4] \times [-12.4, 12.4]$.

INDIVIDUAL ASSIGNMENT:

Prepare a formal graph, using graph paper, for your function. Do not write the equation on the page with the graph. On a second page include the following: (a) the equation you used to create the graph, (b) the linear equation for each segment and the x-interval for which the equation applies, (c) the coordinates of points (if any) where the graph intercepts the y-axis and the x-axis, and (d) an explanation of how each number in the function transformed the graph of $y = |x|$ and affected the appearance of the graph you created.

ACTIVITY

4

THE ABSOLUTE-VALUE GAME

GROUP CHALLENGE:

Each member of your group will choose the most challenging graph that she or he created. Your group will then select one graph from among the graphs created by the members of the group. Your group will present the graph to the class. The class will be given an opportunity to guess the absolute-value function used to create the graph. Following the brief period of guessing, your group will present the information requested above in the individual assignment.

SAMPLE PRESENTATION #1 PAGE 1

Find the absolute-value function used to create the graph in **Figure 1.57**.

SAMPLE PRESENTATION #1 PAGE 2

a) $y = |x - 2| + 3$.

b) For $x < 2$, the slope of the graph is -1, a point on the line is $(2, 3)$, the equation of the line is $y = 3 - 1(x - 2)$ or $y = -x + 5$.

For $x \geq 2$, the slope of the graph is $+1$, a point on the line is $(2, 3)$, the equation of the line is $y = 3 + 1(x - 2)$ or $y = x + 1$.

c) The graph intercepts the y-axis at $(0, 5)$. This graph does not intercept the x-axis.

d) The 2 in $|x - 2| + 3$ causes the graph of $|x|$ to shift two units to the right. This is similar to the shift created by changing the address from 0 to 2 for the one house in Hermitville.

The 3 in $|x - 2| + 3$ causes the graph of $|x|$ to shift three units up.

SAMPLE PRESENTATION #2 PAGE 1

Find the absolute-value function used to create the graph in **Figure 1.58**.

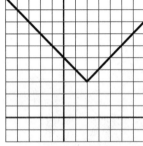

Window: [-5, 7] × [-2, 10]

Figure 1.57.
Graph for sample presentation #1.

Window: [-9, 3] × [-2, 10]

Figure 1.58.
Graph for sample presentation #2.

ACTIVITY

THE ABSOLUTE-VALUE GAME

4

SAMPLE PRESENTATION #2 PAGE 2

a) $y = 2\,|x + 4|$.

b) For $x < -4$, the slope of the graph is -2, a point on the line is $(-4, 0)$, the equation of the line is $y = 0 - 2(x + 4)$ or $y = -2x - 8$.

For $x \geq -4$, the slope of the graph is $+2$, a point on the line is $(-4, 0)$, the equation of the line is $y = 0 + 2(x + 4)$ or $y = 2x + 8$.

c) The graph intercepts the y-axis at $(0, 8)$. The graph intercepts the x-axis at $(-4, 0)$.

d) The 2 in $2\,|x + 4|$ causes the "V" shape of the graph of $|x|$ to become taller as the segments forming the "V" become steeper. It is stretched in the y-direction.

The 4 in $2\,|x + 4|$ causes the graph of $|x|$ to shift four units to the left. This is similar to the shift created by changing the address from 0 to -2 for the one house in Hermitville.

FOLLOW-UP

The graph you created for your individual assignment may be displayed on the wall of the classroom. Other members of the class may want to guess the absolute-value function you used to create your graph (if it was not selected by your group for the group presentation).

SUMMARY

1. Describe how each constant in the equation $y = \frac{1}{2}|x - 3| + 1$ transforms the graph of $y = |x|$.

2. Describe how each constant in the equation $y = 4 - 3|x + 5|$ transforms the graph of $y = |x|$.

3. Describe how each constant in the equation $y = a + b|x - c|$ transforms the graph of $y = |x|$.

INDIVIDUAL WORK 4

The Game Plan

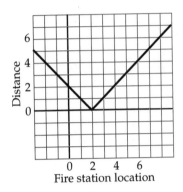

Figure 1.59.
Absolute-value graph for Item 1 (a).

1. a) Write an absolute-value function to match the graph in **Figure 1.59**.

   b) Graph your function on the graphing calculator in the window [–3, 7] × [–2, 10]. Compare your graph to the one in Figure 1.59.

   c) Assuming that the graph in Figure 1.59 shows all relevant information about this situation, how many houses are in this town? State the address of each house, and explain how you know you are correct.

   d) What is the slope of the line for fire station locations to the right of the single house? Explain the meaning of the slope for this part of the graph.

   e) Write a linear equation to represent distance for fire station locations to the right of the single house.

   f) What is the slope of the line representing distance for fire station locations to the left of the single house? Explain the meaning of the slope for this part of the graph.

   g) Write a linear equation to represent distance for fire station locations to the left of the single house.

In Hermitville and Linear Village, the equation $|x - 3| = 5$ represents all points (addresses) whose distance from location 3 is 5 units. Absolute-value notation can also be used to represent intervals between points. The inequality $|x - 3| < 5$ represents all locations within 5 units of 3.

The notation $|x - 3| < 5$ means the "distance between x and 3 is less than 5 units." The solution to this inequality is all values of x between (but not including) –2 and 8 on the number line. In Hermitville or Linear Village, $|x - 3| < 5$ represents fire station locations within 5 units of the house at 3 (see **Figure 1.60**).

Figure 1.60.
Fire station locations within 5 units of 3.

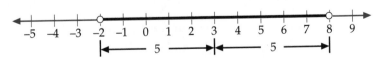

Notice that 3 is the "center" of the interval and 5 is the distance between the "center" and the "edge" of the interval. The diagram above also represents the notation $-2 < x < 8$.

To include the points -2 and 8 and all the points between -2 and 8, write $|x - 3| \leq 5$, which is the same as $-2 \leq x \leq 8$ (see **Figure 1.61**).

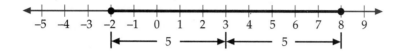

Figure 1.61.
Locations –2 and 8 are filled in to show inclusion.

2. a) Use the context of Hermitville and Linear Village to describe the meaning of the inequality $|x - 8| < 4$.

   b) Use the context of Hermitville and Linear Village and write an inequality to represent all locations within 2 units of 7.

   c) Explain how to use the graph of $y = |x - 8|$ to solve the inequality $|x - 8| \leq 4$.

3. a) Dee Dee is buying a gift to exchange at a Math Party. The party invitation suggests that the price p of the gift in dollars must satisfy the condition: $|p - 25| \leq 6$. What is the minimum amount of money Dee Dee can spend on the gift and meet the conditions of the inequality? What is the maximum amount she can spend on the gift?

   b) On a typical summer day, the air temperature in the town of Arise ranges from a low of 72° F to a high temperature of 94° F. Write an absolute-value inequality to represent all temperatures between the minimum and maximum temperatures.

   c) At a particular beach along the coast of New England, the difference in water levels between high tide and low tide is 16 feet. Assume that the average (or middle) tide level is sea level. Write an absolute-value inequality to represent the tide levels between low and high tide.

   d) The minimum speed allowed on a certain stretch of highway is 45 mph. The maximum speed is 65 mph. Write an absolute-value inequality to represent all speeds between (and including) the maximum and minimum speed.

   e) The average height of students on the women's basketball team is 175 cm (1.75 m). Write an absolute-value inequality to represent all heights within 15 cm of the average height.

f) Jenny writes computer software programs used to manufacture parts for medical equipment. One particular part is specified to have a diameter of 1.125 cm. The amount of error allowed is ±0.005 cm. Write an absolute-value inequality to represent allowable diameters for the part.

g) Normal human body temperature is 98.6° F. Suppose any temperature T that is within 1.2 degrees of this value is considered acceptable. Write an absolute-value inequality to describe the acceptable range of body temperatures.

h) An automobile engine uses several fluids to help it run smoothly or keep it cool. Most of those fluids have a minimum safe level. They also have a maximum level or capacity. The fluid level must be somewhere between the minimum and the maximum. Absolute-value inequalities may be used to describe the range of fluid levels. Suppose the maximum fluid level is six quarts and the minimum fluid level is three quarts. The absolute-value inequality would be $|x - 4.5| \leq 1.5$. The "center" of the interval is 4.5. The distance from the center to the "edge" of the interval is 1.5 quarts.

Name at least three other situations that are familiar to you that have a high and a low mark or a beginning and ending time and write the corresponding intervals using absolute-value inequalities.

4. a) Without plotting points or using a graphing calculator, predict what the graphs of $y = |x - 2|$ and $y = |x - 2| + 3$ will look like.

b) Graph the two functions on a single set of axes. Compare your graph to your predictions.

c) Explain how the 3 in $y = |x - 2| + 3$ transforms the graph of $y = |x - 2|$.

d) What change(s) would you need to make in the equation $y = |x - 2|$ to cause its graph to shift upward 5 units and to the left 3 units?

e) Discuss the meaning of $6 = |x - 2|$ in the context of Hermitville.

f) Use the graph of $y = |x - 2|$ to solve the equation $6 = |x - 2|$. Explain your method.

5. a) Determine an absolute-value function that produces the graph in **Figure 1.62**.

   b) Determine an absolute-value function that produces the graph in **Figure 1.63**.

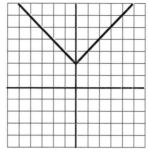

Figure 1.62.
Graph for Item 5(a).

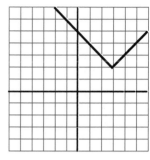

Figure 1.63.
Graph for Item 5(b).

   c) Determine an absolute-value function that produces the graph in **Figure 1.64**.

   d) Determine an absolute-value function that produces the graph in **Figure 1.65**.

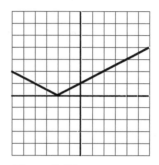

Figure 1.64.
Graph for Item 5(c).

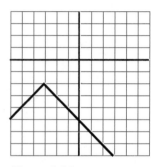

Figure 1.65.
Graph for Item 5(d).

6. Write an equation to produce each of the following graphs. For each graph, one "box" is one unit.

   a) Write an equation for **Figure 1.66**.

   b) Write an equation for **Figure 1.67**.

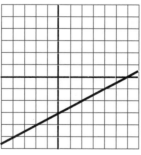

Figure 1.66.
Graph for item 6(a).

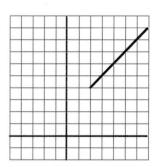

Figure 1.67.
Graph for item 6(b).

   c) Write an equation for **Figure 1.68**.

   d) Write an equation for **Figure 1.69**.

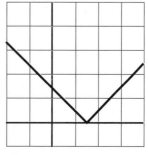

Figure 1.68.
Graph for item 6(c).

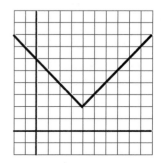

Figure 1.69.
Graph for item 6(d).

INDIVIDUAL WORK 4

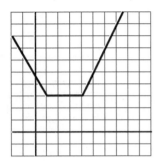

Figure 1.70.
Graph for item 6(e).

e) Write an equation for **Figure 1.70**.

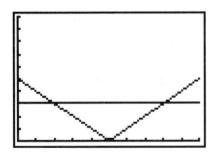

Figure 1.71.
Graph for item 6(f).

f) Write an equation for **Figure 1.71**.

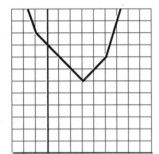

Figure 1.72.
Calculator graph of $y = |x - 5|$
and $y = 3$ on $[0, 10] \times [0, 10]$.

7. You have solved absolute-value equations using the graphing calculator and by using the distance interpretation. Consider the equation $|x - 5| = 3$.

If you solve using the graphing calculator, you could graph $y = |x - 5|$ and trace to find the x-coordinates of the points on the graph where $y = 3$, or you could graph $y = |x - 5|$ and $y = 3$ together and identify the x-coordinates for the point(s) of intersection. As you can see in **Figure 1.72**, to within the accuracy of reading the graph, the solutions appear to be $x = 2$ and $x = 8$. These may be checked by substituting back into the original equation.

If you solve using the distance interpretation, you would understand "solve the equation $3 = |x - 5|$" as meaning "find all points 3 units from 5 on the number line." (See **Figure 1.73**).

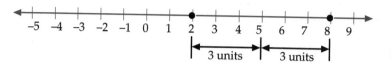

Figure 1.73.
Solving $|x - 5| = 3$ using the number line.

The solutions are $x = 2$ and $x = 8$. Again, you can verify both solutions by substitution.

Thus, graphical methods represent at least two ways to solve this equation. In previous units, however, you have learned symbolic methods for solving linear and quadratic equations. Use arrow diagrams to invent a symbolic method for solving absolute-value equations. The difficult part is reversing the absolute value. Since absolute value transforms two opposite values into the same result, reversing the process must take a single positive value and convert it back into two opposite values, one positive and one negative.

Be sure your method "works" for $|x - 5| = 3$, then apply your method to solve the following equations.

a) $5 = |x - 2|$

b) $|x| + 3 = 13$

c) $2|x + 1| - 3 = 9$

d) $|x - 2| + 7 = 4$

e) Solve $a|x - h| + k = n$

f) The major drawback in using the "graphing calculator" method of solving is that it is only as precise as your reading of the point of intersection on the graph. It can never be better than the precision of the calculator, which, although it is typically 12 or 13 significant figures, is not exact. However, combining a graphing calculator graph with your knowledge of piecewise equations can be at least as powerful in getting exact solutions as the arrow diagram suggested above. Write the piecewise equations for $|x - 5|$, then use them to solve two appropriate linear equations (without absolute values) to get an exact solution to $|x - 5| = 3$.

INDIVIDUAL WORK 4

g) Find the exact solutions to the equation $|x - 5| + |x - 2| = 10$. Explain the method you used and why you selected it.

8. Using one of the methods you developed in Item 7, solve the equation $|x + 3| = |x - 5|$. What might the solution to this equation mean in the context of the problem involving two ATM machines?

9. a) For what numbers x is it correct to write that $|x - 2| = x - 2$? For what numbers x is this wrong? You solved this equation in Individual Work 3 using a guess-and-check method. Now solve this equation by graphing appropriate equations on the graphing calculator.

   b) Conchita says that if x is any number greater than 5, then $|x - 2| + |x - 5| = 2x - 7$. Explain what you discover when you use the graphing calculator to verify or dispute her conclusion.

10. Is $|x + 2| = |x| + 2$? Always? Never? Sometimes? Check and explain your answer using appropriate graphs.

11. a) For what values of x is $|-x| = |x|$? Experiment with several values of x. Explain and justify your final conclusion.

    b) For what values of x is $x^2 = |x|^2$? Experiment with several values of x. Explain and justify your final conclusion.

12. Write an equation involving absolute values that has (a) no solutions, (b) exactly one solution, (c) exactly two solutions, and (d) more than two solutions.

13. a) In the context of linear villages, what is the meaning of the equation $6 = |x - 1| + |x - 7|$? Solve (exactly) the equation $6 = |x - 1| + |x - 7|$. Which of the analytic methods you developed in Item 7 is best here? Explain.

    b) Use the same method you used in part (a) to solve the equation $9 = |x - 1| + |x - 7|$.

    c) Graph the two equations $y = |x - 1|$ and $y = |x - 7|$ and the equation $y = |x - 1| + |x - 7|$ on the same axes in an appropriate window. Describe the relationships among the graphs of these three equations.

    d) What is the meaning in a linear village of the point where the graph of $y = |x - 1|$ intersects the graph of $y = |x - 7|$?

    e) The point (1, 6) is a point on the graph of $y = |x - 1| + |x - 7|$. Explain the meaning of the point (1, 6).

14. a) Without using your calculator, predict what the graph of
$y = |x-1| + |x-2| + |x-3|$ looks like. Sketch a graph to record your prediction. Then use your graphing calculator to check your prediction.

   b) Find the slope of the graph when $x > 3$. Interpret in the context of fire-station locations the meaning of the slope on this interval. Write a linear equation to represent this "piece" of the graph.

   c) What is the slope of the graph for all other intervals? That is, find the slope of the linear piece in each of the following intervals:

   Find the slope of the line for $x < 1$.

   Find the slope of the line for $1 < x < 2$.

   Find the slope of the line for $2 < x < 3$.

   Find the slope of the line for $x > 3$.

   d) Interpret the meaning of the slope for each interval as it relates to the context of distance.

   e) Complete the piecewise description for this function by writing an appropriate linear equation for each linear "piece" of the graph. Check your answers by substituting to verify that the end-points of each piece satisfy its equation.

You know what each of the letters a, h, k in the function $y = a|x-h| + k$ do to transform the graph of $y = |x|$. An arrow diagram may be used to describe a particular transformation. For example, the transformation of $y = |x|$ to $y = |x-5| + 2$ is illustrated with the arrow diagram in **Figure 1.74**.

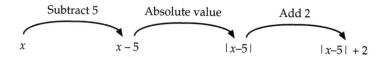

Figure 1.74.
Arrow diagram for transformation from $y = |x|$ to $y = |x - 5| + 2$.

Note that the graph of $y = |x|$, the graph to be transformed, involves the numbers that go "into" and "out of" the absolute function, usually called x and y respectively. However, in the arrow diagram in Figure 1.74, they are called $x - 5$ and $|x - 5|$ instead. The graph of $y = |x - 5| + 2$ has input and output numbers x and $|x - 5| + 2$ so the question is how these values are related to $x - 5$ and $|x - 5|$. Using the arrow diagram, the answers are easy: x is "add 5 to $x - 5$" and $|x - 5| + 2$ is "add 2 to $|x - 5|$" (follow the appropriate arrows). Thus the final graph of $|x - 5| + 2$ versus x is 5 units to the right and 2 units up from the graph of $|x - 5|$ versus $x - 5$, better known as $y = |x|$.

15. Draw and explain an arrow diagram to illustrate each of the following transformations:

   a) from $y = |x|$ to $y = 2|x - 3|$.

   b) from $y = |x|$ to $y = 3|x| + 4$.

   c) from $y = |x|$ to $y = a|x - h| + k$.

16. In Course 1, Unit 2, *Secret Codes*, you studied coding processes with two steps. A typical two-step coding process combines a stretch (multiply by a constant) and a shift (add or subtract a constant).

   a) Does reversing the order of the stretch and the shift make a difference in the output? If so, what is the difference between a coding process that stretches by a factor of 2 followed by a shift of 3, and a coding process that shifts 3 followed by a stretch using a factor of 2? Justify your answer.

   b) You have studied three different transformations for the absolute-value function $y = |x|$. You stretched absolute-value graphs in the vertical direction and you shifted absolute-value graphs in both the vertical and horizontal directions. Does changing the order of the stretch and shift produce a different absolute-value graph? Does it matter if one transformation is horizontal and the other is vertical? Justify your answer.

17. Suppose your linear village has a house at 5 and a duplex at 9. (A duplex is two houses joined together as one.)

   a) How do you find the location for the fire station that minimizes the total distance?

   b) What equation would you use to represent total distance?

c) Describe how the graph for this situation will differ from graphs involving total distance for single-unit houses only.

d) Determine the location(s) that minimize(s) total distance in a linear village with houses at 2, 3, 8 and 12 and duplexes at 5, 7 and 14. Explain your methods.

18. The graph of the absolute-value function resembles a bird in flight. Write a calculator animation program using absolute-value equations to simulate a bird in flight.

19. Use at least seven piecewise-equation descriptions to create a recognizable picture. Draw your picture on graph paper or use the graphing calculator. The equations may be linear, quadratic, or absolute value.

20. What is the average of a number and its absolute value? Experiment with several values to discover a pattern.

21. a) Use a variety of values for x_1 and x_2 to determine what the function $y = \frac{x_1 + x_2}{2} - \frac{x_1 - x_2}{2}$ does.

b) Write a formula that always gives the larger number of two numbers as the answer.

**Chicago Burns
—Bovine Suspected**

Chicago—October 10, 1871

Over one-third of the city of Chicago has been destroyed by a fire that burned out of control since last Sunday evening, October 8. With no rain since July, the summer heat and high winds proved a formula for disaster. The huge conflagration spread quickly, engulfing everything in its path in spite of the efforts of exhausted firefighters. Thousands of people have been left homeless, and the death toll is feared to be in the hundreds.

Fire companies from as far as 300 miles away came to assist. Finally, the blaze was extinguished early this morning when heavy rains poured down on the smoldering city.

Unconfirmed reports indicate that the fire broke out when, as Mrs. O'Leary of Chicago was attempting to milk her cow, the cow kicked a lit kerosene lantern into a pile of hay.

Source: *Fire! How Do They Fight It?*
by Annabel Dean

LESSON FOUR
Minimax Village

KEY CONCEPTS

Absolute-value graphs

Maximum distance in one dimension

Midrange

David Barber

PREPARATION READING

One More Reason

*I*n Lesson 2, you learned that you can minimize the total distance if you build the fire station at the median location or within the median interval in a linear village.

To find the median location or median interval, you begin by listing the addresses representing houses in increasing or decreasing order.

If the number of houses is *odd*, the median address is the address in the middle of the list. The location that minimizes total distance is called the **median location (see Figure 1.75).**

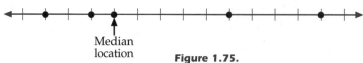

Median location

Figure 1.75.
Median location for odd number of houses.

If the number of houses is *even*, the locations that minimize total distance will be all locations with addresses between and including the two middle houses. The addresses of the two middle houses are the endpoints of the **median interval** (see **Figure 1.76**).

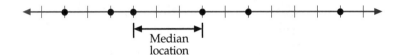

Median
location

Figure 1.76.
Median interval for even
number of houses.

It makes sense to build a fire station at a location that minimizes the total distance. The mathematics does not seem difficult. The median is easy to find. Now, the community must decide if the median location is fair. Does building the fire station at the median location favor some houses more than others? Are there other locations, supported by different criteria, that seem just as fair? Continue the modeling process and look for answers to these questions.

What is the next step in the modeling process? That depends. In this unit you model to seek answers to two main questions:

1. "Which location criteria should I investigate using mathematics?"

2. "How do I find locations that satisfy the criteria?"

So far in this unit, you have discovered how to find the location that minimizes total distance in a linear village. You still need to determine a procedure for finding the location that minimizes total distance in a two-dimensional Gridville.

For your next step you might want to apply what you learned in a one-dimensional town, Linear Village, and find the equivalent of the "median location" or "median interval" in two-dimensional Gridville.

However, you might decide that offering one location based on one criterion is unfair. Perhaps it is better to offer the city leaders a choice, based on competing criteria, and let the community be the judge of what is fair. In that case you might examine another location criterion and determine how to find locations that satisfy the criterion in Linear Village.

This lesson follows the second scenario and examines **minimax location**, another candidate for "best" location. Minimax is another name for the location that **mini**mizes the **max**imum distance a firetruck would have to travel to reach a house in the village. The investigation of minimax begins in Linear Village. Use your mathematical tools and look for ways to find minimax locations in one dimension easily.

Do not forget that Linear Village is a simplified model for Gridville, and Gridville is a simplified model for a real town. Keep track of what you learn in Linear Village so that you can apply it as you move toward more realistic models.

CONSIDER:

1. Name some situations in life where you might want to minimize the total value of some quantities.

2. Name some situations in life where you might want to minimize the maximum value of some quantity.

3. In your opinion, is it more important to minimize the maximum distance a firetruck must travel or to minimize the total distance from the fire station to all of the houses? Give reasons to support your opinion.

CAUGHT IN THE MIDDLE

Where do you build the fire station in Linear Village if you want to minimize the maximum distance a firetruck must travel?

The purpose of this activity is to develop a simple method or rule for finding the fire station location that minimizes the maximum distance in any size linear village.

For example, consider the linear village in **Figure 1.77**. The letters A, B, and C represent locations for three different houses. The letter F represents a possible location for the fire station.

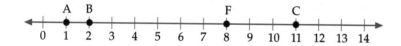

Figure 1.77.
Linear village with three houses.

The distance between F and A is 7 units.

The distance between F and B is 6 units.

The distance between F and C is 3 units.

The maximum distance is the largest of these three distances, or 7 units.

Suppose you build the fire station at location 7 instead of location 8 (see **Figure 1.78**).

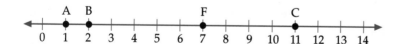

Figure 1.78.
Fire station at location 7.

The distance between F and A is 6 units.

The distance between F and B is 5 units.

The distance between F and C is 4 units.

The maximum distance is 6 units.

ACTIVITY

CAUGHT IN THE MIDDLE

5

In this three-house village, the maximum distance is less if you build the fire station at location 7 than if you build the fire station at location 8.

Where do you build the fire station if you want the maximum distance to be as small as possible? What if the village contains 15 houses? Forty houses? Five thousand houses?

THE TASK:

Develop a rule or strategy for finding the location that minimizes the maximum distance in any size linear village.

THE PROCEDURE:

Develop your own procedure for building and checking possible strategies.

THE REPORT:

Present your procedure and conclusions to the class. Apply your rule to the linear village in **Figure 1.79** by finding the location that minimizes maximum distance.

Figure 1.79.
Linear village to apply your rule.

THE CHALLENGE:

Design a linear village to give to another group to see if they can find quickly the location that minimizes the maximum distance.

INDIVIDUAL WORK 5

Midrange Fair

*I*n the figures in Items 1 and 2, assume the fire station may be built across the street from a house and share the same address on the number line. Remember that houses and the fire station may have addresses anywhere along the line in Linear Village. Sometimes the minimax location is not an integer.

1. For each map below, determine the minimax location. Then decide which house or houses seem to be favored by building the fire station at the minimax location. Explain and defend your decisions.

   a) Refer to the map of Linear Village A in **Figure 1.80**, then copy and complete the table in **Figure 1.81**.

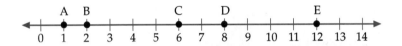

Figure 1.80.
Map of Linear Village A.

Note: This table uses integers to represent addresses. In Linear Village, buildings may be located anywhere along the number line.

Fire Station Location	Farthest House From The Fire Station	Maximum Distance
0		
1		
2		
3		
4		
5		
6		
7		
8		
9		
10		
11		
12		
13		
14		

Figure 1.81.
Table for Linear Village A.

INDIVIDUAL WORK 5

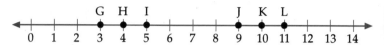

Figure 1.82.
Map of Linear Village B.

b) Refer to the map of Linear Village B in **Figure 1.82** and copy and complete the table in **Figure 1.83**.

Fire Station Location	Farthest House From The Fire Station	Maximum Distance
0		
1		
2		
3		
4		
5		
6		
7		
8		
9		
10		
11		
12		
13		
14		

Figure 1.83.
Table for Linear Village B.

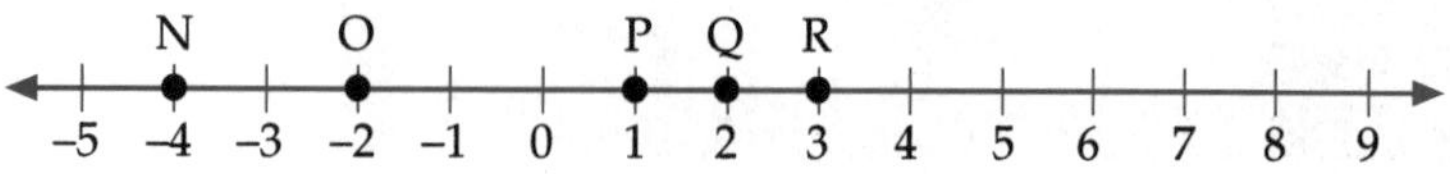

Figure 1.84.
Map of Linear Village C.

c) Refer to the map of Linear Village C in **Figure 1.84** and copy and complete the table in **Figure 1.85**.

Fire Station Location	Farthest House From The Fire Station	Maximum Distance
−5		
−4		
−3		
−2		
−1		
0		
1		
2		
3		
4		
5		
6		
7		
8		
9		

Figure 1.85.
Table for Linear Village C.

Figure 1.86.
Map of Linear Village D.

Fire Station Location	Farthest House From The Fire Station	Maximum Distance
–5		
–4		
–3		
–2		
–1		
0		
1		
2		
3		
4		
5		
6		
7		
8		
9		

d) Refer to the map of Linear Village D in **Figure 1.86** and copy and complete the table in **Figure 1.87**.

Figure 1.87.
Table for Linear Village D.

2. You learned in Lesson 2 that you minimize the total firetruck distance if you build the fire station at the median location or within the median interval.

For each map in Item 1, above, determine all locations that minimize the total firetruck distance. Use inequality symbols (such as, $3 \leq x \leq 5$) to identify intervals. Then decide which house or houses seem to be favored by building the fire station at that location. Explain and defend your decisions.

a) Linear Village A (see Figure 1.80).

b) Linear Village B (see Figure 1.82).

c) Linear Village C (see Figure 1.84).

d) Linear Village D (see Figure 1.86).

3. Compare your answers from Item 1 (for each Linear Village) to your answers for Item 2. In each case, decide which is the best location, and defend your answers.

4. Given a Linear Village with n houses with addresses $x_1, x_2, x_3, \ldots x_n$ (written in increasing order), write a formula for finding the minimax location.

INDIVIDUAL WORK 5

5. Graciela remembers using a graph to find locations that minimized total distance. She is wondering if a graph may be used to determine minimax locations easily. She prepared a graph (see **Figure 1.88**) representing the distances from possible station locations to each of the five houses in a linear village.

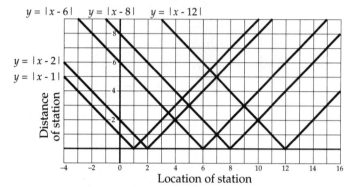

$y = |x - 6|$ $y = |x - 8|$ $y = |x - 12|$

$y = |x - 2|$
$y = |x - 1|$

Distance of station

−4   −2   0   2   4   6   8   10   12   14   16
Location of station

Figure 1.88.
Graph representing distances to five houses separately.

a) Identify the locations for the five houses.

b) Explain a procedure for using the graph to find the minimax location.

6. Suppose there are three houses in Linear Village. House A is located at 1, B at 4, and C at 7. Draw **Figure 1.89** on your own paper.

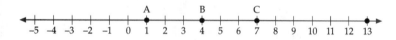

                    A   B   C
−5 −4 −3 −2 −1   0   1   2   3   4   5   6   7   8   9   10 11 12 13

Figure 1.89.
Linear Village with three houses.

a) Mark the median location with the letter "M." Mark the minimax location with the letter "X."

b) Add a fourth house, D, at location 8. How does this change the median location and the minimax location? Explain.

c) Add a fifth house, E, at location 9. How does this change the median interval and the minimax location? Explain.

d) Move house E from location 9 to location 20. How does this change the median location and the minimax location?

e) With five houses at 1, 4, 7, 8, and 9, move the house at location 7 to new location 2. How does this change affect the median location? How does this move affect the minimax location?

7. Suppose there are seven houses in Linear Village. House A is located at −3, B at −2, C at 4, D at 5, E at 7, G at 8, and H at 9. Draw **Figure 1.90** on your own paper.

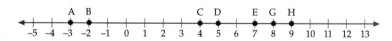

        A B   C D   E G H
−5 −4 −3 −2 −1   0   1   2   3   4   5   6   7   8   9   10 11 12 13

Figure 1.90.
Linear Village with seven houses.

a) Mark the median location with the letter "M." Mark the minimax location with the letter "X."

b) If closeness to the fire station is the criterion, which house or houses in Linear Village might prefer the median location? Which house or houses in Linear Village might prefer the minimax location? Which house or houses might object to both locations? Explain your reasons.

c) Add an eighth house at location 12. Explain how the median location changes. Explain how the minimax location changes.

d) The eight houses are located at –3, –2, 4, 5, 7, 8, 9, and 12. How will the median interval change if you move the house located at 5? How will the minimax be affected?

8. Suppose there are eight houses in Linear Village. House A is located at –1, B at 0, C at 1, D at 2, E at 4, G at 9, H at 11, and I at 14. Draw **Figure 1.91** on your own paper.

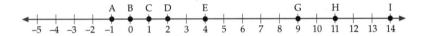

Figure 1.91.
Linear Village with eight houses.

a) Mark the median interval with the letter "M." Mark the minimax location with the letter "X."

b) If closeness to the fire station is the criterion, which house or houses in Linear Village might prefer the median location? Which house or houses in Linear Village might prefer the minimax location? Which house or houses might object to both locations? Explain your reasons.

c) Add a ninth house at location 5. Explain how the median interval changes. Explain how the minimax location changes.

d) Add a tenth house at location –3 and explain what happens to the median location and the minimax location.

9. Suppose there are six houses in Linear Village. House A is located at –6, B at –1, C at 2, D at 3, E at 10, and G at 15. Draw **Figure 1.92** on your own paper.

Figure 1.92.
Linear Village with six houses.

   a) Which houses can be moved without resulting in a change in the median interval?

   b) Which houses can be moved without resulting in a change in the minimax location?

10. Suppose there are five houses in Linear Village. House A is located at –3, B at 5, C at 6, D at 10, and E at 11. Draw **Figure 1.93** on your own paper.

Figure 1.93.
Linear Village with five houses.

   a) How many houses do you need to add to the interior of the village to change the median location? Describe the change in median caused by the change in the number of houses.

   b) How many houses do you need to add to the interior of the village to change the minimax location significantly?

   c) Suppose a four-unit apartment complex is built at location 1. Count each unit as a separate house. Describe the new median location and the new minimax location.

11. Create your own map of Linear Village. Arrange five houses in Linear Village so that the median location is at 4 and minimax location is at –1.

12. Create your own map of Linear Village. Arrange six houses in Linear Village so that the minimax location is –2 and the median interval is represented by all locations x such that $3 \le x \le 6$.

13. a) Develop a spreadsheet, write a calculator program, or write a computer program that uses the definition of maximum distance to find the location that minimizes the maximum distance in any size linear village.

    b) Apply your spreadsheet to the Linear Village in Item 10.

    c) Compare the ease of solution using the definition and technology to using the "by hand" methods you have developed.

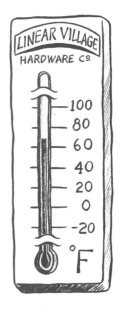

14. Jan prefers to work with the temperature set at 68° F. Susan prefers 76° F. Claudia prefers 71° F, Alfred 71° F, Maria 72° F, Victoria 60° F, and Anton prefers 72° F.

    a) Find the median temperature for this group of employees working in the same office.

    b) Find the midrange temperature. (The midrange temperature is the average of the two extreme temperatures.)

    c) If you had the key to the thermostat, what temperature would you set? Explain your reasoning.

15. Elgin has several deliveries to make in Linear Village. The deliveries are to locations 3, 8, 20, 27, 39, and 50. After each delivery, he must return to the truck to pick up the next package to deliver to the next house.

    a) In front of which house should Elgin park if he wants to minimize the total distance he must walk to make all the deliveries?

    b) Where (at what address) should Elgin park if he wants to minimize the maximum distance he would have to walk from the car to a house?

16. a) Sylvia has been operating a mobile pizza kitchen and now wants to rent a permanent shop location in Linear Village. Her most consistent and dependable customers live at 5, 14, 17, 30, 41, 45, and 62. Location 5 is an apartment complex. Location 45 is a college dormitory. She receives twice as many calls to location 5 as to most other locations. She receives five times as many calls to location 45 as to most other locations. Where should Sylvia locate the pizza shop if she wants to minimize the average distance she must travel to these most consistent customers?

    b) How does Sylvia's question relate to the fire station location problem?

17. Hank is taking nine kids to the movie theater. They all sit in the row labeled "K." They take up most of the row, but leave a few empty seats, one of which will be occupied by Hank. The kids are sitting in seats numbered: 103, 104, 105, 109, 110, 115, 116, 117, and 118. Where should Hank sit? Explain your reasons for choosing this location.

18. Alicia has bowling scores of 120, 145, 192, 160, 155, 138, and 171.

   a) What is the median for her bowling scores?

   b) What is the midrange score?

   c) Alicia bowls two more games. The median is now 160. Write down two possible scores for the two games Alicia bowls.

   d) Suppose the two new games that Alicia bowls increase the midrange score to 171. Write down two possible scores for the two games Alicia bowls.

19. Linear Beach is 500 meters long and 10 meters wide. It is quite narrow. The recreation department will build two lifeguard towers. Where should the lifeguard towers be placed to minimize the maximum distance a lifeguard would travel to reach a place anywhere on the beach? (Assume this Linear Beach can be represented by a straight line.)

20. Houses in Linear Village are located at 3, 6, 7, 9, 15, 17, 18, 19, 24.

   a) Where would you build two mail boxes to minimize the maximum distance to the closer mailbox?

   b) Where would you build three mail boxes to minimize the maximum distance to the closer mailbox?

   c) Describe a method for finding the locations that minimize maximum distance when you are building n mail boxes in a linear village with p houses located at $x_1, x_2, x_3, \ldots x_p$.

21. In the next lesson you will return to Gridville. Practice finding distance by drawing a map of Gridville. Place house A at (4, 5), house B at (7, 10) and the fire station, F, at (1, 6).

   a) What is the distance between F and A?

   b) What is the distance between F and B?

   c) What is the total distance?

   d) What is the average distance?

e) What is the maximum distance the firetruck must travel to reach any house in Gridville if Gridville has only two houses, A and B?

f) What is another location for the fire station that would make the maximum distance less?

g) Find a location for the fire station F in Gridville that minimizes the maximum distance the firetruck would have to travel.

22. **Figure 1.94** shows a map of Gridville when it had only five houses. Apply what you learned in Linear Village. Find the locations that minimize the total distance and the maximum distance. Then, answer the question, "What is the best location to build the fire station in Gridville?" Be prepared to explain and defend the location you select.

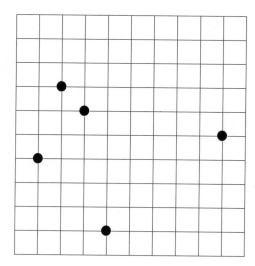

Figure 1.94.
Map of Gridville for Item 22.

LESSON FIVE

Return to Gridville

KEY CONCEPTS

Distance in
firetruck geometry

Circles in
firetruck geometry

Total and average
distance in
two dimensions

Maximum distance
in two dimensions

David Barber

PREPARATION READING

Homecoming

What is the next step in the modeling process? Do you want to investigate another criterion while you are visiting Linear Village? Do you want to return to Gridville and apply what you have learned about finding locations that minimize the total distance or the maximum distance? Are two criteria enough to propose a "fair" solution to the problem of finding the best location for the fire station?

In previous lessons, you used absolute value to calculate distance in one dimension. You discovered that the median location or median interval in Linear Village minimizes both the average and total distance a firetruck must travel. You found that the midrange location minimizes the maximum distance a firetruck would have to travel. You created simple rules and procedures for finding both locations in any size linear village.

Time is running out. Gridville is expecting your recommendation on the best place to build the fire station. In this lesson you will extend what you have learned in the one-dimensional Linear Village to the two-dimensional world of Gridville and will develop procedures for finding locations that minimize the total distance and maximum distance in Gridville. Perhaps, then, you can answer the question, "What is the best location for the fire station in Gridville?"

Remember the assumptions you made about Gridville when you began this unit. Gridville is not a real place. It is a mathematical model representing key aspects of the problem in the real world. It is also a place where circles do not look round. Incredible!

Welcome back to Gridville!

ACTIVITY

6

OPTIMIZATION STATION

It is time to return to Gridville and apply what you learned in Linear Village. Where should this Gridville (see **Figure 1.95**) build its new fire station? Choose the location you think is best. Prepare to present your selection to the class. Be ready to defend your selection using logical reasoning, tables, graphs, and/or equations.

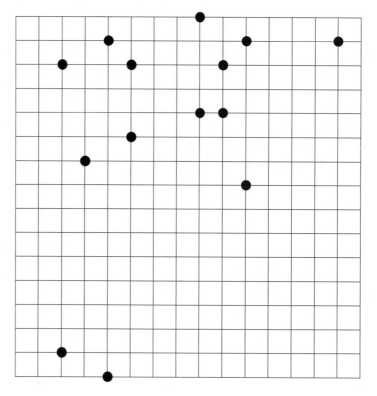

Figure 1.95.
New map of Gridville.

ACTIVITY

MAKING THE RULES

7

If you build the fire station in the median interval or at the median location in Linear Village, then you minimize the total firetruck distance. How do you find the location in Gridville that minimizes total distance? Is there a "median location" in two dimensions?

Find the location or locations that minimize total distance for each map of Gridville in **Figures 1.96–1.105.** Look for patterns as you investigate maps with two houses, three houses, four houses, and more. When you find a location that you believe minimizes the total distance, challenge another person or group to find another location and prove you wrong.

Develop a rule for finding the location that minimizes total distance in Gridville. Explain your rule and prepare to present your rule to the class.

Don't forget that travel is more complicated in Gridville than it is in Linear Village. Firetruck drivers have a choice of routes to take to get to a house. Houses are located at the intersection of grid lines, not inbetween. Locations are identified by two coordinates instead of one. Distance is measured in horizontal and vertical directions.

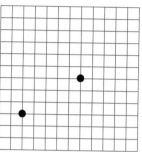

Figure 1.96.

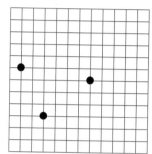

Figure 1.97.

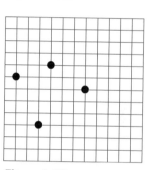

Figure 1.98.

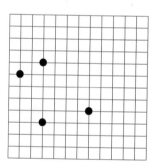

Figure 1.99.

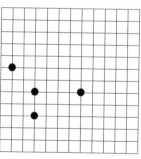

Figure 1.100.

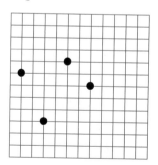

Figure 1.101.

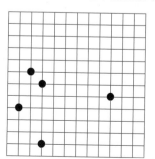

Figure 1.102

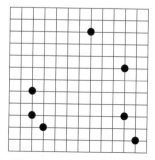

Figure 1.103.

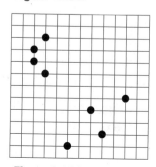

Figure 1.104.

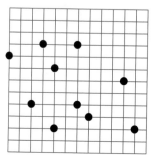

Figure 1.105.

INDIVIDUAL WORK 6

According to the Rules

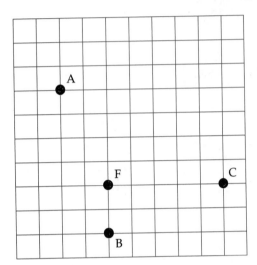

Figure 1.106.
Gridville for Item 1.

1. Suppose there are three houses in Gridville. A city employee claims the fire station F minimizes total distance (**Figure 1.106**). Verify that F is the location that makes the total distance to all three houses smallest. Explain your reasoning.

2. The planning department for Lattice County is proposing three different locations for the new regional mall. The grid lines on the map shown in **Figure 1.107** represent the major connecting roads and highways in Lattice County. The dark dots represent the six communities in Lattice County. Each community has about the same number of residents and wants to be located close to the mall because of the economic benefits, including jobs, revenue from sales tax, and increase in property values.

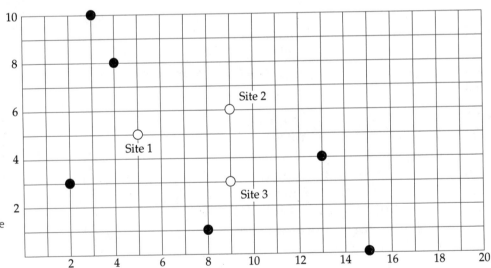

Figure 1.107.
Lattice County with three proposed sites for the new regional mall.

a) Determine the total distance and maximum distance from each proposed site to the six communities.

b) If minimizing total distance is the criterion, then which of the three proposed sites is best?

c) If minimizing the maximum distance is the criterion, then which of the three proposed sites is best?

d) Explain to the planning department how to find a location that produces the lowest total distance.

e) Explain to the planning department how to find the location that produces the lowest maximum distance.

3. Apply your rule or procedure from Activity 6 or Activity 7 and find the median locations for each map of Gridville in Figures 1.108–1.110. Describe each median location or copy each figure on your own paper and mark each median location with the letter "M."

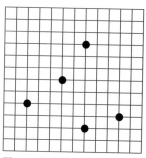

Figure 1.108.
Map of Gridville for Item 3(a).

a) Map A (see **Figure 1.108**)

b) Map B (see **Figure 1.109**)

c) Map C (see **Figure 1.110**)

4. Given the map of Gridville in **Figure 1.111**, describe what happens to the median location(s) when you make the following changes.

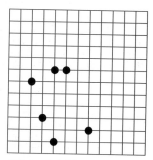

Figure 1.109.
Map of Gridville for Item 3(b).

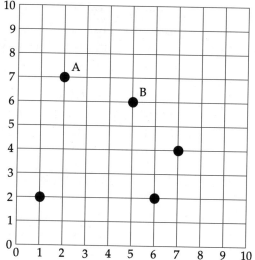

Figure 1.111.
Gridville for Item 3.

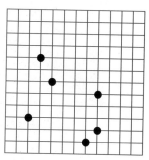

Figure 1.110.
Map of Gridville for Item 3(c).

a) Move house A to (2, 5).

b) Keep house A at (2, 5) and move house B to (4, 5).

c) Keep house A at (2, 5), house B at (4, 5) and add a new house at (6, 6).

d) Keep house A at (2, 5), house B at (4, 5), and the new house at (6, 6), and add two more new houses, one at (1, 4) and the other at (8, 4).

5. There seems to be a systematic way of finding locations where the total distance to all the houses is smallest: take the median location or interval for the *x*-coordinates and the median location or interval for the *y*-coordinates and use only the intersections of grid lines defined by the intervals. Here is a way to verify that this must be true in the map of Gridville in **Figure 1.112**.

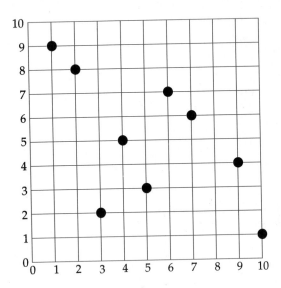

Figure 1.112.
Gridville for Item 5.

a) Suppose you suspect that (4, 6) is the location that minimizes total distance and average distance. If you move one unit to the right of (4, 6), how many distances to houses increase? How many decrease? Explain how you can be sure that (4, 6) is not the location that minimizes total and average distance.

b) Use similar reasoning and explain how you can be sure that (5, 7) is not the location that minimizes the total distance.

c) A median location is (5, 5). If you start at any point other than (5, 5), explain why moving closer to (5, 5) reduces the total distance. How does this show that (5, 5) must be a location that minimizes the total distance?

d) Using **Figure 1.113**, explain why some maps of Gridville have more than one location that will minimize the total distance.

e) Since some Gridville maps have more than one location that minimizes the total distance, how can you tell how many locations produce the same minimum total distance?

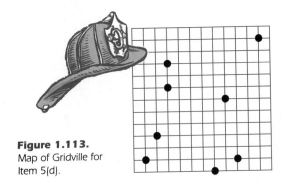

Figure 1.113.
Map of Gridville for Item 5(d).

6. a) Develop a formula for finding the distance between any two points (x_1, y_1) and (x_2, y_2) in Gridville (along the roads, of course).

b) Explain the difficulty in answering part (a) if you were allowed to build houses at locations other than the intersection of grid lines.

7. a) Copy the grid in **Figure 1.114** on your own paper and mark all locations whose firetruck distance from (5, 4) is 3 units. Explain why this figure is called a "firetruck circle."

   b) Make a copy of **Figure 1.115** on your own paper and mark all locations whose firetruck distance from (5, 7) is 2. Mark all locations whose firetruck distance from (5, 2) is 5. Describe all locations that are exactly 2 units from (5, 7) and 5 units from (5, 2).

8. Victoria is the distribution manager for the Gridville Gazette newspaper. She arranges to have the bundles of newspapers dropped at a location that is central for all the delivery people. Copy the map of Gridville on your own paper (see **Figure 1.116**). Find the centers of all smallest firetruck circles that contains the points representing the homes of delivery people. Draw or describe the answer. Is there more than one smallest circle and one center for a smallest circle that contains all the points? Describe the method you used to find the smallest circle and its center.

Note: A house is "contained" in the circle if it lies on the boundary of the circle or inside of the circle. Remember, firetruck circles are not round!

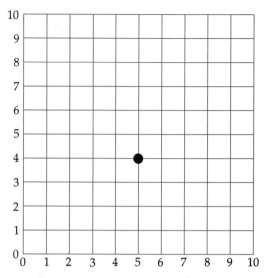

Figure 1.114.
Grid for Item 7(a).

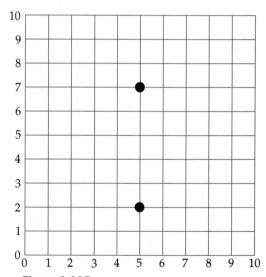

Figure 1.115.
Grid for Item 7(b).

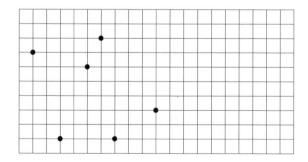

Figure 1.116.
Homes of Gridville Gazette delivery people.

ACTIVITY

8

MINIMAX IN GRIDVILLE

The purpose of this activity is to develop a rule or method for finding the location that minimizes the maximum distance in two dimensions.

1. For each diagram (see **Figures 1.117–1.120**), find the location that minimizes the maximum distance from the fire station to any house. Determine a rule or method for finding the mini-max location in two dimensions.

2. Compare your results with those of another person or group. Find someone or some group with whom you disagree. Convince them that your selected point gives a smaller maximum distance than their point. Compare your methods and determine who has a more efficient or accurate method or rule.

3. When you find another group who agree that your group has accurately found a minimax location, continue on to the next figure. Repeat the process described in (1) and (2) above for each of the four figures.

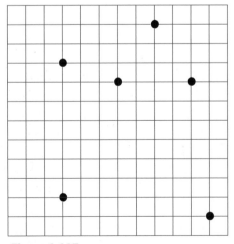

Figure 1.117.

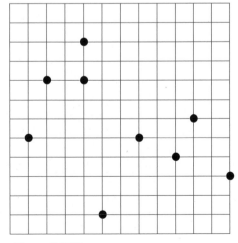

Figure 1.118.

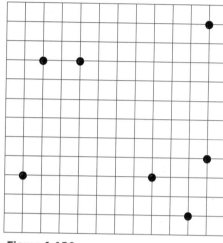

ACTIVITY

MINIMAX IN GRIDVILLE

8

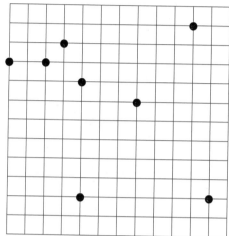

Figure 1.119.

Figure 1.120.

4. Can a group of houses (or points) have more than one mini-max location? Defend your answer. If your answer is "yes," then find an example that shows it is possible for a group of points to have more than one minimax. If your answer is "no," then provide a logical argument explaining why you believe it is impossible to have more than one minimax.

The "King Kamehameha" Mystery Persists

According to legend, the Honolulu Hawaii Fire Department ordered their first piece of fire apparatus around 1840 or 1850. It was a hand-drawn pumper to be dubbed the "King Kamehameha." But the pumper never made it to the shores of Hawaii. It disappeared, thought to be stolen from the docks in San Francisco. Rumors persist to this day that the missing pumper is housed in a museum somewhere in California. Hawaiians would like to solve the mystery and bring the King home, so the search continues.

Source: *Fire Museum Network*

INDIVIDUAL WORK 7

Median v. Minimax

1. Find the minimax location for each map of Gridville (**Figures 1.121 and 1.122**).

   a) Find the minimax location for Figure 1.121.

   b) Find the minimax location for Figure 1.122.

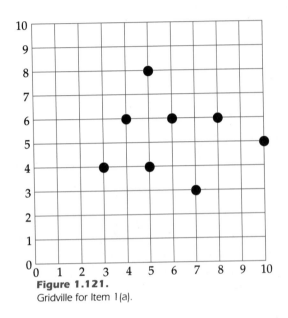

Figure 1.121.
Gridville for Item 1(a).

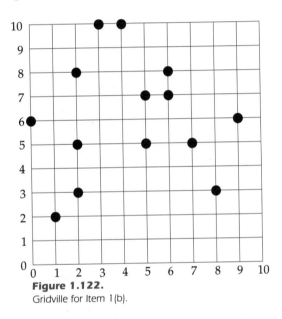

Figure 1.122.
Gridville for Item 1(b).

2. What happens to the minimax location in Figure 1.122 if you:

   a) Remove the house located at (4, 10)?

   b) Add a house at location (10, 9)?

3. Show that there is more than one minimax location in **Figure 1.123**.

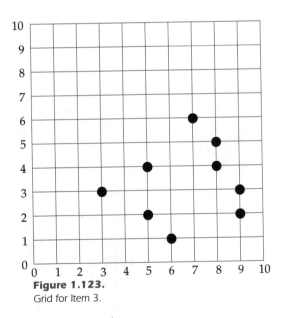

Figure 1.123.
Grid for Item 3.

4. Here is a way to search for minimax in two dimensions. Look at the map in **Figure 1.124**.

   Move line L1 in the direction indicated until it first touches a house location. Repeat for L2, L3, and L4. In this way you "box in" all the houses. Find the shortest firetruck distance from L1 to L3 and L2 to L4. The larger of these two distances determines the "diameter" of the bounding circle. Half of that distance is the "radius" of the bounding circle. (But remember that distances in Gridville are always integers, so you may need to round up after taking half of the larger distance.) Now sketch lines that are this distance inside each of the bounding lines. Every grid point inside or on the resulting figure is a minimax point. Use this method to find the minimax location for the map in Figure 1.124.

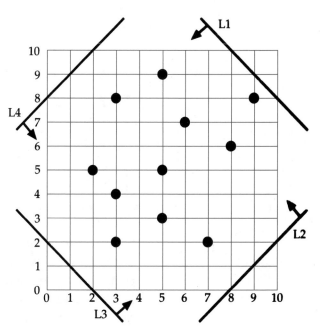

Figure 1.124.
Parallel lines approach for finding minimax locations.

5. Use the method of parallel lines described in Item 4 and identify the radius of the smallest firetruck circle for the map of Gridville in Item 3 (Figure 1.123). Compare your answer using the method of parallel lines with your answer using the firetruck circles.

6. **Figure 1.125** shows a map of Gridville with five houses.

   a) Write the coordinates for the minimax location(s). Describe the method you used.

   b) Write the coordinates for the median location(s). Describe the method you used.

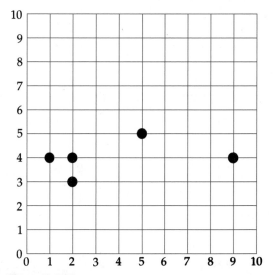

Figure 1.125.
Gridville for Item 6.

   c) Where could you add a sixth house and seventh house so the minimax location(s) and median locations remain the same? Justify your answer.

INDIVIDUAL WORK 7

7. a) Find the locations that minimize the total distance for each Gridville map in **Figures 1.26–1.29**. Copy each figure and label each median location by writing the letter "M" on the map.

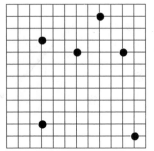

Figure 1.126.

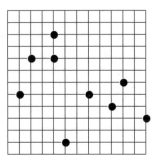

Figure 1.127.

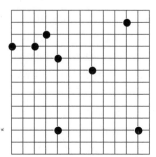

Figure 1.128.

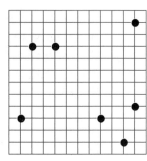

Figure 1.129.

b) Compare the minimax location with the median location for each of the four Gridville maps. For which maps does the median location seem to be the most fair? For which maps does the minimax location seem most fair? Explain your criteria for fairness and defend your choice based on your fairness criteria.

8. Develop a spreadsheet, write a calculator program, write a computer program, or create a sketchpad script to find locations that minimize total distance for the fire station in any size Gridville. Use any suitable method as the basis of your work.

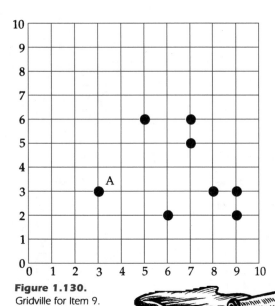

Figure 1.130.
Gridville for Item 9.

9. In the map of Gridville in **Figure 1.130**, A is an apartment house with six units. The other houses in Gridville are single dwellings. Assuming that a fire at A is six times as likely as at any of the other houses, where should you build the fire station? Explain your reasoning and defend your choice.

10. For each of the following situations, create your own map of Gridville with five or six houses.

    a) The median location is a single point.

    b) The median locations are the grid points along a strip or segment.

    c) The median locations form the corners of a unit square.

    d) The median locations are grid points along the sides of, and in the interior of, a rectangle.

11. Draw your own map of Gridville with five houses. Place the houses so the minimax location and median location are the same.

12. Real estate agents find that houses are more valuable if they are situated close to schools and certain services. Suppose one buyer wants to buy a house that is within (less than) 4 blocks of the school S and within (less than) 5 blocks of the fire station F. Copy **Figure 1.131**, the map of Gridville, and mark all locations that enjoy both of these advantages.

13. a) Copy **Figure 1.132** and draw a firetruck circle of radius 3 passing through the points A and B. Write the coordinates of the center.

    b) Find a location that is the same distance from A, B, and C. Use the location as the center and draw a firetruck circle that passes through A and B and C. (Do not connect the points on the firetruck circle.)

14. a) Explain why there is no firetruck circle that passes through B, C, and D in Figure 1.132.

    b) In Gridville, houses are located at the intersection of grid lines. Suppose you temporarily remove that requirement and allow houses to be built along the grid lines between the intersections, or you add more streets halfway between the current streets. Now, can you construct a firetruck circle that passes through C and D? Explain your answer.

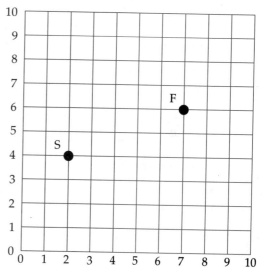

Figure 1.131.
School and fire station.

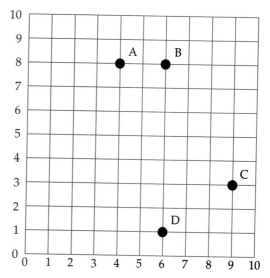

Figure 1.132.
Grid for Item 13(a) and 13(b).

15. Explain how to use firetruck circles to find the minimax location in two dimensions.

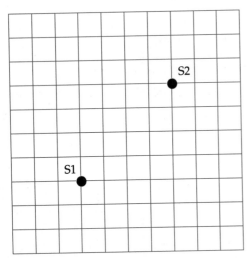

Figure 1.133.
Map of Gridville with two schools.

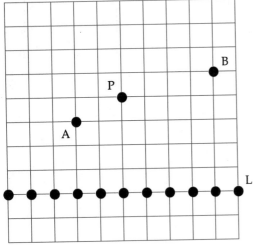

Figure 1.134.
Map of Gridville for Item 17.

16. In Individual Work 1, Item 13(b), you marked all the points that formed the boundary line between two schools. Students attend the school closest to their homes, in firetruck distance. Suppose the two schools are located at opposite corners of a square. Draw the map in **Figure 1.133** on your own paper and mark on the grid all the points that are the same distance from S1 and S2.

17. In the grid in **Figure 1.134**, point A is at the same firetruck distance from point P as it is from the nearest point that lies along a line L. (Both distances are 3.) Point B is another point at the same firetruck distance from P as it is from the nearest point along line L. (Both distances are 5.)

   a) Why is it reasonable to say that the distance from point B to line L is 5?

   b) Mark at least 9 other points on the grid that are the same distance from the point P as they are from line L.

18. a) Find the firetruck perimeter for circles with radius 2, 3, and 5. In firetruck distance, the perimeter of a figure is the length of a shortest path that begins at one point on the figure, passes through all the points on the figure, and returns to the starting point, traveling only along grid lines.

   b) The radius of each circle is the distance the set of points are from P, the center. The circumference of each circle is the firetruck perimeter. Refer to your answers to part (a) and describe the relationship between the radius and circumference of a firetruck circle.

19. The town of Gridville decides to build two fire stations. Using **Figure 1.135,** where do you think they should be located? Explain.

20. Gridville decides to build two fire stations. Using **Figure 1.136,** answer parts (a) and (b).

    a) Where should the two fire stations be placed to minimize the maximum distance to the closer fire station?

    b) Where should the two fire stations be placed to minimize the total distance?

21. The Grid River flows through Gridville, separating the town into eastern and western sections (**Figure 1.137**). There is only one bridge across the river, and the town can only afford one fire station. Where should they build the fire station?

22. Mark a grid on a square piece of cardboard. Glue five objects (all the same) at different locations on the grid. Determine the median and minimax locations for the objects. Find the balance point by placing your finger under the cardboard until the "village" balances. Draw your map of Gridville on a grid and mark the locations for the median, minimax and balance points. What is the "best" location for the fire station: median, minimax, or balance point? Explain your reasoning.

23. For any size Gridville, how do you determine where to build two fire stations if you want to minimize maximum distance? Explain your answer.

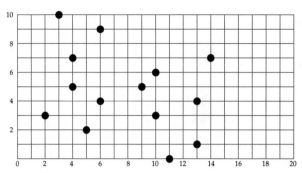

Figure 1.135.
Gridville for Item 19.

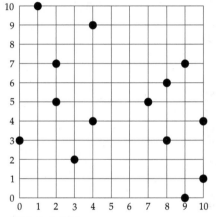

Figure 1.136.
Map of Gridville for planning two fire stations.

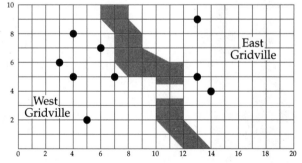

Figure 1.137.
Gridville for Item 21.

Wrapping Up Unit One

1. a) House A is located at 3. Write an absolute-value equation to represent fire station locations x that are a distance of 4 units from A.

   b) House B is located at –2. Write an absolute-value equation to represent fire station locations x that are within 5 units of B.

2. Linear Village has two pizza places. Dumbo's is located at 1 and Jumbo's is located at 20. Dumbo's will deliver up to a distance of 12 blocks. Jumbo's will deliver up to a distance of 15 blocks. Describe all locations in Linear Village where you can have pizza delivered from both. Write inequalities to describe the locations.

3. Slopeville is a linear village on a hillside. Going uphill takes the firetruck twice as long as going downhill. (**Figure 1.138.**)

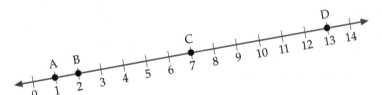

Figure 1.138.
Slopeville

a) Where do you build the fire station to minimize the average time it takes the firetruck to reach each house?

b) Where do you build the fire station to minimize the maximum time it takes the firetruck to reach each house?

4. Slopeville has grown (see **Figure 1.139**). Remember that the firetruck takes twice as long to travel uphill as it does to travel the same distance downhill. Without computing, give a convincing logical argument to show that the average time to get to a fire is smallest at position G.

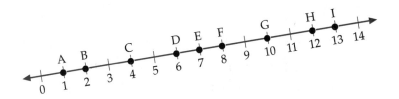

Figure 1.139.
Slopeville revisited.

5. Cross Corners is an interesting town. It has only two streets that meet at right angles in the center of town. From an airplane the town looks like a giant "X." Horizontal Highway runs east and west. Vertical Boulevard runs north and south. The center of town at coordinates (0, 0) is the intersection of Horizontal Highway and Vertical Boulevard. There are 11 houses in Cross Corners. The houses are located at: (–5, 0), (1, 0), (2, 0), (3, 0), (6, 0), (8, 0), (0, –3), (0, –2), (0, 1), (0, 2) and (0, 4).

a) Where should the fire station be built to minimize total distance? Explain your reasoning.

b) Where should the fire station be built to minimize the maximum distance? Explain your reasoning.

6. A helicopter does not follow the path of a grid. Compare the helicopter distance between house A at (3, 5) and the fire station F at (7, 8) to the firetruck distance.

a) What is the firetruck distance from F to A?

b) Estimate the helicopter distance from F to A. Discuss how you might find the exact helicopter distance from F to A.

c) When, if ever, is helicopter distance the same as firetruck distance? Explain.

d) Is it possible for firetruck distance between two locations to be twice the helicopter distance between two points? Explain your answer.

7. Explain the meaning of each of the following in the context of Linear Village or Gridville.

   a) $y = |x - 5|$.

   b) $y = |x + 1| + |x - 3|$.

   c) $y = |x - 8| + 3$.

   d) $y = 2|x - 4| + |x + 2|$.

8. Danielle is a candidate for mayor in her small town. She has strong views about a particular issue. She realizes that her personal views are not favorable to a majority of the community so she adjusts her stated opinions so she will appear as a more favorable candidate to a greater number of voters.

   Compare the politics of running for office to finding the best location for the fire station in Gridville. Use ideas from this unit to illustrate the similarities.

9. a) In Linear Village, you have solved the problems of finding firetruck locations that minimize both the total distance and the maximum distance between it and the village's houses. State as clearly as possible how you would solve these two kinds of problems.

   b) In your search for solutions to the two kinds of problems mentioned in part (a), you have used tables, graphs, equations, and arrow diagrams to model relationships in Linear Village. Explain how to use at least two of these representations to verify the locations that minimize the total distance in Linear Village.

   c) Which of the various representations that you have used and studied do you find most helpful in understanding why your answer for minimum total distance in part (a) is correct? Explain.

   d) Explain how to use at least two of these representations to verify locations that minimize the maximum distance in Linear Village.

   e) Which of the various representations that you have used and studied do you find most helpful in understanding why your answer for minimum maximum distance in part (a) is correct? Explain.

Mathematical Summary

Mathematics is useful when you investigate criteria involving quantities such as distance. In this unit you use multiple representations to understand and apply distance relationships in one and two dimensions.

One candidate for the best location for the fire station is the location that minimizes the total or average distance from the fire station to each house in the village.

Another candidate for best location for the fire station is the location that minimizes the maximum distance from the fire station to any house in the village.

Both criteria focus on an aspect of distance.

Firetruck geometry involves looking at distance in two dimensions in a different way. The firetruck distance between two points is the length along grid lines of a shortest path between them. The total distance associated with a fire station location is the sum of all the lengths of the shortest paths from the fire station to all of the houses. The average distance is equal to total distance divided by the number of houses.

Simplified models of real towns aid the study of important properties. Simplifications may be made in the number of buildings to examine and in the number of dimensions in which the town is laid out. Distance on the number line serves as a simplified model for studying distance in the plane.

The absolute value of a number is defined as the distance, on the number line, between the given number and 0, and is written as $|x|$. The distance between the numbers A and B on the number line is $|A - B|$, which is also $|B - A|$.

The absolute-value function $y = |x|$ is piecewise linear, with a graph that forms a characteristic "V" shape. When $x \geq 0$, $y = |x|$ is $y = x$. When $x < 0$, then $y = |x|$ is $y = -x$.

The graph of the absolute-value function $y = |x|$ can be transformed by changing the control numbers a, h, and k in the equation $y = a|x - h| + k$. Transformations of the function $y = |x|$ are similar to transformations of other functions you have studied in earlier units.

The control number **h** translates the graph to the right h units. If h < 0, then moving "to the right h" is a negative move, so it's really a move to the left.

The control number **a** stretches the graph of $y = |x|$ in the y-direction if a > 1 or a < –1. When a < 0 the graph is also reflected across the x-axis. If –1 < a < 1, the result is a shrink in the y-direction.

The control number **k** translates the graph up k units. If k < 0, then the "negative move" is really a downward move.

When distance is measured along a line, the general equation $|x - 3| = y$ means the "distance between 3 and x is y units." The specific equation $|x - 3| = 4$ means "the distance between 3 and x is 4 units." Since the numbers –1 and 7 are each 4 units from point 3, they are the values of x that make the equation true (see **Figure 1.140**). Therefore, we say $x = -1$ and $x = 7$ are solutions to the equation $|x - 3| = 4$.

Figure 1.140.
Number line solution to an absolute-value equation.

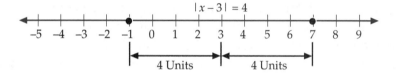

Absolute-value inequalities represent intervals on a number line.

$|X - A| < d$ means all values X within d units of A. See **Figure 1.141**.

"All temperatures T within 5°F of 80°F" is written: $|T - 80| < 5$.

"All scores within 15 points of 78" is written: $|S - 78| < 15$.

Figure 1.141.
Values within "d" units of "A."

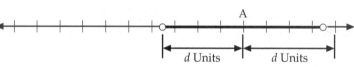

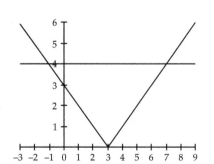

Figure 1.142.
A two-dimensional graph to solve an absolute-value equation.

The equation $4 = |x - 3|$ may be solved using symbolic methods based on using arrow diagrams to reverse the effect of the absolute value, producing two equations to solve: $x - 3 = 4$ and $x - 3 = -4$.

An absolute value equation such as $4 = |x - 3|$ may also be solved using the graph of $y = |x - 3|$ and piecewise equations. (See **Figure 1.142**) The solutions are the x-coordinates of all points (x, y) where $y = 4$. The corresponding piecewise equations are $4 = x - 3$ and $4 = -x + 3$, whose solutions give the two points on the graph where $y = 4$, namely (–1, 4) and (7, 4).

When the number of houses in Linear Village is more than one, total distance may be represented using the sum of absolute-value expressions.

The graph of the total distance function for any linear village has a very distinctive pattern of slopes. For locations to the left of the left-most house, the slope is the negative of the number of houses. Moving along the graph from left to right, for each house the graph reaches, the slope increases (becomes more positive) by exactly 2. To the right of the right-most house, the slope is equal to the number of houses. In fact, if you write the total distance function using its piecewise-defined equation, the first and last pieces will always have exactly opposite equations.

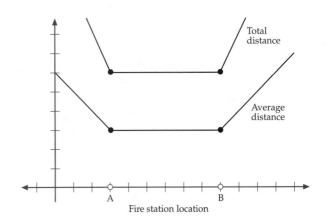

The graph of the average distance function is similar to total distance. For locations to the left of the left-most house, the slope is –1. To the right of the right-most house, the slope is +1. Locations that minimize total distance also minimize average distance. See **Figure 1.143**.

Figure 1.143.
Graph of total distance and average distance for a two-house linear village.

In general, locations that minimize total distance in a linear village may be found using the median rule. If the number of houses is odd, build the fire station at the median; if the number of houses is even, build the fire station anywhere in the median interval (see **Figure 1.144**). This rule extends directly to two-dimensional Gridvilles.

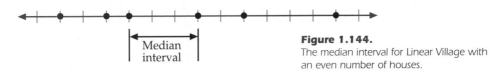

Figure 1.144.
The median interval for Linear Village with an even number of houses.

Inequalities such as $A \leq x \leq B$ may be used to identify a median interval.

Minimizing the maximum distance is another criterion for optimizing fire station location. The minimax location is the midrange for the addresses in a linear village. Although a bit more difficult to implement in two dimensions, solutions based on the midrange may be determined in Gridvilles using sets of parallel lines or sets of concentric firetruck circles.

Fire station locations that minimize total distance are not sensitive to changes in locations of houses "near the edges," but changes near the central houses may affect the optimal location.

The minimax location is completely insensitive to changes in house locations between the highest and lowest locations. However, the minimax location can change considerably if the house with the highest or lowest address changes; the minimax location is sensitive to outliers.

The two-dimensional models studied required that buildings be located at the intersections of grid lines. Locations and distance in Gridville are discrete.

The set of all houses that are the same distance from a given fire station form a firetruck circle. **Figure 1.145** shows a firetruck circle with center, or fire station, at P and a "radius" of 3 units. Each house on the "circle" is a distance of 3 units from P. The points on the "circle" are not connected. The points of the "circle" form a diamond-shaped pattern.

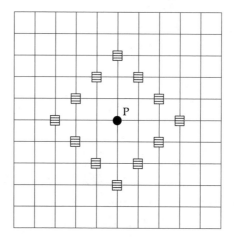

Figure 1.145.
Firetruck circle with a radius of 3 units.

Using the mathematics of medians, midrange, and firetruck circles allows you to optimize (minimize) total distance or maximum distance in one and two dimensions. The location that minimizes total distance may favor some houses and not others. The location that minimizes maximum distance may favor still other houses. Mathematics does not determine fairness. The criteria for fairness will be decided by community values. However, once criteria for judging fairness are established, mathematics provides tools to help find or verify locations that meet the criteria. Mathematics also permits "trying out" criteria before selecting one.

Glossary

ABSOLUTE VALUE:
The non-negative number representing the distance between a given number and zero on the number line. $|A - B|$ represents the distance between A and B.

AVERAGE DISTANCE:
The total distance divided by the number of houses.

DISTANCE:
Between a house and a fire station in a linear village, the non-negative number representing the difference in their coordinates on a number line.

FIRETRUCK CIRCLE:
The set of locations that are the same grid-line distance from a given location.

FIRETRUCK DISTANCE:
The shortest path between two points.

HELICOPTER DISTANCE:
The direct distance between two points.

MEDIAN INTERVAL (ONE DIMENSION):
With an even number of houses, the locations between (and including) the two middle houses in a linear village. The addresses must be listed in increasing or decreasing order.

MEDIAN LOCATION (ONE DIMENSION):
With an odd number of houses, the location representing the "middle" house in a linear village. Addresses must be listed in numerical order.

MEDIAN LOCATION (TWO DIMENSIONS):
All locations (at the intersection of grid lines) whose x-coordinates represent the median (or median interval) for the x-coordinates of all houses in Gridville and whose y-coordinates represent the median (or median interval) for the y-coordinates of all houses in Gridville.

MIDRANGE:
The number or address that is halfway between the highest or lowest coordinate or address in Linear Village. The midrange location is found by taking the average of the highest and lowest coordinates or addresses.

MINIMAX LOCATION (ONE DIMENSION):
The minimax location minimizes the maximum distance a firetruck would travel to any house in Linear Village. It is found at the midrange location.

MINIMAX LOCATION
(TWO DIMENSIONS):
All locations (at the intersections of grid lines) that minimize the maximum distance to any house in Gridville.

PIECEWISE-DEFINED FUNCTION:
A combination of two or more functions, each defined over specific (but different) values of x, joined together to form one function. The function defined: $y = x - 3$ for $x \geq 3$ combined with $y = -x + 3$ for $x < 3$ is a piecewise-defined function.

POINT:
The place on a number line representing a house in a linear village.

ROUND-TRIP DISTANCE:
Twice the distance between a house and the fire station.

SLOPE:
The slope of a line is the ratio of the difference in y-coordinates divided by the difference in x-coordinates between two points on the line. If (x_1, y_1) and (x_2, y_2) are points on the line, the slope of the line is
$$\frac{y_2 - y_1}{x_2 - x_1}.$$

TOTAL DISTANCE:
The sum of the one-way distances between the fire station and each of the houses.

TOTAL ROUND-TRIP DISTANCE:
Twice the total distance.

UNIT

Strategies

People have been making decisions throughout history. This unit examines one kind of decision making and introduces an important (and fairly new) area of mathematical thought — game theory. First developed during the 1940s by mathematician John von Neumann and economist Oscar Morganstern, the study of game theory has already led to at least one Nobel prize. Although theoretical work can get complicated, all you need to understand in order to get started in this fascinating subject are a few concepts from algebra, such as matrices and systems of linear equations, and some elementary probability. So dig in and enjoy this look at one of the important ideas of the 20th century!

The Image

EVERYDAY STRATEGIES

Strategic situations occur all around us every day. They involve ordinary people, businesses, and even governments. Often, you don't even characterize them as strategic situations, and you certainly don't think about using mathematics to analyze them. However, after working through this unit, you might find yourself approaching such situations quite differently.

Are you a bit skeptical about the previous statement? Here's a test.

Suppose that you are on a crowded train and you think that you saw your friend board a compartment further down from yours. As it turns out, your friend thinks that he caught a glimpse of you. Is it worth giving up your seat to try and meet your friend? If it is, should both of you get up from your seats and go look for each other? Should just one of you get up, and if so which one? How do you decide?

Can you identify the mathematics in the train situation? After you have worked through part of this unit, you should return to this introduction and think about how your interpretation of the situation described above has changed.

LESSON ONE
Decisions

KEY CONCEPTS

**Strategic situations
(games)**

**Elements of a game:
players, strategies,
payoffs (or rankings)**

Opposing interests

Payoff matrix

The Image Bank

PREPARATION READING

Endless Decisions

Decisions, decisions. You make many of them every day. Should you take your umbrella when you go for a walk? Should you skip putting a quarter in a parking meter and risk getting a ticket? In a game of Monopoly™, should you buy Boardwalk when you land on it? Should you skip your homework tonight and risk a low grade on a possible quiz tomorrow?

Making decisions is one aspect of life that we all have in common. Many decisions involve another person whose interests conflict with yours. For example, if you don't buy Boardwalk, another player may do so and bankrupt you with rental charges. If you don't put a quarter in the parking meter, the police department may give you a $25 ticket.

There are many situations in which you must choose from one of several courses of action while one or more other parties are trying to do the same. In this lesson you will learn how mathematics can help you organize the key features of the situation. Later, you will discover that mathematics can help you select the best strategy.

In your own life, what are some situations in which you could benefit by choosing the right strategy?

INDIVIDUAL WORK 1

Conflicts, Crises, and Games

Game theory can be viewed as a tool for analyzing strategic situations. A **strategic situation** exists whenever you must make a decision that affects one or more other parties who must also make decisions that affect you. The parties involved in a strategic situation are called **players** because strategic situations often resemble games. In fact, strategic situations are commonly called **games**. The actions the players can choose to take are called their **strategies**. Here is one example of a strategic situation.

SCENARIO #1:
THE ZAIRIAN CONFLICT OF 1997

In 1997, Zaire was involved in a conflict. At one point, rebel troops had captured three-quarters of Zaire and were advancing closer to the capital, Kinshasa. The rebel leader, Laurent Kabila, demanded that President Mobutu Sese Seko step down and leave the country. Otherwise, Kabila declared, his troops would take the capital by force and capture President Mobutu. The president proclaimed he would die in office rather than live in exile and demanded that the rebels cease their hostilities. The international community, aided by South African President Nelson Mandela, began pressuring the two leaders to meet and negotiate a peaceful resolution to the conflict.

Certainly, the Zairian conflict is more complicated than the description in the preceding paragraph. However, in beginning to model any situation, it is generally necessary to simplify the real world situation by identifying and isolating the most influential factors.

1. Who do you think are the players in the Zairian conflict? What are their choices or strategies? What do you think might be the consequences of their strategies?

The mathematician John von Neumann is credited as the founder of game theory.

INDIVIDUAL WORK 1

SCENARIO #2:
SIMPLE MATCHING GAME

A simple matching game is also a good example of a strategic situation. It requires that one player be the even player and one player be the odd player. Each player has a choice of two strategies shown in **Figure 2.1**.

Figure 2.1.
Strategies for matching game.

Strategy #1: show one finger.   **Strategy #2:** show two fingers.

The even player wins a point from the odd player if the sum is even; the odd player wins a point from the even player if the sum is odd. If a player wins a point, the **payoff** to that player is 1; if the player loses a point, the payoff to that player is –1.

Mathematics is helpful in organizing the important features of a strategic situation. You can organize the players, their strategies, and the consequences of their strategies, which are sometimes called **payoffs,** in a **payoff matrix**, as shown in **Figure 2.2**.

Figure 2.2.
Payoff matrix for simple matching game.

	Column player	
Row player	**One finger**	**Two fingers**
One finger	1	–1
Two fingers	–1	1

The even player has been arbitrarily assigned to the rows of the matrix. It is customary to write the payoffs to the player associated with the rows of the matrix and to call this player the row player. Sometimes, the payoffs to each player are listed as an ordered pair in the matrix. In the case of the simple matching game, the ordered pairs should be in the form of (row player's payoff, column player's payoff). A partial payoff matrix written in this form appears in **Figure 2.3**.

Column player

		One finger	Two fingers
Row player	One finger	(1, −1)	(−1,)
	Two fingers	(−1,)	(1,)

Figure 2.3.
Payoff matrix for simple matching game.

2. What are the missing payoffs to the column player in the matrix shown in Figure 2.3?

Quite often players have **opposing interests**.

3. What do you think it means to say that the row player and column player in the simple matching game have opposing interests?

4. The players in a simple matching game may play it more than once. Do you think the game in Item 1 will be played more than once? Explain.

SCENARIO #3:
THE CUBAN MISSILE CRISIS OF 1962

An example of a strategic situation in which the interests of the players were not completely opposed is the Cuban Missile Crisis of 1962. The U.S. discovered that the USSR was establishing a missile base in Cuba, and President Kennedy ordered a naval blockade of the island. The USSR had to decide whether to challenge the blockade or back down. The U.S. would then have to decide how to respond to the Soviet strategy.

It can be very hard to attach payoffs to the outcomes of a situation like the Cuban Missile Crisis. An alternative to attaching payoffs is to rank the outcomes. The United States, for example, felt that escalation on its part (continuing the blockade) and withdrawal of the missiles (backing down) on the part of the Soviets was the best outcome, and ranked it first. Both sides probably ranked joint escalation last. Suppose you decide to use the following ranking convention: rank the outcomes from 1–4 with 1 as the best outcome and 4 as the worst outcome. Using this convention, **Figure 2.4** shows possible ranks from the row player's (U.S.'s) perspective.

USSR

		Escalate	Back down
U. S.	Blockade	(4,)	(1,)
	Back down	(3,)	(2,)

Figure 2.4.
Payoff matrix with rankings.

5. Complete the payoff matrix in Figure 2.4 by adding the column (USSR) player's rankings. Justify your answer.

6. Suppose, instead, you wanted to use the following ranking convention: rank the outcomes from 0 to 3 with 0 the worst outcome and 3 the best. (Using this convention, more preferable outcomes are associated with higher rankings.)

   a) What is your matrix for the Cuban Missile Crisis using this convention?

   b) How are the entries in this matrix related to the entries for your matrix in Item 5?

   c) Which ranking convention do you prefer, the one used in this item or the one used in Item 5? Explain.

7. a) When rankings are used instead of payoffs, what do you think it means for the players to have opposing interests?

   b) Based on your completed matrix in Item 5, do you think that the U.S. and the USSR had opposing interests? Explain.

Unlike the simple matching game that can be played repeatedly, the Cuban Missile Crisis was a situation that did not repeat. So, neither side had the benefit of previous similar situations on which to base its choice of strategy.

Furthermore, the structure of the payoff matrices for the simple matching game and the Cuban Missile Crisis differ. Look at the payoff matrix for the simple matching game (your answer to Item 2). Notice that for each set of row and column player strategies, the sum of the payoffs to the row and column players is zero. Whenever this occurs, the game is said to be a **zero-sum game**, that is, the payoffs to one player are the opposite of the payoffs to the other player. Now look at the matrix of rankings for the Cuban Missile Crisis. Clearly, this is not a zero-sum game since the sum of the rankings is always positive.

8. a) Return to the Zairian Conflict. Organize the elements of this situation into a payoff matrix. Justify your choice for payoffs (or rankings). If you choose to use rankings, be sure to specify the ranking convention.

   b) Whenever you are modeling a real-world situation, you must first decide which factors are most influential and create your model based on those factors. If later you find additional information that you think is highly influential, you can adjust your model accordingly. In the case of the Zairian conflict, here are two new pieces of information: (1) President Mobutu is dying of cancer and (2) Laurent Kabila predicts his rebel troops will reach the capital within two weeks. In light of this new information, do you want to make any changes to your payoff matrix in part (a)?

In each of the strategic situations in Scenarios #1–#3, you could identify two players. However, there are many strategic situations that do not involve two "true" players. An example follows.

SCENARIO #4:
THE UMBRELLA DECISION

Should you take your umbrella with you on a walk?

Here, there is no second player in the conventional sense. However, since rain may or may not happen, nature can be considered the second player.

9. a) How is nature like a conventional player? How is nature different?

   b) Organize the information in Scenario #4 into a matrix. Explain your choice of payoffs or rankings.

   c) Does it make sense to list the payoffs or rankings for each outcome as an ordered pair? Explain why or why not.

   d) Is the umbrella situation a one-time deal or a situation that is likely to be repeated many times?

In this assignment you have seen that information about key features of strategic situations—players, strategies, and payoffs—can be organized into a matrix. In some examples, such as the Zairian conflict, the payoffs are not clearly identified and you must choose payoffs (or rankings) based on your interpretation of the situation. Determining reasonable payoffs can, at times, be the most challenging part of key-feature identification. Were the students in your class, for example, able to agree on the payoff matrix for the Zairian conflict?

In the next activity, you will create payoff matrices for a number of different strategic situations. In each case, you should ask yourself: Is this new situation more like the simple matching game, the Cuban Missile Crisis, the Zairian conflict, or the Umbrella problem?

In each of Items 1–7, prepare a payoff matrix that shows the players, their strategies, and the payoffs (or rankings) for the row player.

If your matrix shows payoffs, be sure to label the units (dollars, for example). If you choose rankings, be sure to specify your ranking convention.

Remember that reaching agreement on payoffs or rankings is often the hardest part of the analysis of a strategic situation. The "right answer" is not always obvious, and often does not exist. (A model seldom describes reality perfectly.) Allow members of your group who have dissenting opinions to express and defend them. However, in the end, your group must reach consensus on the model (the payoff matrices) for each situation.

In addition, answer each of the following after you have finished your matrix:

(a) Is this a situation in which both players are true players, or is one of them like nature in the umbrella situation?

(b) Do the players have opposing interests, or would it be to their advantage to cooperate?

(c) Is this a one-time situation, or can the players repeat it over and over again?

1. A baseball player is about to bat against a pitcher who throws only fast balls and curve balls. If the batter guesses the correct pitch, he has a 40% chance of getting a hit; if he guesses the wrong pitch, he has only a 10% chance.

2. Sarah has $1,000 that she earned from a summer job and intends to invest toward her college education. She can place the money in a savings account or the stock market. If she invests it in the stock market and the economy does well during the coming year, her investment will gain about 20% in value, but if the economy does poorly, her investment will lose about 10% of its value. A savings account, on the other hand, will pay her about 7% interest if the economy does well, and 5% if it does poorly.

EXPLORING STRATEGIC SITUATIONS

3. You and a friend have decided it would be "cool" if you both showed up at school tomorrow with really weird haircuts. After dinner, you leave your house to fulfill your part of the bargain, but you begin to worry. It would be extremely embarrassing to show up at school with a wild haircut if your friend chickens out.

4. Maka Deel sells cars for a dealership that advertises it will reduce the price of a car by $200 if the customer presents one of its competitor's business cards before the transaction ends. Maka is close to an agreement with a customer that will earn her a $500 commission. However, her boss has said that any further reduction in price, including $200 if the customer has a business card from a competitor, will come out of the commission, and the customer is still undecided. Maka feels that if she drops the price another $100, the customer will agree, but will the customer have a business card?

5. Suppose that two competing grocery chains are each planning to open a new store in a city that has two major malls. Potential monthly profits are $20,000 at the larger mall, and $15,000 at the smaller mall. If the companies choose different malls, each store will capture all of the profits from the mall at which it is located. If they are at the same mall, suppose that Store A can expect 60% of the total profits from that mall, and Store B 40%.

6. A week into a hot, dry spell, Droughtsville has announced that there is a severe water shortage and all citizens should practice water-conserving measures. But you have been planning to clean the house, wash cars, and do several other water-intensive activities. Will it matter that much if you go ahead with your plans, or should you conserve? What if you're a water hog and so is the rest of the community?

ACTIVITY

EXPLORING STRATEGIC SITUATIONS

1

7. Here is another example of a "one-finger, two-finger" matching game. If on any play the two players show an even number of fingers, then the even player wins; otherwise the odd player wins. The winning player is awarded points equal to the total number of fingers showing; the losing player loses as many points.

8. If a game has the property that the sum of the payoffs to the row and column players is the same regardless of the row and column player's choice of strategies, then the game is a **constant-sum game**. (Zero-sum games are examples of constant-sum games because the sum of the row and column player's payoffs is always zero.) Which of the games in Items 1–7 are constant-sum games?

9. A **pure strategy** is a player's choice of play on a single game.

Pair off with another student so that you can play the game in Item 7. Within each pair, decide who will be the even player and who will be the odd player. Before playing the game, each player should mentally select a pure strategy: decide whether you will show one finger or two.

a) Play the game once using the strategy that you selected. Did the even player or the odd player win? What are the players' scores after one play?

b) If you were to play this game 9 more times using your pure strategy (that is, you can't switch from strategy to strategy in different games), what would the players' scores be after you had completed the 10 games?

c) Next, assume that you don't have to stick with a single pure strategy but are allowed to mix strategies. Play the game for several minutes or until your teacher tells you it's time to stop. What are the players' scores?

CONSIDER:

1. Do you think the game that you have just played is fair?

2. What do you think it means for a game to be fair?

3. Which students in the class were the best players? How did you decide?

4. Do you think the best players were just lucky or did they use a winning strategy? How can you find out?

INDIVIDUAL WORK 2

Show Me the Money

*I*n Course 1, Unit 7, *Imperfect Testing*, tree diagrams were useful in calculating probabilities when two factors (or characteristics) were involved. For example, suppose you wanted to know if the percentage of female students with driver's licenses at your school differed from the percentage of male students with driver's licenses. After collecting data on gender and driving status, you might organize your data in a tree diagram such as the one in **Figure 2.5**.

1. Two players each begin with a pile of 50 pennies. To play the game, each player takes either one or two pennies from her pile and hides them in her hand. Simultaneously, both players open their hands and show their pennies. If the total number of pennies is even, the even player gets to keep both of them; if the total number is odd, the odd player gets all three pennies.

   a) Write a payoff matrix for this game.

   b) Represent the same information that is in your payoff matrix using a tree diagram.

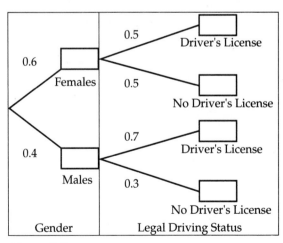

Figure 2.5.
Tree diagram of gender and driving status.

   c) Do you think this game is fair? If not, would you rather be the row or the column player? Why?

INDIVIDUAL WORK 2

d) Sasha and Joel played this game 25 times. Jan and Dana played the game 50 times. A summary of the results from their plays appears in **Figures 2.6 and 2.7**, respectively. Which of the four players do you think is the better player? Explain.

Joel

Sasha		One penny	Two pennies
	One penny	5	7
	Two pennies	9	4

Figure 2.6.
Number of times each outcome was observed in 25 plays.

Dana

Jan		One penny	Two pennies
	One penny	12	12
	Two pennies	8	18

Figure 2.7.
Number of times each outcome was observed in 50 plays.

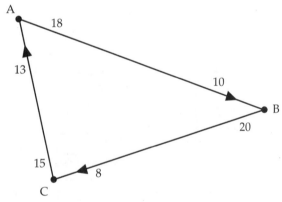

Figure 2.8.
Graph made up of edges and vertices.

In Item 1, you saw that you could represent the information contained in a payoff matrix using a tree diagram. You have used graphs made up of edges and vertices as models in other units. For example, if you used Course 1, Unit 1, *Pick a Winner: Decision Making in a Democracy,* you saw a graph like the one in **Figure 2.8**.

The arrow on the edge from A to B means that A beats B in an election. The 18 and 10 indicate that A received 18 votes and B received 10. Similarly, B beats C by 20 to 8, and C beats A by 15 to 13.

2. Suggest a way in which a graph could be used to model the important features of one of the strategic situations described in Items 1–7 of Activity 1. What useful information could you get from your graph?

In this lesson, you have learned to organize key features of a game–the players, the strategies, and the payoffs–into a matrix. As you can see, you can represent information from a payoff matrix using tree diagrams and graphs, too. Sometimes it can be helpful to have a visual representation of a game.

LESSON TWO
Changing Your Strategy

KEY CONCEPTS

Average points per play (expected payoff)

Zero-sum game

Strategy in repeated play

Random strategy

Optimal strategy

Equilibrium

Fair game

The Image Bank

PREPARATION READING

Your Loss is My Gain

*I*n Lesson 1, you saw many examples of games. Now you will explore them in more detail in order to understand players' strategies. Players' strategies generally depend on the payoffs associated with the outcomes.

Some games have payoffs that are easily measured. Often, a gain to one player is a loss to the other—the interests of the players are completely opposed. The simple matching game discussed in Lesson 1 is one example. Competitive business situations often are very similar to the simple matching game.

In other games, one player's gain is not the other player's loss. A loss for one may also be a loss for the other. Warfare between two nations, for example, can result in extensive losses for both sides.

The Cuban Missile Crisis is unique in history; however, some strategic situations are repeated many times. Games like the simple matching game are played more than once, and competitive situations between businesses often occur on a regular basis.

To begin your study of games, you will look at those in which one player's gain is the other's loss. You'll also focus on situations that are repeated.

What's the best strategy when your opponent's gain is your loss? Is it possible to take advantage of your opponent's mistakes? What exactly does "strategy" mean?

PLAYING THE GAME

In Lesson 1, you created models for strategic situations by identifying and organizing the key features of the situations into matrices. Now it's time to investigate how to play such games intelligently.

CONSIDER:

1. What do you think would be a good strategy for playing the simple matching game with matrix shown in **Figure 2.9**?

2. What is meant by "strategy" anyway?

3. What makes a strategy good?

Figure 2.9.
Payoff matrix for a simple matching game.

		Column player	
		One finger	**Two fingers**
Row player	**One finger**	1	–1
	Two fingers	–1	1

SIMULATION 1

In this activity you will get a chance to design your own strategy for the simple matching game in Figure 2.9. You will play the game to test your strategy against an opponent. Check the success of your strategy after 60 plays of the game by comparing your **average points per play** (your total points divided by the number of plays) with your opponent's and then with those of the other members of your group.

Remember that the payoffs in Figure 2.9 are payoffs to the row (even) player. The column (odd) player's payoffs are the opposites of the row player's payoffs.

ACTIVITY

PLAYING THE GAME

2

DIRECTIONS FOR SIMULATION 1:

1. You and your partner (opponent) will play 60 rounds of the game in Figure 2.9. Before you begin, decide which of you is the row player, and which is the column player. Next, take a moment to think about what strategy you will use. Then, play the game using your strategy. Keep a record of each play on Handout H2.2, "Game Record for Simulation 1." Use the small square at the top right of the record sheet to tally the number of each type of outcome.

2. When you finish, calculate the average points per play both for the row and column player. How is the row player's average related to the column player's average?

3. Compare the various strategies used by members of your group. Did some do better than others? Do you think that's because the strategies were good, or was it just due to luck?

As you probably have seen in Simulation 1, the strategies used by players in a strategic situation (or a game) have a great impact on the outcome. One way to measure the success (or failure) of a strategy is by calculating the average points per play over a *large* number of games. In the previous simulation, you played 60 games. If you could imagine extending the number of times the game is played to infinity, then the long-range average points per play to the row player is also called the row player's **expected payoff**. If a large (but not infinite) number of games are played, then a player's actual average points per play should be *close* to the theoretical expected payoff.

The expected payoff depends not only on the payoff matrix but also on the row and column players' strategies.

4. a) Suppose the row player uses the pure strategy of showing two fingers, and the column player uses the pure strategy of showing one finger. What is the row player's expected payoff? What is the column player's expected payoff?

   b) Are either the row or column player playing this game wisely? Explain.

ACTIVITY

2

PLAYING THE GAME

5. A **fair game** is a zero-sum game in which neither player has the advantage.

   a) Do you think that the game in Figure 2.9 is a fair game? Why or why not?

   b) If you think the game is fair, what strategies should the players use so that neither has the advantage? If you don't think the game is fair, which player has the advantage?

6. **Figures 2.10–2.14** present five payoff matrices for the simple matching game. Imagine that your teacher assigns one of Games 1–5 to each of five pairs of students from your class. She tells each pair to play their assigned game 100 times. Suppose that each player decides to use the same strategy of flipping a coin—if it lands heads up, choose 1; otherwise, choose 2.

Column player

Game 1		One finger	Two fingers
Row player	**One finger**	−3	3
	Two fingers	3	−3

Figure 2.10.
Payoff matrix for Game 1.

Column player

Game 2		One finger	Two fingers
Row player	**One finger**	−3	4
	Two fingers	1	−2

Figure 2.11.
Payoff matrix for Game 2.

Column player

Game 3		One finger	Two fingers
Row player	**One finger**	2	−1
	Two fingers	1	−2

Figure 2.12.
Payoff matrix for Game 3.

ACTIVITY

PLAYING THE GAME

2

Column player

Game 4		One finger	Two fingers
Row player	One finger	3	–6
	Two fingers	–1	2

Figure 2.13.
Payoff matrix
for Game 4.

Column player

Game 5		One finger	Two fingers
Row player	One finger	3	–5
	Two fingers	–5	3

Figure 2.14.
Payoff matrix
for Game 5.

For each of the games in Figures 2.10–2.14:

a) Estimate the average points per play for the row player in the 100 plays.

b) Do you think that by tossing coins the row and column players were playing this game wisely or foolishly? If you think they were playing foolishly, suggest wiser strategies for the row and column players.

c) Do you think this is a fair game? If not, who do you think has the advantage, the row player or the column player?

Did you have difficulty in deciding whether the players using the coin tossing strategy were playing Games 1–5 wisely or foolishly? How certain are you that the wiser strategies that you suggested are better than the coin tossing strategy? In the next activity, you will have a chance to play one or more of Games 1–5 using your suggested strategies. Then you can see for yourself how well your wiser strategies work.

INDIVIDUAL WORK 3

It's Not Fair!

A fair game is a zero-sum game in which neither player has an advantage.

1. a) Write a payoff matrix for a game in which the row player does have an advantage.

   b) How do you know that the row player has the advantage in this game?

   c) Would the coin toss strategy be a good strategy for the row player? Explain why or why not.

2. a) Write a payoff matrix for a game in which the column player has the advantage.

   b) How do you know that the column player has the advantage in this game?

   c) Would the coin toss strategy be a good strategy for the column player? Explain why or why not.

Nobel Prize for Game Theory

October 11, 1994

John F. Nash, Jr., John C. Harsanyi, and Reinhard Selton were awarded the Nobel Prize in Economic Science for their work in game theory. The three shared $930,000 for their achievements. The basis of Nash's award was a Ph.D. thesis he had produced in 1949 at the age of 21; essentially, it focused on rivalries in which mutual gain is possible.

ACTIVITY

THE SCRIPT OF THE PLAY

3

1. If your teacher does not assign your group a specific game, select one of the Games in Figures 2.10–2.14 (Activity 2) for which you think the coin tossing strategy is not a wise strategy for at least one of the players. Split your group into row players and column players. One of the row players should use the coin toss strategy and another player should choose the wiser row player strategy suggested for that game by your group. Similarly, one of the column players should use the coin toss strategy and another the wiser column player strategy.

   a) Design an experiment that tests each row player strategy against each column player strategy. Conduct your experiment and report your results.

   b) Do you think the wiser strategies that your group suggested for the row and column players are better than the coin toss strategy? If you are unsure, what additional experiments would help you decide? Explain.

   c) What do you think it means for a player to play wisely?

Were you able to find wise strategies for the row and column players for each of the games in Figures 2.10–2.14? If not, in the spirit of modeling, start with a simpler situation. Examine how changing the row player's strategy can affect the average points per play for one specific game in which the column player's strategy is preassigned.

SIMULATION 2

The design of this experiment is very similar to that of Simulation 1 in Activity 2. This game uses the same payoff matrix (Figure 2.9) and you will again play 60 games. However, in this game, the column player must follow a script. The row player should try to determine what kind of script the column player is using and take advantage of this knowledge to increase his total points.

ACTIVITY

3

THE SCRIPT OF THE PLAY

Keep a record of the plays and the outcomes on Handout H2.3, "Game Record Sheet for Simulation 2." Your teacher will give each pair of players an envelope containing the script, which only the column player should see.

DIRECTIONS FOR SIMULATION 2:

Decide which of you is the row player and which is the column player. The column player should open the envelope containing the script and must play according to the script. The row player may use any strategy he or she wants for the first 30 times. After playing the first 30 games, stop. At this time, the row player should study the column player's first 30 moves. During the last 30 plays, the row player should, if possible, take advantage of the knowledge gained during the first 30 plays. The column player must not deviate from the script even if the row player seems to have detected what is being done.

When you have finished the game, discuss each of the following items with your opponent.

2. What strategy did the row player think the column player's script was following? Was the row player right, or close to being right?

3. Was the row player able to take advantage of the column player's scripted strategy during the last half of the game? What evidence can you give to support your answer?

4. After you have finished Items 2 and 3, your teacher will distribute Handout H2.4, which reveals the actual column players' scripts. Each script is read from left to right beginning on the top row.

   a) Describe the column player's strategy for each of the scripts.

   b) For each script, decide on a wise strategy for the row player. Explain.

THE SCRIPT OF THE PLAY

c) What would be the best (or a better) strategy for the column player? Why? How could you decide if you are not sure?

A **mixed strategy** ia a blend of pure strategies over repeated plays. By now, you should have discovered that it is foolish to use a mixed strategy (sometimes choosing 1, and other times 2) that is patterned. Even if the pattern is fairly complicated, a smart opponent can discover the pattern and then capitalize on your strategy. It is more difficult to capitalize on a random mixed strategy. When using a **random strategy**, your choice of what to play on any given play is determined by a random device such as the toss of a coin or your calculator's random number generator. However, at least for this game, it is sometimes possible to capitalize on an opponent's random strategy, particularly if there is a noticeable imbalance between 1s and 2s. In Individual Work 4, you will take a closer look at random mixed strategies.

INDIVIDUAL WORK 4

Is This Strategy Random?

*S*ome players do not use the best strategy and lose to smart opponents who can discover the first player's strategies and adjust their own play to take advantage of the first player's poor choice. Suppose, for example, that in the simple matching game with payoff matrix in Figure 2.9, the row player alternates 1s and 2s: 1 2 1 2 1 2 . . . . Before long, the column player will realize this and play in such a way that the sum is always odd. The column player has taken advantage of the opponent's strategy.

Some players mix their strategies well, maybe even using a random strategy, but do not use the numbers of 1s and 2s in the proper proportions. In some games, such as the simple matching game in Figure 2.9, a player may be able to capitalize on this imbalance. If, for example, the row player plays 1s much more often than 2s, the column player can capitalize by playing 2s all the time and win much more often than not. For other games, a strategy that produces some other proportion of 1s and 2s may be a wise strategy.

It is important to understand the difference between a patterned and a random strategy. A random strategy does not contain a predictable pattern. It does not have to contain an equal mix of 1s and 2s. A random strategy is often described by the intended ratio of 1s and 2s—for example, 30-70 or 50-50. On the other hand, a patterned strategy is one for which, once the pattern is discovered, the next play is predictable.

1. Suppose each of the following is a string of 1s and 2s played by your opponent (the column player) in a simple matching game. (Read the lists from left to right, row by row, starting with the top row.) In each, tell whether or not you think your opponent is playing a random strategy or a patterned strategy, and explain your reasoning. If you think your opponent is playing a random strategy, tell whether you think it's a 50-50 random strategy, and again explain your reasoning.

   a)
   ```
   1 2 1 2 1   1 1 2 2 2   2 2 1 1 2   2 1 2 1 1   1 2 2 1 2
   2 1 1 1 2   2 2 1 2 1   1 2 1 2 1   1 1 2 1 2   2 2 1 2 1
   ```

   b)
   ```
   1 2 1 2 2   2 1 2 2 1   1 2 1 1 2   2 2 1 1 1   1 2 2 1 1   2 1 2 2 1
   1 1 1 1 1   1 1 1 1 2   1 1 1 1 1   1 1 2 1 2   1 1 1 1 2   2 2 2 1 2
   2 1 2 2 2   1 2 2 1 2   2 2 1 1 2   2 1 1 2 2   2 1 2 1 2   2 1 2 2 1
   ```

c)
```
11221   12211   22112
21122   11221   12211
```

d)
```
11111   11122   11112   11112   22112
11111   11122   11221   11112   12111
```

e)
```
22211   21112   12111   21121   12112
11212   12121   12121   21111   11111
```

2. Generate a list of 100 1s and 2s by flipping a coin. (Choose 1 if the coin lands heads up and 2 if the coin lands tails up.) When you have finished, answer each of the following questions.

   a) How many 1s are in your list?

   b) How many 2s are in your list?

   c) A "run" is several 1s or several 2s in a row. What is the longest run in your list?

   d) Now, compare your list to the lists in Item 1. Do you still agree with your answers in Item 1?

3. Suppose that you were to play the simple matching game with the payoff matrix in **Figure 2.15**.

Column player

		One finger	Two fingers
Row player	**One finger**	2	–2
	Two fingers	–2	2

Figure 2.15.
Payoff matrix for simple matching game.

   a) For each of the scripts in Item 1, describe the strategy you would use to capitalize on the scripted strategy of your opponent. Assume that you are the row player and your opponent is the column player.

   b) Against which scripted strategy or strategies do you think you would be most successful?

   c) If both the row and column players are playing wisely, what strategies do you think they will use? Explain.

INDIVIDUAL WORK 4

d) Do your think this game is fair? If not, who has the advantage?

e) How is this payoff matrix related to the payoff matrix in Figure 2.10? How are the row and column players' wise strategies for these two games related?

4. Suppose that you were to play the simple matching game with the payoff matrix in **Figure 2.16**.

Figure 2.16. Payoff matrix for simple matching game.

		Column player: One finger	Column player: Two fingers
Row player	**One finger**	4	–2
Row player	**Two fingers**	2	–4

a) For each of the scripts in Item 1, describe the strategy you would use to capitalize on the scripted strategy of your opponent. This time, assume that your opponent is the row player and you are the column player.

b) Against which scripted strategy or strategies in Item 1 do you think you would be most successful?

c) If both row and column players are playing wisely, what strategies will they use? Explain.

d) Do your think this game is fair? If not, who has the advantage?

e) How is this payoff matrix related to the payoff matrix in Figure 2.12? How are the row and column players' wise strategies for these two games related?

As you have discovered, sometimes you can capitalize on another player's strategy. But then, what is to keep your opponent from attempting to capitalize on your strategy? After all, if you play against really good opponents, they have a good chance of figuring out your strategy, even if you use a random mix.

5. You and an opponent are playing the simple matching game in Figure 2.15. You are the row player, and your opponent the column player.

    a) Suppose you notice that your opponent is playing many more 1s than 2s. To capitalize on his strategy, you decide to play mostly 1s. Once your opponent notices what you are doing, how will he change his strategy?

    b) How will you react to the change in your opponent's strategy?

    c) Once your opponent notices your reaction, how will he counter?

Situations like the one described in Item 5 are, in a sense, unstable. In an attempt to improve your scores, both you and your opponent are playing risky strategies and hoping the other player doesn't notice. In the long run, the resulting gains and losses are very likely to average out. So, why take the risk? That's the idea behind an **optimal strategy**—a strategy that provides security so that you don't have to worry about being a victim of a clever opponent.

ACTIVITY

4

EXPLORING STRATEGIES

CONSIDER:

An optimal strategy prevents even the very best opponent from being able to capitalize on your strategy. It is reasonable to assume that your opponent will be able to figure out your strategy (at least approximately) after you play for a bit. A really good opponent will "put herself in your place" and know what you are planning before you even begin play! How do you play *that* kind of opposition?

1. What do you think it means to say that an opponent can't capitalize on your strategy?

2. If you are playing a game using an optimal strategy, do you think you should switch your strategy from time to time so that your opponent doesn't discover what is being done? Explain.

If you fail to use the best strategy, your opponent may be able to adjust his strategy and end up the winner. This will certainly happen if you use a patterned strategy and plan to play the game repeatedly with a smart opponent. However, it can also happen if you chose the wrong random strategy. Is there an optimal strategy on which an opponent cannot capitalize?

One way to find out is to carry out another experiment by having a "contest" among several random strategies and see how they do. If one always seems to do well, regardless of the opposing strategy, then you may be well on your way to finding an optimal strategy. That's what Simulation 3 is all about.

ACTIVITY

EXPLORING STRATEGIES

4

SIMULATION 3

In this simulation, you will test row and column player strategy combinations in two different games—Games 1 and 2—with payoff matrices shown in **Figures 2.17 and 2.18**.

Column player

Game 1		One finger	Two fingers
Row player	One finger	1	−1
	Two fingers	−1	1

Figure 2.17.
Payoff matrix for Game 1.

Column player

Game 2		One finger	Two fingers
Row player	One finger	−3	4
	Two fingers	1	−2

Figure 2.18.
Payoff matrix for Game 2.

DIRECTIONS FOR SIMULATION 3:

For this simulation, you and a partner will test six different combinations of row and column player random strategies: 50-50 v. 50-50, 20-80 v. 50-50, 70-30 v. 50-50, 20-80 v. 20-80, 70-30 v. column player's choice, row player's choice v. column player's choice. In these strategy combinations, the row player's strategy is listed first, followed by the column player's strategy.

Before you begin the simulation, all the row players and column players should get together in separate groups. Each group should decide what strategy or strategies they want to use as the "choice strategies." Remember, the average points per play are more reliable when based on large numbers of plays. So, you probably don't want each pair to use a different "choice strategy."

ACTIVITY

EXPLORING STRATEGIES

4

Play each matching game 20 times using each of the six strategy combinations. Thus, each payoff matrix will be used in 120 plays. Keep track of your outcomes on Handout H2.5,"The Random Strategy Record Sheets."

To play the game, use the random-number function on your calculator to generate a random number and then use this number to choose option 1 or option 2. For example, if your random strategy is 20-80 then you should choose option 1 if the random number is 0.2 or smaller, and option 2 otherwise. (So, for a 20-80 random strategy, if the random number is 0.7123456, you should choose option 2.)

Now, play the games. When you have finished and have recorded your results on Handout H2.5, answer as much of Items 1 and 2 as time allows.

1. Use the results from your simulations to answer the following questions.

   a) For each of the first four strategy combinations, what is the row player's average points per play in Game 1?

   b) For each the first four strategy combinations, what is the row player's average points per play in Game 2?

   c) Have your answers to parts (a) and (b) helped you determine an optimal strategy for Games 1 and/or 2?

More data may produce clearer patterns. After everyone has played their six strategy combinations, combine the results for the entire class. Record the class results on Handout H2.6 (Figures 1–3). You will need to make separate entries if different groups used different choice strategies for the last two strategy combinations (Handout H2.6, Figure 3).

2. On Handout H2.6, complete the third columns of Figures 4 and 5.

ACTIVITY

EXPLORING STRATEGIES

4

3. Suppose Game 1 is being played.

   a) If the column player is playing a 50-50 random strategy, what is the row player's best strategy? How well do you think the row player will do with this strategy?

   b) If the column player is playing a 20-80 random strategy, what do you think is the row player's best strategy? How well do you think the row player will do against the column player's 20-80 strategy?

   c) Is there a column-player strategy against which the row player is unable to capitalize? Explain.

   d) Based on results from Simulation 3, do you think this game is fair? Explain why or why not.

4. Suppose the players decide to play Game 2.

   a) If the column player is playing a 50-50 random strategy, what is the row player's best strategy? How well do you think the row player will do using this strategy?

   b) If the column player is playing a 20-80 random strategy, what do you think is the row player's best strategy? How well do you think the row player will do against the column player's 20-80 strategy? Are the results from the class simulation helpful in answering this question?

   c) Based on the class data from Simulation 3, can you identify a column player strategy against which the row player is unable to capitalize? Explain.

   d) Based on results from Simulation 3, do you think this game is fair? Explain why or why not.

When either player uses a strategy that keeps the average winnings per play near some constant value regardless of the strategy used by the other player, the game is said to be in **equilibrium**. A strategy that keeps a game in equilibrium is the most common kind of optimal strategy. The expected payoff for a game in which one player uses the optimal strategy is called the **value of the game**.

ACTIVITY

4

EXPLORING STRATEGIES

At this point, your results from Simulation 3 have probably given you a good understanding of what can happen when you play Game 1. Game 2, on the other hand, is a bit more complicated and may require further study. So, the next item deals only with Game 1.

5. Assume that players are playing Game 1.

   a) What is the optimal strategy? What does it mean that this strategy keeps the game in equilibrium? Approximately what is the value of this game?

   b) Suppose that you are the row player and that you have discovered that your opponent is playing a 70-30 random strategy. Would you choose to use the optimal strategy? Why or why not?

   c) Why do you think that a player might want to use an optimal strategy? When does it make sense not to use an optimal strategy?

6. Describe an experiment that would help you discover the row player's best strategy against a column player's 70-30 strategy.

INDIVIDUAL WORK 5

Play the Calculator

R eturn again to the simple matching game with payoff matrix in Figure 2.17. In Activity 4, you discovered that if you play this game using a 50-50 random strategy, you will, on average, break even, and your opponent can do nothing to lower your average payoff. However, if your opponent plays a random strategy other than a 50-50 mix of 1s and 2s, and you can discover her strategy, you may be able to capitalize on her strategy and do much better. Again, the best way to see how this can work is to try it out.

In this activity, your opponent is the calculator. The calculator will play a random strategy characterized by Q, the portion of times 1s are played— for example, if the calculator chooses to play a 20-80 random strategy, $Q = 0.2$. You will not know, however, the exact proportion of 1s and 2s. Instead, you will need to study the calculator's plays to learn something about its strategy and then respond accordingly.

You will need Handout H2.7, "Record Sheet for 'Me Against the Calculator'," and a calculator with the program GAME1.

DIRECTIONS:

Decide on a strategy to use for the first 25 games. Before the calculator reveals its choice of one finger or two, it will display a random number. If you plan to use a random strategy, use this number to help you decide whether to play one or two fingers. After you enter your choice, 1 or 2, the calculator will respond with its play.

Run GAME1 on your calculator. Enter 50 for the number of games (so the calculator will use the same strategy for the second 25 games as for the first 25).

Use the first chart on Handout H2.7 to record the first 25 plays and the second chart for the second 25 plays. For each play, in the column labeled "Me," write 1 or 2, according to whether you played one finger or two. In the column labeled "Calculator," do the same for the calculator's play. In the column labeled "Payoff," write 1 or –1, according to whether or not you (as the row or even player) gained or lost one point.

When 25 plays are completed, calculate your average points per play.

Before proceeding, study the first chart and guess the calculator's strategy. Then, continue the game for another 25 plays and use what you have learned from the first 25 plays to try to beat the calculator. Record the choices and payoffs in the second chart.

At the end of the 50 games, the calculator will give you a summary of the outcomes and display the random strategy that it actually used. However, if you want to play it again, it will use a different strategy next time!

Now answer the following questions about your experiment.

1. What strategy did you use for the first 25 plays? Why did you use this strategy?

2. After studying the calculator's first 25 plays, what was your guess for its strategy? How did you determine your guess? How close was your guess to the actual strategy?

3. What strategy did you use for the second 25 games? Why did you choose this strategy? Were you able to beat the calculator?

4. Suppose that the GAME1 program was set so that the calculator, as the column player, always used a 30-70 random strategy. Gina and George both decide to take on the calculator. George decides to match the calculator's strategy and uses a 30-70 strategy. Gina decides to use a pure strategy and always plays 2s.

   a) About what percentage of the time do you think Gina will win against the calculator?

   b) About what percentage of the time do you think that George will win against the calculator? (A tree diagram may be helpful in determining this percentage.)

   c) Which player do you think does better against the calculator? Explain.

5. Suppose the payoff matrix in **Figure 2.19** (this was the matrix for Game 2 in Activity 2) was used to calculate your average points per play for the second group of 25 plays on Handout H2.7. What would your average points per play be in this case?

<table>
<tr><td></td><td></td><td colspan="2" align="center">Column player</td></tr>
<tr><td></td><td></td><td align="center">**One finger**</td><td align="center">**Two fingers**</td></tr>
<tr><td rowspan="2">Row player</td><td>**One finger**</td><td align="center">−3</td><td align="center">4</td></tr>
<tr><td>**Two fingers**</td><td align="center">1</td><td align="center">−2</td></tr>
</table>

Figure 2.19.
Payoff matrix .

6. Use the outcomes from the second set of 25 plays on Handout H2.7 to complete the following items.

   a) Make up a payoff matrix for which your average points per play for the second 25 plays on Handout H2.7 would be greater than 2.

   b) Make up a payoff matrix for which your average points per play would be less than –2.

   c) Do you think that either of your payoff matrices in parts (a) and (b) represent fair games? If not, which player is favored?

7. Suppose you plan to play Game 2 with payoff matrix in Figure 2.19. You will be the row player and you would like to find an optimal strategy for this game. Design an experiment using the GAME1 program that would help you decide if either a 30-70 or a 70-30 random strategy might be an optimal strategy for the row player. Be sure to explain how you would decide if a particular strategy is an optimal strategy for the row player based on your experiment.

8. Vapor and Evacuate are competing chemical companies. They each do best financially if they dispose of wastes as cheaply as possible. This means that they practice no pollution control, because pollution controls are expensive.

   a) In the preceding scenario, identify the players and their strategies, and organize the results into a payoff matrix.

   b) Do the players have opposing interests?

   c) What would it mean for Vapor to play a mixed strategy?

   d) Is this a fair game?

LESSON THREE
Changing the Payoffs

KEY CONCEPTS

Fairness

Equilibrium

Optimal strategies

Strictly determined games

Dominant strategy

Value of a game (determined empirically)

The Image Bank

PREPARATION READING

What's the Payoff?

You have seen several competitive situations in which strategies are important. Some of these situations are games; others are not. However, mathematicians use the technical term **game** when referring to any of these competitive situations, thus the topic you are now studying is called game theory. These situations, or games, that you have been studying have certain features in common.

In all of the games you've been studying:

- There are two players.

- Each player chooses a strategy.

- The strategies determine a payoff.

- Payoffs can be measured numerically.

- The strategies and payoffs can be organized in a matrix.

- Both players know all possible payoffs.

The simple matching game in which the even player wins a point if there is a match and otherwise loses a point is one in which it is wise to mix your plays in a random way if you plan to play the game repeatedly. This is because neither pure strategy, one finger or two, is better. The value of each of your choices depends on what your opponent does. When you use a random mix, your opponent cannot detect a pattern in your plays.

An optimal strategy is one that does not let your opponent capitalize on what you are doing. An optimal strategy provides a sense of security, because with it your average payoff cannot be decreased by your opponent, even by an opponent who knows your strategy!

But the simple matching game in which you either gain a point or lose a point depending on the outcome is almost too simple. Most strategic situations in the real world are not so simple. The payoff may be a gain of 400 points with one outcome, but a loss of 7 points with another.

What happens when the payoffs change from those in the simple matching game? Will the game still be a fair game? Is there still an optimal strategy? Remember how things worked (or didn't) with Game 2 in your experiment.

ACTIVITY

5

LET THE GAMES BEGIN!

In Lesson 2, you were introduced to the concept of a fair game. One example of a fair game is the simple matching game with payoff matrix in **Figure 2.20**.

Column player

		One finger	Two fingers
Row player	**One finger**	1	–1
	Two fingers	–1	1

Figure 2.20.
Payoff matrix for simple matching game.

CONSIDER:

1. What makes this game fair?

2. Next, modify the matrix in Figure 2.20 by changing the 1 in the upper left to a 2. Suppose two players plan to play the modified game. The even (row) player, reasoning that she wins more points when she shows one finger than when she shows 2, decides to use a 60-40 random strategy. Will the odd (column) player be able to capitalize on her strategy? If so, how?

3. How sure are you of your answer to the previous question? How could you check that your answer is best?

Remember, by definition, if you use an optimal strategy your opponent can't capitalize on your strategy. No matter what strategy he tries, your average points per play will not decrease. On the other hand, if you know that your opponent is not using an optimal strategy, and you are sure that he's not smart enough to change strategies, then you can capitalize on his strategy and

LET THE GAMES BEGIN!

increase your average points per play. But how do you determine what strategy is optimal for a particular game? How do you decide which strategy is best for capitalizing on another?

You tried a few strategies in the games in Lesson 2. Ideally, data from all possible strategy combinations in all possible games would give you perfect information, but that would take a *long* time. Instead, in Simulation 4 you will collect data on outcomes for the one-finger, two-finger game using a variety of row and column player strategy combinations. You can't try all possible strategy combinations, but you can collect systematic data that might help you determine optimal strategies, as well as strategies that capitalize on an opponent's non-optimal strategy.

SIMULATION 4

Directions:

Record the outcomes of your games on Handout H2.8a or Handout H2.8b. On the handout, each square with a dark border is divided into four smaller squares. The four smaller squares correspond to a game in which the rows and columns represent playing either one or two fingers.

Run the calculator program GAME2 with the values of P and Q and the number of trials that your teacher assigns you. For this program, P determines the row player's strategy, and Q determines the column player's strategy. P and Q are between 0 and 1 and are the decimal equivalents of percentages. For example, if $P = 0.25$, then the row player is using a 25-75 random strategy. After running the program, the calculator will display a matrix of results. Record them in the appropriate four small squares of your handout.

After all groups have completed their results, they will be combined with the results of other groups to produce the class results. Copy the class results onto Handouts H2.9a or H2.9b. Save the class results. You will need them again in this lesson and in Lesson 4.

ACTIVITY

LET THE GAMES BEGIN!

5

Using data recorded on Handout H2.9a or H2.9b:

1. In the row labeled 0.6 and the column labeled 0.4, what do the four entries represent?

2. Select a square with a dark border at one of the corners of the table. Explain the four entries contained in this corner square.

3. Study the completed table from Handout H2.9 for patterns. Then, discuss with other members of your group the patterns that you notice. What patterns did your group find?

Notice that the data you have collected are independent of pay-offs. Therefore, they will be useful in analyzing any strategic situations in which each player has a choice of two strategies, not just the game shown in Figure 2.20.

4. Return to the modified game you examined in Consider questions 2 and 3.

a) Complete Figure 1 on Handout H2.10 by entering the row player's average points per play for this game for each strategy combination used in Simulation 4.

b) Do the data that you have collected indicate that the column player can capitalize on a row player who uses a 60-40 random strategy? If so, what strategy would be best for the column player? Explain how you determined the column player's "best" strategy in this situation.

c) What seems to be the optimal strategy for the row player?

d) Is this game fair? Explain.

In Individual Work 6, you will revisit Games 1–5 introduced in Activity 2 (see Figures 2.10–2.14, Lesson 2). Using the data from Simulation 4, you will again try to determine wise strategies for playing these games. Also, you will try to answer the question, "Which of these games are fair?"

INDIVIDUAL WORK 6

Revisiting Games 1–5

1. Suppose the payoff matrix for a matching game is shown in
   Figure 2.21.

Column player

		One finger	Two fingers
Row player	One finger	4	–11
	Two fingers	–1	9

Figure 2.21.
A new payoff matrix for the matching game.

a) Use your data from Simulation 4 (Handout H2.9a or H2.9b) to
   make an average points per play chart for this game. Write your
   answer on Figure 2 of Handout H2.10. Save this chart for use in
   Activities 7–9.

b) Examine the entries in the average points per play chart. Describe
   any patterns that you notice.

c) Based on your data, what would happen if the row player played
   one finger 20% of the time and the column player played one
   finger 40% of the time?

d) If the column player keeps records and realizes that the row
   player is playing one finger about 20% of the time, what should
   the column player do?

e) Suppose the column player keeps accurate records of what the
   row player is doing. After every few plays, he studies his records
   with the hope of discovering the row player's strategy so that he
   can capitalize. In this case, what is the best course of action for the
   row player?

Recall that if one player has a strategy that keeps his average winnings essentially constant (so that the other player cannot capitalize),
then that strategy is an optimal strategy. The value of the game is the
expected payoff to the row player (the row player's long range average points per play) that results when one of the players uses an
optimal strategy.

f) Based on your experimental data, what is the optimal strategy for the row player? What is the optimal strategy for the column player?

g) Approximately what is the value of the game? Is this a fair game? If not, who has the advantage?

You first encountered the five matching games with payoff matrices shown in **Figures 2.22–2.26** in Activity 2. At that time, you were asked to suggest wise strategies for playing these games and whether these games were fair. Next, armed with the class results from Simulation 4, you will revisit these questions.

2. For each game in Figures 2.22–2.26 that you are assigned, complete the following items. Be prepared to discuss your results during the next class.

   a) Make a chart showing the row player's average points per play for each strategy combination used in Simulation 4.

   b) What strategy do you think might be optimal for the row player? Explain.

   c) What strategy do you think might be optimal for the column player? Explain.

   d) Based on your chart in part (a), do you think this game is fair? Why or why not?

3. Remember, by using an optimal strategy a player can keep his opponent from taking advantage of his play.

   a) Suppose that the column player uses her optimal strategy. After playing 1000 games, the row player's average winnings are –1.5 points per play. The row player did not use an optimal strategy. Could he have done better if he had? Explain.

   b) Are the players in part (a) playing a fair game? Explain.

Column player

Game 1		One finger	Two fingers
Row player	**One finger**	−3	3
	Two fingers	3	−3

Figure 2.22.
Payoff matrix for Game 1.

Column player

Game 2		One finger	Two fingers
Row player	**One finger**	−3	4
	Two fingers	1	−2

Figure 2.23.
Payoff matrix for Game 2.

Column player

Game 3		One finger	Two fingers
Row player	**One finger**	2	−1
	Two fingers	1	−2

Figure 2.24.
Payoff matrix for Game 3.

Column player

Game 4		One finger	Two fingers
Row player	**One finger**	3	−6
	Two fingers	−1	2

Figure 2.25.
Payoff matrix for Game 4.

Column player

Game 5		One finger	Two fingers
Row player	**One finger**	3	−5
	Two fingers	−5	3

Figure 2.26.
Payoff matrix for Game 5.

WHAT'S THE WORST THAT CAN HAPPEN?

CONSIDER:

Return to the simple matching games with payoff matrices in Figures 2.22–2.26.

1. For which of these games were you able to determine optimal strategies? What are the optimal strategies? How did you find them?

2. Which of these games do you think are fair games? Which do you think are unfair? For which of these games aren't you sure? Explain.

Recall that the reason for looking for an optimal strategy is to avoid being taken advantage of by your opponent. Remember what happened when you were able to guess your opponent's strategy? You improved your payoffs—at your opponent's expense! Now assume that your opponent is equally shrewd and will do as well as possible against whatever strategy you select. How will you come out in such a situation?

1. Return to the game with payoff matrix shown in Figure 2.21. In Item 1 of Individual Work 6, you should have found optimal strategies for both the row and column players.

   a) Imagine that you are the row player. Use the data from your average points per play chart in Item 1(a) of Individual Work 6 to help you complete a chart similar to the one in **Figure 2.27**. Your completed chart should show the outcomes for each row player strategy against the best possible opposing play.

WHAT'S THE WORST THAT CAN HAPPEN?

6

Row player strategy, P	Average points per play possible with this strategy against best opposition
0.0	
0.2	
0.4	
0.6	
0.8	
1.0	

Figure 2.27.
Worst outcomes corresponding to various row player strategies.

b) Which strategy has the best average points per play against the best opposition? What would happen if the row player used this strategy against the column player?

c) Repeat parts (a) and (b), this time examining the column player's strategies and payoffs.

d) The average payoff per play against the best opposition represents the "worst" that can happen to a player's selected strategy. Thus, the "best of the worst," is a "maximin" strategy for the row player and a "minimax" strategy for the column player. How do these row and column players' best-of-the-worst strategies compare to the row and column players' optimal strategies found in Individual Work 6?

e) Which row player strategies involve the greatest risk? Which row player strategies involve the least risk? What does "risky" mean?

ACTIVITY

6

WHAT'S THE WORST THAT CAN HAPPEN?

2. In Item 1, the optimal strategies and the best-of-the-worst strategies are the same. Why do you think this is the case? (The patterns in the average points per play charts may be helpful.)

3. Suppose in a new game the payoffs are doubled from those in Figure 2.21.

   a) What is the payoff matrix for the new game?

   b) Complete a "worst outcomes" chart similar to Figure 2.27 for the game in part (a). How is this chart related to your chart in Item 1? Explain.

   c) What is the best-of-the-worst strategy for the row player? What about for the column player? What effect did doubling the payoffs in Figure 2.21 have on the best-of-the-worst strategies?

   d) Is this a fair game? If not, who has the advantage?

Computer Defeats World Champ Chess Player
May, 1997

What game strategy can defeat a 1.4-ton IBM computer in a chess match? World champion chess player, Garry Kasparov, did not have the correct answer to that question. He was defeated by "Deep Blue"—a computer with a system capable of evaluating 200 million chess positions each second. Kasparov won the first game of the match, but Deep Blue then won two games. The three other games of the match were draws.

INDIVIDUAL WORK 7

Mixing It Up

1. Recall that a zero-sum game is fair if, when both players are playing wisely, neither has the advantage.

   a) What does it mean for the players to play wisely? Explain.

   b) If players are playing a fair game wisely, what can you say about their average points per play after many games? Explain.

2. **Figure 2.28** presents another payoff matrix for the matching game.

Column player

Row player		One finger	Two fingers
	One finger	6	−4
	Two fingers	−14	1

Figure 2.28.
A payoff matrix for the matching game.

Figure 2.29 shows an average points per play chart for the game in Figure 2.28 constructed by another class using their Simulation 4 data.

Q

P	0	0.2	0.4	0.6	0.8	1.0
0	1.0	−1.8	−4.6	−7.7	−10.9	−14.0
0.2	0.1	−1.7	−4.4	−5.9	−7.8	−10.1
0.4	−1.0	−2.0	−3.2	−4.2	−5.5	−6.4
0.6	−2.0	−1.9	−2.1	−1.9	−1.4	−2.1
0.8	−3.0	−1.7	−1.2	0.4	1.4	2.1
1.0	−4.0	−1.9	0.0	2.1	4.0	6.0

Figure 2.29.
Chart of average points per play for the game shown in Figure 2.28.

a) What is the optimal strategy for each player? Explain your methods.

b) What is the approximate value of this game?

c) Is this game fair? If not, whom does it favor? Explain.

3. Suppose the payoff matrix for the matching game were changed to the one in **Figure 2.30**.

Column player

		One finger	Two fingers
Row player	One finger	1	2
	Two fingers	−1	1

Figure 2.30.
A payoff matrix for a matching game.

a) Without creating an average points per play chart for this game, what strategy do you think the row player play should play? How should the column player respond? Explain.

b) Is this game fair? Explain.

4. For the game in Figure 2.30, the row player's strategy of always showing one finger is called a **dominant strategy**. Using this strategy, the row player's payoffs are greater than the corresponding payoffs for showing two fingers, regardless of the column player's strategy. Does either player have a dominant strategy in the game in Figure 2.28? Explain.

Games in which the players prefer one pure strategy to all others are called **strictly determined games.** When a game is strictly determined, you should play the dominant strategy rather than mixing strategies. Strictly determined games are boring because, if players play wisely, they will always choose a pure strategy. That means that the outcome of the game will be the same play after play after play.

5. Which of the games in Figures 2.22–2.26 are strictly determined? What are the respective dominant strategies?

6. Can a strictly determined game ever be fair? If so, give an example of a payoff matrix for a fair game that is strictly determined.

The simple matching game (in which the row player wins a point if there is a match and loses a point if there is no match) is one in which players are wise to mix their strategies—sometimes showing one finger, sometimes showing two. It has no dominant strategy. In many other strategic situations, mixing strategies over the course of time also makes sense.

7. Go back to the six situations you encountered in Activity 1 (Lesson 1). Discuss whether you think mixing strategies makes sense for each of the players.

   a) The baseball batter and pitcher (Item 1).

   b) Sarah's investment (Item 2).

   c) The cool haircuts (Item 3).

   d) Maka Deel's car sales (Item 4).

   e) Store A and Store B (Item 5).

   f) Conserving water in Droughtsville (Item 6).

8. Suppose that Maka Deel, in the car-sales situation from Lesson 1, views the payoffs as shown in **Figure 2.31**.

Customer

		Card	No card
Maka Deel	**Drop price**	200	400
	Don't drop price	300	500

Figure 2.31. Payoffs from Maka's perspective.

   a) What strategy do you think Maka should pursue? Explain.

   b) Suppose, on the other hand, that Maka fears that the customer will feel the price is too high unless it drops a bit more. She views the payoffs as shown in **Figure 2.32**.

Customer

		Card	No card
Maka Deel	**Drop price**	200	400
	Don't drop price	300	0

Figure 2.32. Maka reassesses her payoffs.

Does this change the strategy you'd recommend? Explain.

LESSON FOUR
Optimal Strategies

KEY CONCEPTS

Strictly-determined game

Dominant strategy

Linear equations

Strategy lines

Solving systems of equations

The Image Bank

PREPARATION READING

Testing Strategies

The optimal strategies in a game can be found by testing various strategies repeatedly. In Lesson 3, you built a large table of results by trying the strategies 0, 0.2, 0.4, 0.6, 0.8, and 1 for both players. It is a fairly rough table: you did not test common strategies such as 0.5 and 0.25. Even with the aid of technology, building this rough table can take quite a bit of time, particularly if you use calculators rather than computers. Simulating a thousand games with a particular strategy may seem like a lot, but the expected payoffs you calculated from your table would be more reliable if you had done many more than a thousand trials of each game.

How can you find optimal strategies more easily and more accurately? You can learn much by observing the effects on the game when you hold one strategy constant and allow the opponent's strategy to vary. Scanning your tables of data provides some good information, but how can you "see" the patterns even better?

VISUALIZING THE EFFECTS OF STRATEGIES

7

The purpose of this activity is to examine graphically patterns in the average points per play charts completed in Lesson 3. Is it possible to visualize the equilibrium you identified in these charts? To answer this question, go back to your average points per play charts from Individual Work 6 (Lesson 3).

In Item 1 of Individual Work 6, you created an average points per play chart for the matching game with payoff matrix in **Figure 2.33**. Your results are on your completed Figure 2 of Handout H2.10.

Column player

		One finger	Two fingers
Row player	**One finger**	4	−11
	Two fingers	−1	9

Figure 2.33.
The payoff matrix for a matching game used in Individual Work 6.

Although you have already searched this chart for numerical patterns, you may not yet have tried graphing the data. That's next.

Look at the chart on your completed Handout H2.10. The data in your chart involve three variables: E, the expected payoff (or average points per play); P, the row player's strategy; and Q, the column player's strategy. Visualizing relationships involving three variables can be difficult because your calculator's screen (or a piece of graph paper) is two-dimensional. So you'll need a method that compresses three-dimensions' worth of data into two dimensions. One technique for doing this is to look at how two of the quantities vary while keeping the third constant.

1. Notice that the numbers in the first column of your average points per play chart show the expected payoffs that result when the column player never plays one finger (in other words, when the column player's strategy is $Q = 0$). Using expected payoffs in this column as the response variable and the corresponding row player strategies as the explanatory variable, graph these data on Handout H2.12.

2. What pattern do these points form? Draw it. Label it $Q = 0$.

VISUALIZING THE EFFECTS OF STRATEGIES

3. a) Notice that your chart does not contain the expected payoff when the row player uses a 50-50 random strategy and the column player plays all 2s. Use your graph to estimate this expected payoff.

   b) If the column player always plays 2s, what row player strategy will result in an expected payoff of 3 points per play?

4. Next, on the same graph, repeat Items 1 and 2 for $Q = 0.2$. Then do the same for $Q = 0.4$, $Q = 0.6$, $Q = 0.8$, and $Q = 1$. (Save this graph for use in Activity 8 and again in Lesson 5.)

5. When you have all of the drawings on the same grid, what do you notice? Why is this significant?

6. Describe how the graph corresponding to $Q = 0.7$ should look and where you think it should be placed.

7. Illustrate on your graph the set of worst-possible expected payoffs for the row player. Explain how the graph helps the row player visualize a maximin strategy.

8. a) Suppose that you are the row player and that you discover that the column player is using a 20-80 random strategy. Based on your graph from Item 4, what is your best strategy? Explain.

   b) Suppose the column player notices that you are playing all 2s and counters by playing a 60-40 random strategy. How should you respond? Explain.

   c) Explain how to see in this graph the column player's optimal strategy, the one of which you *cannot* take Advantage.

9. The graphs in Item 4 were created by holding the column player's choice of strategy fixed and then plotting the expected payoffs versus the row player's choice of strategy. The row player's optimal strategy can easily be approximated from this graph. Devise a similar plan for creating such a graph from the column player's perspective. Draw this graph on your other copy of Handout H2.12.

After finishing this activity, save both copies of Handout H2.12 for use in Activity 8.

Strategy Lines

*I*n Individual Work 3 (Lesson 2), you encountered five games. Their matrices are shown again in **Figures 2.34–2.38**. You revisited these games in Individual Work 6 (Lesson 3). In that assignment you created average points per play charts for one or more of them. Using your charts, you tried to determine optimal strategies for the players. Most likely, this was more successful for some of the games than for others.

Column player

Game 1		One finger	Two fingers
Row player	One finger	−3	3
	Two fingers	3	−3

Figure 2.34.
Payoff matrix for Game 1.

Column player

Game 2		One finger	Two fingers
Row player	One finger	−3	4
	Two fingers	1	−2

Figure 2.35.
Payoff matrix for Game 2.

Column player

Game 3		One finger	Two fingers
Row player	One finger	2	−1
	Two fingers	1	−2

Figure 2.36.
Payoff matrix for Game 3.

Column player

Game 4		One finger	Two fingers
Row player	One finger	3	−6
	Two fingers	−1	2

Figure 2.37.
Payoff matrix for Game 4.

Column player

Game 5		One finger	Two fingers
Row player	One finger	3	−5
	Two fingers	−5	3

Figure 2.38.
Payoff matrix for Game 5.

Now you will apply the graphical methods that you learned in Activity 7 to find optimal strategies for players of these games. However, you won't have to do all the work yourself. Instead, you'll have help from Sarah, a student who is also working through the Strategies Unit.

Sarah began with Game 1. She used the data collected by her class to make an average points per play chart for this game. Then, she plotted the expected payoffs in each column versus the corresponding values for P, the row player's strategy. Her graph appears in **Figure 2.39**.

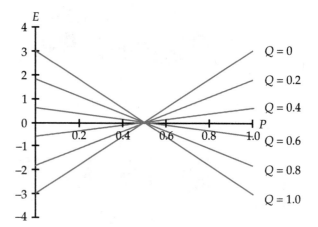

Figure 2.39.
Graphs of strategy lines for Game 1.

Each of these represents the relationship between the expected payoff and the row player's strategy when the column player uses a particular strategy ($Q = 0$, 0.2, 0.4, 0.6, 0.8, or 1.0). Such lines are called **strategy lines**. Since each of the strategy lines in Figure 2.39 is associated with a particular value of Q, these lines will also be referred to as Q-lines. The Q-lines in this family appear to be concurrent because they appear to intersect at the same point.

1. Use Sarah's graph to answer the following questions.

   a) What, if anything, can you say about the spacing of the lines on her graph?

   b) Where would the line corresponding to $Q = 0.3$ fall? Where do you think it would cross the vertical axis?

c) What is the row player's optimal strategy? How can you tell from Sarah's graph?

d) What is the column player's optimal strategy? How can you tell from Sarah's graph?

e) Is this game fair? Explain how you can tell from Sarah's graph? If the game is not fair, which player has the advantage?

2. **Figure 2.40** shows the average points per play chart that Sarah made for Game 2.

		Q					
		0	0.2	0.4	0.6	0.8	1.0
	0	-2.0	−1.4	−0.9	−0.3	0.4	1.0
	0.2	−0.9	−0.7	−0.3	−0.3	−0.1	0.2
	0.4	0.4	0.1	0.0	−0.1	−0.3	−0.5
P	0.6	1.6	0.9	0.4	−0.2	−0.9	−1.4
	0.8	2.8	1.7	1.0	−0.3	−1.2	−2.2
	1.0	4.0	2.5	1.2	−0.3	−1.6	−3.0

Figure 2.40.
Sarah's average points per play chart for Game 2.

a) Use Sarah's chart (or one that you made for Lesson 3) to draw the strategy lines that correspond to $Q = 0.0, 0.2, 0.4, 0.6, 0.8,$ and 1.0.

b) What are the row and column players' optimal strategies? How can you tell from your graph?

c) Is this game fair? How can you tell from your graph? If the game is not fair, which player has the advantage?

INDIVIDUAL WORK 8

3. The average payoff chart that Sarah made for Game 3 appears in
Figure 2.41.

	Q					
	0	0.2	0.4	0.6	0.8	1.0
0	-2.0	−1.4	−0.9	−0.3	0.4	1.0
0.2	−1.8	−1.2	−0.6	−0.1	0.6	1.2
0.4	−1.6	−1.0	−0.4	0.1	0.8	1.4
0.6	−1.4	−0.8	−0.1	0.4	1.0	1.6
0.8	−1.2	−0.5	0.0	0.6	1.2	1.8
1.0	−1.0	−0.4	0.2	0.8	1.4	2.0

P (row labels)

Figure 2.41.
Average points per play
chart for Game 3.

a) Use the data from Figure 2.41 to draw the strategy lines corresponding to $Q = 0.0, 0.2, 0.4, 0.6, 0.8$, and 1.0 for Game 3.

b) What is different about your strategy lines for Game 3 and the
strategy line graphs for Games 1 and 2? What do you think
accounts for this difference?

c) What is the "best-of-the-worst" strategy for the row player?
Explain how you get this information from your graph.

4. For a change of pace, Sarah decided
to draw P-lines instead Q-lines for
Game 4. Corresponding to each
row-player strategy, she graphed the
expected value (to the row player) versus
the column player's strategy. Her graph
appears in **Figure 2.42**.

a) What does her graph tell you about
the optimal strategies for this game?
Explain.

b) Do you think this game is fair? Why or
why not?

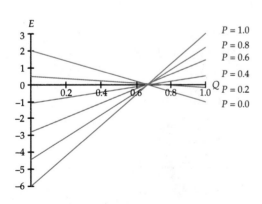

Figure 2.42.
P-lines for Game 4.

5. Sarah's average points per play chart appears in **Figure 2.43.**

P \ Q	0	0.2	0.4	0.6	0.8	1.0
0	3.0	1.5	0.0	–1.6	–3.3	–5.0
0.2	1.5	0.6	–0.8	–1.3	–2.3	–3.4
0.4	–0.2	–0.4	–0.9	–1.3	–1.7	–1.9
0.6	–1.9	–1.4	–1.3	–0.8	–0.3	–0.2
0.8	–3.4	–2.3	–1.7	–0.3	0.6	1.4
1.0	–5.0	–3.3	–1.8	–0.1	1.4	3.0

Figure 2.43.
Average points per play chart for Game 5.

a) Use the data from Sarah's chart to graph the *P*-lines corresponding to $P = 0.0, 0.2, 0.4, 0.6, 0.8$, and 1.0.

b) What does your graph tell you about the row and column players' optimal strategies?

c) Do you think this game is fair? Why or why not?

6. Sketch the graph of a hypothetical set of at least two strategy lines that illustrate an unfair game. From your graph, identify the row player's optimal strategy and the value of the game. Which player is favored in your game?

By now you should be comfortable using strategy lines to find at least one player's optimal strategy and the value of the game. However, you may also be tired of plotting simulation data. Remember, you generated all those data from information in the payoff matrix. What if you could find optimal strategies and the value of the game exactly, without such plots?

FINDING OPTIMAL STRATEGIES

The purpose of this activity is to find a better procedure for determining optimal strategies than the graphing method that you used in Activity 7. You will again need your average points per play chart from Item 1, Individual Work 6 (Figure 2 of Handout H2.10) as well as your graph from Item 4, Activity 7 (Handout H2.12).

In Activity 7, you graphed a family of six strategy lines from data generated for the matching game with payoff matrix in **Figure 2.44**. You observed that the strategy lines appear to intersect at a point. Because of the experimental nature of the data in the average points per play charts from which these lines were generated, the point of intersection may be difficult to find accurately and may not even look like a single point. In fact, the lines may not even really be lines! However, in this activity you will assume that the strategy lines are, in fact, **concurrent lines**—three or more lines that have a single point in common—and you will use this assumption to solve for a player's optimal strategy and the value of the game. In Lesson 5, you will return to this situation and prove that these assumptions are correct.

Column player

Row player		One finger	Two fingers
	One finger	4	−11
	Two fingers	−1	9

Figure 2.44.
A new payoff matrix for the matching game.

1. Study your graphs of the strategy lines (*Q*-lines) that you drew for Item 4, Activity 7. Compare your graphs to the payoff matrix in Figure 2.44.

   a) What does the −11 in the matrix mean? How can you "see" this same information in your graph?

   b) There are two *Q*-lines that you can graph using "special points" from the payoff matrix, without using any of the simulated data from your average points per play chart.

FINDING OPTIMAL STRATEGIES

Which two are they? Why is it possible find them without the simulation data?

c) Find the equations of the two lines identified in part (b) using only the entries in the payoff matrix. Do not refer to your average points per play chart.

d) Solve the equations of the lines to find the coordinates of the point of intersection of the two lines.

e) Now, go back to the previous activity and compare the Q-lines for $Q = 0$ and $Q = 1$ to the Q-lines for $Q = 0.2$, $Q = 0.4$, $Q = 0.6$, and $Q = 0.8$. What do you notice?

f) The Q-line for $Q = 0.8$ is approximately horizontal. Explain the significance of this.

g) Explain why the point associated with $P = 0.4$ ($x = 0.4$) represents the optimal strategy for the row player.

The geometric representation of finding an optimal strategy is different for each player. When inspecting the graph of the family of Q-lines, the row player is looking for a point, but the column player is looking for a line. In either case, however, the "other player" has no control over the average payoff if the choice of strategy is made optimally. The roles would be reversed in a family of P-lines.

2. Next, consider the payoff matrix in **Figure 2.45**.

Column player

		One finger	Two fingers
Row player	**One finger**	4	–2
	Two fingers	–3	1

Figure 2.45.
A payoff matrix for the matching game.

a) Apply the method used in Item 1 to determine the equations of two strategy lines for the column player.

ACTIVITY

8

FINDING OPTIMAL STRATEGIES

b) Use these lines to determine the row player's optimal strategy and the value of the game.

c) Is this a fair game? If not, whom does the game favor? If so, explain how you know.

You have found the optimal strategies and associated expected payoffs for games with two different matrices. The process was the same for both games: determine the equations for two strategy lines, then find the point of intersection. Each time the payoff matrix changes, you will have to repeat this process, unless you can generalize the process for all payoff matrices. That's what the next item is all about.

3. Use the matrix in **Figure 2.46** to generalize the process of finding the row player's optimal strategy and the value of the game.

Column player

		One finger	Two fingers
Row player	**One finger**	a	b
	Two fingers	c	d

Figure 2.46.
A general payoff matrix.

a) Determine equations for the Q-lines corresponding to $Q = 0$ and $Q = 1$ for the general payoff matrix in Figure 2.46.

b) State the conditions under which the Q-lines in part (a) will intersect.

c) Use your answer to part (a) to find a formula for the row player's optimal strategy. State the conditions under which this solution makes sense in the context of game theory. Then find a formula for the value of the game.

d) Test that your general solution is valid for the matrix in Figure 2.45. Compare your answer using the general formula with your answer to Item 2.

FINDING OPTIMAL STRATEGIES

4. If a game is fair, what must be true about the strategy lines corresponding to $Q = 0$ and $Q = 1$?

5. Suppose the strategy line corresponding to $Q = 0$ can be described by the equation $y = -5x + 4$ and that the strategy line corresponding to $Q = 1$ has y-intercept -2. If the game is fair, what is the payoff matrix?

6. Consider the matching game with payoff matrix in **Figure 2.47**.

Column player

		One finger	Two fingers
Row player	One finger	5	−2
	Two fingers	2	−5

Figure 2.47.
Another payoff matrix for the simple matching game.

   a) Using the techniques used in this activity, find the row player's optimal strategy and explain your results.

   b) Is this a fair game? If not, whom does the game favor?

   c) Suppose you are allowed to change two of the payoffs in Figure 2.47. What payoffs would you change to make this a fair game? Explain.

   d) Can you create a fair game by changing only one of the payoffs in Figure 2.47? What payoff would you change to make this a fair game? Explain.

After completing this activity, you have learned much about playing the matching game with a variety of payoff matrices. In the next assignment, you will apply what you've learned here to other strategic situations. Because the payoff matrices for these strategic situations have the same basic structure as the ones for the matching game, you can use the same methods of analysis for finding optimal strategies.

INDIVIDUAL WORK 9

Pure or Mixed?

*I*n Activity 8, you were able to find the optimal strategy for the row player in a matching game by using two lines; one representing all pairs of the form (row player's strategy, expected payoff for row player) when $Q = 0$, and the other when $Q = 1$. The intersection of these lines gave you the row player's optimal strategy and the value of the game, provided neither player had a dominant strategy.

1. Adapt the procedure that you used in Activity 7 to find the equations of two *P*-lines (instead of *Q*-lines) for the game with payoff matrix in **Figure 2.48**. Then use these equations to determine the column player's optimal strategy. Explain your method and interpret the results.

Figure 2.48.
The payoff matrix for a matching game.

Column player

		One finger	Two fingers
Row player	**One finger**	4	−11
	Two fingers	−1	9

2. **Figure 2.49** shows still another variation of the simple matching game.

Figure 2.49.
Another variation.

Column player

		One finger	Two fingers
Row player	**One finger**	6	−4
	Two fingers	−14	1

a) Write a pair of equations to find both the row player's optimal strategy and the solution. Illustrate your solution using a graph.

b) Now write a pair of equations to find both the column player's optimal strategy and the solution.

c) Compare and interpret your results.

3. Suppose two players were playing the simple matching game with payoff matrix in **Figure 2.50**.

Figure 2.50.
A payoff matrix for the matching game.

Column player

		One finger	Two fingers
Row player	**One finger**	6	0
	Two fingers	−1	−5

a) Use the methods from Activity 8 to determine the row player's optimal strategy. Does your answer make sense in the context of this game? Explain.

b) Illustrate your solution for part (a) by drawing two of the *Q*-lines for this game. Be sure you scale your axes so that you can observe the point of intersection. What is different about the intersection of these *Q*-lines from the ones that you drew for Item 2(a)?

c) Use the idea of "maximin" to find the value of the game and the best strategies for the players, and determine whether the game is fair. Explain your method.

4. **Figure 2.51** shows a matrix summary for a strategic situation from Lesson 3. Write a pair of equations that will find Maka's optimal strategy and the solution (the value of the game). Discuss your results.

Customer

		Card	No card
Maka Deel	**Drop price**	200	400
	Don't drop price	300	0

Figure 2.51.
Payoff matrix from Maka's point of view.

5. **Figure 2.52** shows a different interpretation of the payoffs in the same strategic situation that you examined in Item 4. Again, write a pair of equations that will find Maka's optimal strategy. Discuss your results.

Customer

		Card	No card
Maka Deel	**Drop price**	200	400
	Don't drop price	300	500

Figure 2.52.
A different interpretation of Maka's payoffs.

6. Suppose you are a legislator faced with an upcoming vote on a campaign reform bill. You think that if the bill passes, you will lose about $50,000 in campaign funds in the next election. However, if you vote against it, you estimate that it will cost you an additional $30,000 in campaign funds (beyond the $50,000) to undo the anger of the voters, who are known to favor reform. If you vote for the bill and it fails, there will be no additional cost or gain to you in terms of campaign funds.

a) Organize this strategic situation into a matrix that shows the players and their strategies.

b) Is this a situation that is likely to repeat itself? What is your best strategy?

7. Suppose you operate a sandwich shop near a beach and have only one competitor, whose business is somewhat smaller than yours. She occasionally drops her prices slightly to lure customers away from you. If you drop your prices too, a price war often develops and your profits drop considerably. If you both drop your prices, your daily profits drop about $500. If you drop yours and she doesn't, your profits go up by about $100. If she drops her prices and you don't, your profits drop about $200. Of course, if neither of you drop your prices, your profits remain unchanged. (Assume that the total daily profits for the two shops together is constant.)

a) Organize this strategic situation into a matrix that shows the players and their strategies.

b) What is your best strategy?

c) Is this a zero-sum game? If so, is it fair? Explain.

8. Imagine that you are the dictator of a small country and that your totalitarian practices have offended a neighboring superpower. You are negotiating with a representative of the superpower when she informs you that her country's president will launch an immediate air strike against you unless you agree to leave the country. Is the president bluffing? You can stay and risk a war that you will almost certainly lose (you may even lose your life), or you can give up your dictatorship, agree to leave the country, and live a reasonably comfortable life elsewhere.

a) Draw a payoff matrix that shows the players, their strategies, and the payoffs (or rankings). If your matrix shows payoffs, be sure to label the units. Otherwise, write the word "rankings" and state your ranking convention.

b) Is this a zero-sum game? If so, is it fair? Explain.

c) Is this game likely to be repeated? Would mixed strategies be appropriate here?

LESSON FIVE
Optimal Strategies Revisited

KEY CONCEPTS

Probability trees

Zero-sum and constant-sum games

Expected payoff and optimal strategies (theoretically determined)

Mixed strategies and equilibria

Probability distributions and expected value

Linear equations and systems of linear equations

Weighted averages

PREPARATION READING

Multiple Approaches

Some zero-sum games are strictly determined and some are not. When a game is not strictly determined, players should mix their plays in a random way so that the other player cannot detect a pattern in repeated play and capitalize on it. Each player has an optimal strategy—one for which long-term results are unchanged regardless of what the other player does.

Your initial approach to finding optimal strategies required that you simulate the game with a calculator or computer. The large amount of data that resulted contained valuable patterns. Keeping one player's strategy constant, the other's expected payoff seems to vary linearly with the second player's strategy. Moreover, the lines associated with each of the first player's strategies appear concurrent. In Lesson 4, you assumed that these indeed are lines and that they are concurrent. In that case, their intersection gives the optimal mixed strategy (if there is one) and corresponding expected payoff (also called the value of the game).

Finding this intersection requires only two lines, and two lines require only two points each. The necessary four points can be found directly from the game's payoff matrix without the aid of simulated data and can be used to determine optimal strategies exactly.

But your calculator or computer-generated data are not perfect. Experimental data hardly ever are. Although the data indicated that certain properties hold (such as linear patterns in the average points per play charts and a common intersection of the "strategy lines"), the "noise" (variability) in the data could hide actual non-linearity or non-concurrence. Exact methods (not dependent on randomly-generated data) can *prove* that the "strategy lines" really are lines and that they really are concurrent!

ACTIVITY

9

STRATEGIES IN TREES

If you used Unit 7, *Imperfect Testing*, and Unit 8, *Testing 1, 2, 3,* of Course 1, you have studied tree diagrams. In Lesson 1 of this unit, you were able to use tree diagrams to describe the play of games. This representation is helpful in developing models for games and in examining players' strategies. More importantly, it can confirm that the assumptions about strategy lines are correct!

In previous lessons you have generated repeated plays by using a random device (a coin or your calculator's random number generator) to select a choice for each play. Therefore, the portion of the plays in which a player chooses an option can be thought of as a probability. For example, if a player in the simple matching game displays one finger 70% of the time, then the probability the player will show one finger on any given play is .7.

For this activity, you will need Handout H2.9 (the class results for Simulation 4 in Lesson 3) and Handout H2.10 or H2.11 (your average points per play chart from Item 1, Individual Work 6, also in Lesson 3). You will also need Handout H2.12 (your graph from Item 4, Activity 7 in Lesson 4). Have these available before starting this activity.

1. Suppose two players are playing the one-finger, two-finger matching game.

    a) Draw a tree diagram to model the situation in which the row player uses a 20-80 random strategy and the column player uses a 60-40 random strategy.

    b) Use your tree diagram and your knowledge of probabilities to determine the probability that the row player shows one finger and the column player also shows one finger. Compare your answer to the simulation results for that combination of strategies on your experimental data recorded on Handout H2.9.

    c) Using the simulation data from Handout H2.9, estimate the probabilities associated with all other combinations of

ACTIVITY

STRATEGIES IN TREES

9

plays by the row player and the column player using the probabilities from part (a). Then, use tree diagrams to calculate the exact probabilities. Compare the theoretical probabilities to your estimates. Explain any discrepancies.

2. Suppose the payoff matrix for this version of the simple matching game is the one shown in **Figure 2.53**.

Column player

Row player		One finger	Two fingers
	One finger	4	−11
	Two fingers	−1	9

Figure 2.53.
A payoff matrix for the simple matching game.

The probabilities you determined in Item 1 can be summarized in a table, like that in **Figure 2.54**.

Payoff	4	−11	−1	9
Probability				

Figure 2.54.
Table summarizing payoffs and their probabilities.

a) Complete a table similar to the one in Figure 2.54 for the situation in Item 1. Use the exact probabilities obtained from your tree diagram.

The table you have just completed is called the **probability distribution for the payoffs to the row player**. If you used Course 1, you may recall that in Unit 8, *Testing 1, 2, 3,* you calculated the expected values for other probability distributions by multiplying the values by the probabilities and then summing the results. For example, in a problem dealing with a hypothetical lottery you determined the distribution shown in **Figure 2.55** for the possible payoffs to the person purchasing lottery tickets.

Amount won or lost	−$1	$19	$299	$999
Probability	.947	.050	.002	.001

Figure 2.55.
Distribution table for a hypothetical lottery.

ACTIVITY

9

STRATEGIES IN TREES

Given the probability distribution table in Figure 2.55, you calculated the expected winnings as follows:

(–$1)(0.947) + ($19)(0.050) + ($299)(0.002) + ($999)(0.001) ≈ –$0.20.

Thus, players in this lottery lose an average of approximately 20 cents per ticket.

This same process used to determine the average winnings per lottery ticket can be used to find the average points per play or expected payoff of a strategic game.

b) Find the expected value of the row player's payoffs for the distribution in part (a). Interpret your answer in the context of the game in Figure 2.53.

c) For Item 1 in Individual Work 6, Lesson 3, you created an average points per play chart for the game in Figure 2.53. Your results are recorded in your Figure 2 of Handout H2.10. Compare the expected value that you computed in part (b), above, to the corresponding experimental results in Handout H2.10. Explain any discrepancy.

d) Your simulations in Lesson 3 indicated that the optimal strategy for the row player in this game is to play one finger 40% of the time. The optimal strategy for the column player seems to be to play one finger 80% of the time. Construct the exact probability distribution of the row player's payoffs using these particular strategies. Then, use your probability distribution to calculate the expected payoff to the row player in this situation.

e) Rework parts (c) and (d), but this time choose a different strategy for the column player. (Leave the row player's strategy as stated.) Then, find the expected payoff. Discuss the result.

ACTIVITY

STRATEGIES IN TREES

9

3. Recall that the decision in Lesson 4 to examine data involving three variables by keeping one variable fixed and watching the other two is what first allowed you to graph strategies. For example, in Item 4 of Activity 7, Lesson 4, you used data from an average points per play chart to make a graph of strategy lines for the game in Figure 2.53 by holding Q fixed and graphing E as P varied. Your graph should have looked similar to the one in **Figure 2.56**.

Can you find a way to use tree diagrams to get the equations of the Q-lines shown in your graph? What are the variables in this graph? Be prepared to share your method with the others in your class.

Notice that for a fixed value of Q, your work with trees shows that the expected pay-off to the row player really *does* vary linearly as the row player's strategy changes. Thus, *one* of the assumptions has been verified, at least for this game. What about for other games?

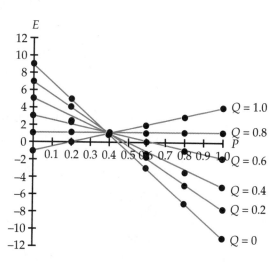

Figure 2.56.
A family of strategy lines for the game in Figure 2.53.

4. Suppose two players are playing the simple matching game with payoff matrix **in Figure 2.57**.

Column player

		One finger	Two fingers
Row player	One finger	7	−8
	Two fingers	−2	8

Figure 2.57.
A payoff matrix for a simple matching game.

STRATEGIES IN TREES

a) Use your method from Item 3 to determine the equations of Q-lines corresponding to $Q = 0$, $Q = 0.2$, $Q = 0.4$, $Q = 0.6$, $Q = 0.8$ and $Q = 1.0$. (To save time, divide the work among members of your group.)

b) Graph this set of Q-lines. (Use an interval for the domain that makes sense in the context of this game.) Do they appear to be concurrent? If so, how could you prove it?

c) What is the value of this game? Explain how you determined your answer.

d) Study the equations of the strategy lines that you have determined in part (a). What patterns do you notice in these equations?

e) Are any of the lines on the graph that you have drawn for part (b) halfway between the lines associated with $Q = 0$ and $Q = 1$? What do you think "halfway" means? If no such line is already drawn, can you suggest a way to find the equation of a line that is halfway between these two Q-lines?

f) Now, look at the line associated with $Q = 0.8$ in your graph. Is it closer to the line associated with $Q = 0$ or to the line associated with $Q = 1$? How much closer? How could your answer be used to find the equation of the $Q = 0.8$ line from the equations of the $Q = 0$ and $Q = 1$ lines?

INDIVIDUAL WORK 10
Sum Games

1. Most tennis players have better forehand
   returns than backhand. When an opponent is serving, a tennis player can get an
   edge by anticipating whether the serve
   will be to her forehand side or her backhand side and reacting accordingly.
   Suppose that a tennis player returns a
   serve to her forehand side 90% of the time
   when she guesses it is coming to that side
   and 50% of the time when she guesses it is
   coming to her backhand side. She returns
   a serve to her backhand side 70% of the
   time when she guesses the correct side
   and 40% of the time when she doesn't.

   a) Summarize this strategic situation in a payoff matrix.

   b) This game is not zero sum because the opponent's payoffs are not
      the opposite of those you wrote in your matrix. For example, if
      the tennis player has a 90% chance of returning the serve, then the
      opponent has a 10% chance of getting the serve by her. The pay-
      offs of the player and the opponent always total 100% (or 1 if you
      use probabilities instead of percentages). When the sum of the
      payoffs is some constant (not necessarily equal to zero), the game
      is called constant sum. Do you think mixed strategies are appro-
      priate in constant-sum games like this one? Explain.

   c) Regardless of whether you think mixed strategies are appropriate
      for this situation, find the optimal mixed strategies and the value
      of the game. Interpret your results.

2. Mr. Fender just scraped a parked car with no apparent damage to his own vehicle, and he is tempted to drive away. He doesn't notice any witnesses, but can't be sure that there are none because the area is heavily populated. If he leaves the scene and a witness reports him, he could sustain about $1,000 in fines and insurance costs. If he reports the accident, the costs will be somewhat less—about $500. Of course, if he leaves the scene and no witness reports him, there are no costs to him except feelings of guilt.

   a) Summarize this situation in a payoff matrix.

   b) Do you think this is a zero-sum game? Do you think mixed strategies are appropriate here? If mixed strategies are appropriate, find the optimal mixed strategies. If not appropriate, explain why not.

   c) It can be difficult to attach a dollar value to feelings, but estimate a dollar value on the guilty feelings a person would experience by leaving the scene. Does this change your analysis of the game?

3. The equations for Q-lines corresponding to $Q = 0$ and $Q = 1$ for a particular game are as follows:

   $Q = 1$: $y = 25x - 4$

   $Q = 0$: $y = -5x + 8$

   where x is the row player's strategy (given as a probability) and y is the expected payoff to the row player.

   a) Sketch a graph of these strategy lines. (Use a window that makes sense in the context of game theory.)

   b) What is the row player's optimal strategy and the value of this game?

   c) What is the column player's optimal strategy? How can you tell? If the column player uses his optimal strategy, what will his average payoff be?

   d) Is this a fair game? If not, which player has the advantage? How can you tell from your graph?

   e) Suppose that the row player decides to play a 50-50 random strategy no matter what happens, and the column player finds out. How should the column player respond? How does your graph in part (b) help you determine the column player's best strategy in this situation?

   f) What if the row player decides to use a 20-80 random strategy?

SEEING THE FOREST THROUGH THE TREES

From a modeling point of view, there are still a couple of nagging questions remaining. You have verified the linearity of strategy lines in some specific cases, but what about all games? And what about the assumption of concurrence? How can that be checked?

You can often learn much about a situation by first using variables to represent the situation and then applying algebra. In this activity, you will apply the method of generalization you may have first studied in Course 1, Unit 7, *Imperfect Testing*, to games of strategy. Remember that the key feature of this method is using variables instead of specific values for important quantities.

1. Use p to represent the probability that the row player shows one finger, and use q to represent the probability that the column player shows one finger, in the version of the simple matching game shown in **Figure 2.58**.

Column player

		One finger	Two fingers
	One finger	4	−1
Row player	**Two fingers**	0	2

Figure 2.58.
A payoff matrix for the matching game.

a) Make a tree diagram and a probability distribution for this game.

b) Use the probability distribution from part (a) to write an algebraic expression for the expected payoff to the row player in the game, in terms of the strategies, p and q, of the respective players.

c) Once you have the algebraic expression for the expected payoff to the row player in the game when the row player uses strategy p and the column player uses strategy q, discuss it with members of your group. How can you use it to find the value of the game (the expected value when either player is using an optimal strategy)?

ACTIVITY

SEEING THE FOREST THROUGH THE TREES

10

d) Suppose the row player decides to play 10% 1s, and the column player 80% 1s. Use your algebraic expression to determine the expected payoff to the row player. Should the row player be happy with how this game is being played? Explain.

e) Suppose the row player knows that the column player will use a 70-30 random strategy. What is the best strategy for the row player to use? What is the expected payoff to the row player if he uses this strategy? Explain your reasoning.

2. **Figure 2.59** shows a matrix in which the payoffs are represented by the letters A, B, C, and D.

Column player

		One finger	Two fingers
Row player	**One finger**	A	B
	Two fingers	C	D

Figure 2.59.
A general payoff matrix for the matching game.

a) Follow the procedure you used in the previous item to find the expected payoff when the row player plays strategy p, and column player strategy q. The resulting general solution can be used to write an expected payoff function for any game quickly.

b) Use your equation from part (a) to find general expressions for the slopes of the $Q = 0$ and $Q = 1$ strategy lines.

c) Based on your answer to part (b), under what conditions do the $Q = 0$ and $Q = 1$ strategy lines never intersect?

SEEING THE FOREST THROUGH THE TREES

d) Use your expected payoff equation to find the row and column players' optimal strategies and the value of the game. Comment on the validity of your solution.

e) If the denominator in your expression for p is zero, what can you say about the Q-lines corresponding to $Q = 0.0$ and $Q = 1.0$? What can you say about this game?

f) How can you identify a strictly determined game directly from the payoff matrix? What do your results in part (d) tell you about the row player's optimal mixed strategy in strictly determined games?

g) Compare the conclusions from this item to those you obtained in Lesson 4's Activity 8, Item 3. What is the major new insight?

3. In Item 2, you found a general formula for the expected payoff to the row player for any combination of strategies for the two players in a two-by-two zero-sum game. That formula allows finding the value of a game and the corresponding optimal mixed strategies in all non-strictly-determined games, provided the assumption of concurrence is valid for all strategy lines.

a) Describe a method that you could use to prove from your formulas that the concurrence assumption is valid.

b) Carry out your proof.

INDIVIDUAL WORK 11

Optimal Strategies Your Way

I n the Items that follow, you can use any means that you prefer for finding optimal strategies. Select the methods that seem easiest for the given situations. Be sure, however, to show your method.

1. A group of law enforcement agents frequently gets tips about smugglers operating in the area. The smugglers typically try to bring their contraband into the country by sea or by air. Using a plane is more risky than using a ship because the rough terrain makes landing difficult, but the same terrain also makes detection less likely. When the law enforcement agents receive a tip, they can concentrate their search on the coast or inland. If they concentrate their search on the coast, there is a 90% chance that they will catch the smugglers if the smugglers use a ship, but only a 10% chance if the smugglers use a plane. If the search is primarily inland, there is a 70% chance that they will catch the smugglers if the smugglers use a plane, but only a 10% chance if the smugglers use a ship.

   a) Prepare a matrix that summarizes the important characteristics of this strategic situation.

   b) Find the value of the game and the optimal strategy for each of the players.

   c) How do the optimal strategies of the two players compare? Why?

   d) Did you write your payoffs as decimals or percentages? Would the optimal strategies or value of the game change if you wrote them the other way?

2. Charlie Brown is kicking a football that Lucy is holding. Charlie Brown can either kick or run up to the football and stop. Lucy can either leave it in place or pull it back at the last second. Charlie Brown takes greatest delight in a successful kick, but Lucy prefers to see him fall flat.

   a) Prepare a matrix summary that shows how each of the two players ranks the various possibilities. Use the ranking convention of 1 for the most preferred outcome and 4 for the least preferred outcome. Do Charlie and Lucy have opposing interests? Is this a constant-sum game? Explain.

b) Although rankings are not really payoffs, frequently people use a ranking convention in which the top choice is assigned the highest rank similar to the situation when using payoffs. Suppose that you change your ranking convention in part (a). This time use 4 for the most preferred outcome and 1 for the least preferred. What is your payoff matrix based on this ranking convention? How are the entries in this matrix related to the ones in part (a)?

c) Suppose Charlie Brown attaches 10 points to the result he most prefers, 7 to his second choice, 3 to his third and 0 to his last. Suppose Lucy does the same. Redo your matrix to reflect this ranking convention.

d) Use your payoff matrix from part (b) to find the optimal strategies for Charlie and Lucy and the value of the game.

e) Do you get the same results as in part (d) if you use the ranking matrix in part (a)?

f) What about if you use the ranking matrix in part (c)?

3. Is either of the games in Items 1(a) or 2(a) a zero-sum game? Is either a constant-sum game? How are the expected payoffs to the two players related in each item?

4. Identify characteristics of a payoff matrix that indicate that the optimal strategy is to choose both options equally. In other words, is there a quick check that tells you whether the row or column player should pursue a 50-50 strategy? Interpret your answer geometrically. (In other words, what can you say about the Q-lines or P-lines in this situation?)

5. Find a quick check that tells whether a matrix represents a fair game—one in which the expected payoff is 0. (Remember that the term "fair" refers only to zero-sum games.)

6. Find a quick check that tells whether a matrix represents a strictly determined game. How could you tell that a game was strictly determined from graphs of strategy lines?

7. In all the games you have examined in this unit, each player has had only two choices. What would happen if one player had more than two? Consider a game represented by the payoff matrix in **Figure 2.60**. (In this game, the row player can chose to play options A or B, and the column player options C, D, or E.)

INDIVIDUAL WORK 11

Column player

		C	D	E
Row player	**A**	2	1	−1
	B	−2	−3	2

Figure 2.60.
Payoff matrix for game where one player has three strategies.

What do you think the players should do in this game?

8. Suppose in an odds-evens matching game, the even player can choose to show one or two fingers but the odd player can choose to show one, two, or three fingers. On any given play, the even player wins if an even number of fingers is showing and the odd player wins if an odd number of fingers is showing. On each play, the players win or lose as many points as the number of fingers showing.

   a) What is the payoff matrix for this game?

   In Lesson 3, you generated strategy lines by holding the column player's strategy fixed and then plotting the row player's expected payoff versus the row player's strategy.

   b) Adapt this approach to this situation by assuming that the column player will play the pure strategy of always showing 1s and that the row player can mix her strategies by sometimes choosing 1s and other times 2s. Write an equation that describes the relationship between the row player's expected payoff and the row player's strategy.

   c) Repeat part (b) above for each of the column player's other pure strategies.

   d) Graph the three equations from parts (b) and (c).

   e) Based on your graph, can you find an optimal strategy for the row player? What about a "best-of-the-worst" (maximin) strategy? If, in either case, you can find such a strategy, what is it? Explain.

LESSON SIX
Games That Are Not Zero Sum

KEY CONCEPTS

Games that are neither zero sum nor constant sum

Dilemmas

Cooperation and defection

Tit-for-tat strategy

The Image Bank

PREPARATION READING

When One Player's Loss is Not the Other Player's Gain

Games of strategy are very common in everyday life. Some of these games are strictly determined because there is really only one option that makes sense. In others, however, the best way to play is to use a strategy that mixes plays in a random way. Sometimes a 50-50 mix is best, but in other cases it is best to play one option more often than another. If you know the optimal mix, you can prevent your opponent from capitalizing on your strategy.

You have spent a lot of time in this unit learning to find the optimal strategies in a wide variety of situations. But all these situations have one thing in common—one player's gain is the other's loss. In other words, these situations are zero sum or constant sum.

Many real-world situations, however, are **non-zero sum;** the sum of the corresponding payoffs is not constant. Often the players have something to gain by cooperating with each other; the interests of the players are not completely opposed. How do you tell if the interests of players are not completely opposed? What strategies are best in games in which interests are not completely opposed?

ACTIVITY

11

THE HAIRCUT

The purpose of this activity is to examine two somewhat different versions of the same game, the haircut game from Lesson 1.

Your job is to write a payoff matrix for each player. The interests of the players might not be complete opposites, so you should write the pay-offs for both players. You can do this either by writing a separate matrix for each player or by writing one matrix with a pair of numbers in each cell.

In the past, you have often used rankings in games like these. There is one problem with rankings: ranking something number 1 gives it the lowest number. Most people like to use the highest number for the best ranking because higher usually means better. So, do this: Use 3 for your highest ranking, 2 for second, 1 for third, and 0 for last.

1. Once again, imagine that you and a friend have agreed to get similar wild haircuts. You agree to get your haircuts at separate shops near your homes, then meet at school the next day. Of course, it would be extremely embarrassing to be the only one with the haircut. Prepare a payoff matrix showing the players, their options, and the payoffs. Assume that both you and your friend have the same first choice: to show up together with the haircuts. Also assume that both of you value your friendship quite highly.

2. Now change the scenario slightly. Suppose the nature of your friend-ship is quite different. Each of you is fond of playing pranks at the other's expense. Both of you take delight in embarrassing the other person. No doubt you would rank things differently. Revise your previous payoff matrix.

3. Are these situations zero sum, constant sum, or neither? Explain.

4. In situations like these, the terms "cooperate" and "defect" are often used. What do you think they mean in this haircut context?

5. Discuss the similarities and differences in the two games described in Items 1 and 2. In particular, do you think either person would be likely to defect in either game?

6. In which situation, the one in Item 1 or Item 2, do the players face a dilemma? Why?

INDIVIDUAL WORK 12

Non-zero-sum Games

*I*n the previous activity, you created a game based on a hypothetical agreement between yourself and a friend to get wild haircuts. **Cooperation** in this setting means that each of you abides by the agreement; **defection** occurs if one of you breaks the agreement. You considered two different payoff matrices for the haircut situation. In the first situation (Item 1), the game turned out to be strictly determined, and mutual cooperation was the best strategy. However, for the second situation (Item 2), mutual cooperation was ranked low by both players; this game posed a **dilemma**. On one hand, if you defected and your friend did not, you increased your payoff; on the other hand, if both you and your friend defected, both of your payoffs decreased.

CONSIDER:

Suppose that you are about to play a game that poses a dilemma. Your choices are to defect or to cooperate. Would you play this game differently if you plan to play the game only once as opposed to repeatedly? Explain.

In each of the games described below, use a payoff matrix to summarize the players, strategies, and payoffs. Give 3 points to the most preferred outcome, 2 to the outcome preferred second, 1 to third, and 0 to last. Write a short explanation of why you rank your choices the way you do.

1. For nearly half a century, the United States and the Soviet Union engaged in a nuclear arms race in which each side could choose to continue building up its arsenals or to disarm them.

   a) Prepare a matrix summary of this situation and then answer the remaining questions based on your matrix.

   b) Which haircut situation does this most resemble—the game that was strictly determined or the dilemma? How is it similar? How is it different?

   c) Do the players face a dilemma in this situation?

   d) Which strategy is similar to cooperation? Which is similar to defection?

   e) Is this a one-time situation or one that is repeated? Do you think this makes a difference? Why or why not?

2. Suppose that you are one of two children in a family. Your parents have assigned the two of you the task of taking out the garbage, but left it to the two of you to divide the work. You can choose to take out the garbage or not, and your sibling can do the same. Each of you dislikes the task, but not nearly as badly as you dislike the stench from garbage standing in the house (or angering your parents).

   a) Prepare a matrix summary, then answer the questions that follow.

   b) Which haircut situation does this most resemble? How is it similar? How is it different?

   c) Do the players face a dilemma in this situation?

   d) Which strategy is similar to cooperation? Which is similar to defection?

   e) Is this a one-time situation or one that is repeated? Do you think this makes a difference? Why or why not?

3. This game is similar to one in Lesson 1. Two competing chain stores are planning franchises in a city with two malls—one large, the other small. Each chain knows it will make the most money if it gets the large mall and the other gets the small mall. Both also know that they run a risk of financial failure if they choose the same mall, but the risk is greater if both choose the small mall than if both choose the large one.

   a) Prepare a matrix summary, then answer the questions that follow.

   b) What do you think cooperation means here?

   c) Do the players face a dilemma in this situation?

   d) Which haircut situation does this most resemble? How is it similar? How is it different?

   e) Is this a one-time situation or one that is repeated? Do you think this makes a difference? Why or why not?

4. A company's labor and management are engaged in negotiations. The union has been on strike for several weeks. Both sides are suffering because operations have been shut down. Each side is trying to decide whether to agree to the other's offer or to continue with its demands.

   a) Prepare a matrix summary; then answer the questions that follow.

b) Which haircut situation does this most resemble? How is it similar? How is it different?

c) Do the players face a dilemma in this situation?

d) Which strategy is similar to cooperation? Which is similar to defection?

e) Is this a one-time situation or one that is repeated? Do you think this makes a difference? Why or why not?

5. To understand various strategies in dilemma situations, it is a good idea to try some of them. Your class will soon be conducting a mini-tournament in a dilemma situation (like the second haircut problem), whose matrix summary appears in **Figure 2.61**.

Friend

		Haircut	No haircut
You	**Haircut**	(2, 2)	(0, 3)
	No haircut	(3, 0)	(1, 1)

Figure 2.61.
Matrix showing rankings in haircut situation.

Plan a strategy to try in your class tournament. Here are some examples of strategies:

• Always cooperate. (Always get the haircut.)

• Always defect. (Never get the haircut.)

• Play a 50-50 random mix of cooperate and defect.

• Cooperate on the first play, then alternate between cooperating and defecting.

Plan your strategy now. Write a description of it and be ready to use it in the tournament!

STRATEGIES IN DILEMMAS

The best way to get a feel for a game and to evaluate possible strategies is to play the game. That approach proved extremely useful with

constant-sum games. Now's your chance to try it again, this time with dilemma situations.

1. Use the strategy you devised in Item 5 of Individual Work 12 (or one that your teacher assigns) to play the game with matrix shown in **Figure 2.62**. Play 25 times with each of eight opponents. Record your plays on Handout H2.14, "Strategies in Dilemmas Record Sheet," which your teacher will provide. You will record 200 trials of your strategy versus a variety of other strategies. Do not deviate from your strategy regardless of how it seems to be working. Remember, this is an experiment among strategies, not a contest among people.

Opponent

		Cooperate	Defect
Me	**Haircut**	(2, 2)	(0, 3)
	No haircut	(3, 0)	(1, 1)

Figure 2.62.
Rankings for cooperation and defection.

2. After you have completed eight rounds of 25 plays each, find your average points for all 200 rounds. You'll be comparing it with similar results from the other people in your class.

3. Next, test the strategy that you used in Part I against the calculator's strategy used in program PD1 or PD2. (Your teacher will assign the program against which you should play. If the program is not already stored in your calculator, you will need to transfer a copy of the program to your calculator.)

   a) Describe the strategy that you think the calculator is using.

   b) How did you and the other members of your group do against the assigned strategy?

INDIVIDUAL WORK 13

Four Kinds of Games

One good strategy in games with dilemmas is tit-for-tat, which you saw in the calculator program PD2. In the **tit-for-tat strategy,** you cooperate on the first move, then for each successive move repeat whatever your opponent did on the previous move.

1. What happens if both players use the tit-for-tat strategy?

2. The tit-for-tat strategy can be changed slightly so that defection is used on the first play rather than cooperation. What happens if both players use this version of tit-for-tat?

3. What happens if one player uses tit-for-tat with cooperation on the first move and the other uses tit-for-tat with defection on the first move?

Zero-sum and constant-sum games can be either strictly determined or not. Similarly, games that are neither zero sum nor constant sum can present a dilemma or not. Thus you can classify two-person games into the following categories:

- zero sum that is strictly determined,

- zero sum that is not strictly determined,

- non-zero sum with a dilemma, or

- non-zero sum without a dilemma.

In each of the games in Items 4–9, prepare a matrix summary and explain why you rank the choices as you do. Then tell which of the above four types of games your payoff matrix represents and give any advice you can to the players on their best ways to play.

4. Two people have been arrested by police and charged with a robbery. The police have isolated the two in separate jail cells and are trying to get each to confess. The police tell each person that if neither confesses, they will both be tried on a concealed-weapons charge, which will result in a year in prison if they are convicted. If both confess, they will be convicted of robbery and be sent to prison for five years. If, however, only one agrees to confess and implicates the other, the prosecutor will reduce the charges against the one who confesses and throw the book at the other. The first will get a very light sentence— no more than six months in jail, but the second will probably go to prison for seven years.

5. Two gas stations located close to each other compete for the same business. Each can choose to set its prices high or low. Together, they sell a total of 10,000 gallons of gas in a week. Each of them sells about half that amount if their prices are similar. However, if one sets prices high and the other low, the one with the low price sells about 80% of the gas. A high price makes a profit of 30¢ a gallon, and a low price a profit of 15¢ a gallon.

6. Change the last situation a bit. If both stations set their prices high, many customers go elsewhere and together they sell only 6,000 gallons per week. If both set their prices low, they attract extra business and sell a total of 12,000 gallons per week. Everything else remains the same.

7. You may have heard stories in which two foolish drivers play a game called "chicken." They agree to drive toward each other at a high rate of speed and see which of them "chickens out" and swerves before the other does.

8. A retail store is preparing for another holiday season. The decision is whether to order large stocks of merchandise or small stocks. A large stock in a good shopping season means a gain of $5,000 over normal seasonal profits, while a large stock in a bad season means a loss of $3,000 over normal seasonal profits. On the other hand, a small stock in a good season means a loss of $1,000 over normal profits, and a small stock in a bad season means a gain of $2,000 over normal profits.

9. Many animals travel in groups. Fish, for example, often swim in schools. One reason for this is protection—there is strength in numbers, as it is often said. An individual fish in a school can either swim in front or lag behind, and others can do the same. Being one of the first fish to advance is risky due to exposure to predators, but advancing itself is necessary in order to find food.

In Item 4, you saw an example of a dilemma in which both players did not rank mutual defection the lowest. Because of the situation in Item 4, a dilemma in which mutual defection does not rank lowest is commonly called a prisoner's dilemma. In Item 7, you saw an example of a dilemma in which both players ranked mutual defection the lowest of all possible outcomes. Because of the situation in Item 7, dilemmas in which mutual defection does rank lowest are often called chicken.

10. List as many different payoff matrices using the 3-2-1-0 ranking scheme as you can. How many of these present dilemmas to the players?

Wrapping Up Unit Two

1. The row player and the column player are playing the basic version of the simple matching game with payoff matrix shown in **Figure 2.63**.

Column player

		One finger	Two fingers
Row player	**One finger**	1	−1
	Two fingers	−1	1

Figure 2.63.
Payoff matrix for simple matching game.

After quite a few plays, both players' winnings are near 0. The row player suggests that the payoffs be changed to those in **Figure 2.64**.

Column player

		One finger	Two fingers
Row player	**One finger**	3	−2
	Two fingers	−2	1

Figure 2.64.
A new payoff matrix for the matching game.

The row player hopes to take advantage of the column player by creating a larger payoff in the upper-left corner of the matrix and then playing one finger more often.

a) Do you agree with the row player's analysis? Explain.

b) If the row player proceeds with the strategy of playing one finger frequently by using a 60-40 random strategy, what advice would you give the column player?

2. You own a business and have a single competitor. The weekly revenues for the two of you together total $50,000. Each week you must decide whether to purchase a large advertisement in the local newspaper. The advertisement is expensive—it costs $5,000. If both of you advertise, or if both of you do not, you split the weekly revenues equally. If one advertises and the other does not, the one who advertises takes about 65% of the week's revenues. Use your knowledge of game theory to analyze the situation.

3. It is common for people to complain about their taxes, yet if no one paid taxes, everyone would suffer from the loss of government services. Suppose that you are not required by law to pay taxes, but do so voluntarily. Use your knowledge of games of strategy to analyze the situation.

4. Public transportation is a heated issue in the state of Confusion, and it appears that the next elections will be decided mainly on the transportation issue. The state's two political parties, the Republicars and the Democabs, have done extensive polling to determine the effects of taking a position for or against public transportation or remaining neutral. (Each district's party chapters are free to adopt their own positions.) **Figure 2.65** shows the percentage of districts in which the Republicars can expect to take control of the district commissioner's office.

Democab position

		Favor	Oppose	Neutral
Republicar position	**Favor**	35	25	60
	Oppose	35	50	55
	Neutral	40	30	65

Figure 2.65.
Party Positions.

What strategy would you recommend to each party? Explain your answer.

5. You and another person are guests on a television game show called *The Cartesian Challenge.* You must pick either 1, 2, or 3 for an x-coordinate and the other person must do the same for a y-coordinate. If the point (x, y) falls on or below the line with equation $y = 4 - x$, each player wins $1000 times the number picked by that person. What should you do?

6. In the previous question, suppose the line is $y = 7 - 2x$, but all other circumstances remain the same. How does your answer change?

7. A football team's statistics show that against their current opponent they gain an average of ten yards when they pass and the opponent uses a running defense, and they lose an average of five yards when they pass and the opponent uses a passing defense. When they run and the opponent uses a passing defense, they gain an average of five yards, and when they run against a running defense, they lose an average of one yard. What should the teams do?

Mathematical Summary

A strategic situation exists whenever you must make a decision that affects one or more other parties who must make decisions that affect you. Mathematicians use the technical term "game" when referring to competitive situations, and they model games with matrices that show the players, their pure strategies, and their payoffs.

The type of game is determined by the game's matrix. In zero-sum and constant-sum games, the player's interests are completely opposed. A loss to one player is a gain to the other. In non-zero-sum games, players' interests are not completely opposed.

Zero-sum and constant-sum games themselves occur in two varieties: games that are strictly determined and games in which the players' optimal strategies are mixed. In a strictly-determined game, mixed strategies are not appropriate because at least one of the players has a dominant strategy.

A player's expected payoff against a fixed opposing strategy varies linearly with that player's own mixed strategy. A strategy is optimal if the average payoff to the player using that strategy is the best that it can be without being decreased by a wise opponent. One such optimal strategy is one that creates a constant payoff line for the opponent; such an optimal mixed strategy keeps the game in a state of equilibrium. An alternate interpretation of the "not decreased by a wise opponent" definition is to maximize the minimum average payoff for yourself— a "maximin" strategy. Graphically, this is typically at the intersection of your opponent's strategy lines.

These methods may be implemented by finding equations for two of the strategy lines directly from the payoff matrix. Using probability, you can write an algebraic expression for the expected payoff in terms of the row and column players' optimal strategies. This, in turn, can provide an algebraic solution to the game.

Not all games are zero-sum games. Non-zero-sum games also occur in two distinct varieties: those that pose a dilemma and those that don't. In dilemma situations, it is not possible to find optimal strategies. However, strategies such as tit-for-tat have been studied in simulations and found to do well. The search for better strategies in dilemma situations is still the subject of mathematical research.

Glossary

AVERAGE POINTS PER PLAY:
The total points divided by the total number of plays.

CONCURRENT LINES:
Three or more lines that have a single point in common.

CONSTANT-SUM GAME:
A game in which the corresponding payoffs to the row player and the column player always have the same sum.

COOPERATION:
A strategy pursued by both players to their mutual benefit.

DEFECTION:
A noncooperative strategy. Usually, a player defects in order to receive a higher payoff than that realized in the cooperative strategy, while leaving the opponent with a lower payoff than that realized in the cooperative strategy.

DILEMMA:
A game in which each player realizes a higher payoff by defecting, but a lower payoff if both players defect.

DOMINANT STRATEGY:
A strategy that produces payoffs greater than the corresponding payoffs for another strategy, regardless of the opponent's strategy.

EQUILIBRIUM:
A state that results when one player is able to keep the average winnings per play near some constant, no matter what strategy the opponent uses.

EXPECTED PAYOFF, OR THE EXPECTED VALUE OF THE PAYOFF:
An alternate term for the long run average points per play. It represents the amount, on the average, that a player can expect to win per play of a game, and it depends on the strategies of both players as well as on the game matrix.

FAIR GAME:
A game in which neither player has an advantage. Theoretically, the average winnings per play in a fair game is 0 provided at least one of the players uses an optimal strategy.

GAME:
Any situation in which one party must make a decision that affects one or more other parties who must also make decisions that affect the first party. In some games the second player is considered to be nature (the weather for example). Mathematical games resemble common games, so the term "game" is often used in place of "strategic situation."

MIXED STRATEGY:
A blend of pure strategies over repeated plays. For example, a player who sometimes shows two fingers and other times shows one finger in repeated plays of a simple matching game is using a mixed strategy. Mixed strategies may be random or patterned.

NON-ZERO-SUM GAME:
A game in which the sum of the corresponding payoffs is not constant.

OPPOSING INTERESTS:
The interests of the players are considered to be opposed whenever one player does better at the expense of the other.

OPTIMAL STRATEGY:
A strategy that produces the best expected payoff to its player that wise opposition cannot decrease.

PAYOFF MATRIX:
A matrix used to summarize key features of a game. It shows the players, the pure strategies available to each player, and the payoffs that result. Instead of payoffs, a payoff matrix can show the rankings that players attach to the outcomes.

PAYOFFS:
The amounts that a player in a game can win or lose on a single play. Losses are represented by negative values.

PLAYERS:
The parties involved in a game.

PROBABILITY DISTRIBUTION OF PAYOFFS:
A table showing each payoff in a game and the portion of the plays in which each payoff can be expected to occur.

PURE STRATEGY:
A player's choice of play on a single play.

RANDOM STRATEGY:
A strategy in which the option for each play is determined by a random device.

STRATEGIC SITUATION:
A game.

STRATEGIES:
The actions that the parties in a game may choose. A strategy may be a choice of options for a single play or a plan for choosing options in repeated play.

STRATEGY LINE:
A line showing possible expected payoffs against a specific strategy, consisting of E versus p when q is fixed, or of E versus q when p is fixed.

STRICTLY DETERMINED GAME:
A game in which the players prefer one pure strategy to all others.

SYSTEM OF EQUATIONS:
A group of equations that must all be true. In this unit, a system of equations is composed of the equations of strategy lines for either the row player or the column player.

TIT-FOR-TAT STRATEGY:
A strategy used in repeated play: the player cooperates with his opponent on the first move and then, on every move thereafter, repeats his opponent's play from the previous move.

VALUE OF A GAME:
The expected payoff to the row player that occurs when one of the players pursues an optimal strategy, thereby keeping the game in equilibrium. Sometimes called the equilibrium value.

ZERO-SUM GAME:
A game in which the payoffs to one player are the opposites of the payoffs to the other player, and the sum of the payoffs to the players is zero.

UNIT

3

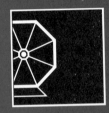

Hidden Connections

Often in your study of mathematics you have seen that the way a problem is represented can be important to the solution of the problem. For example, when you analyze a relationship that you have described with a mathematical function, a graph of the function helps you visualize the relationship and answer questions about it.

However, it isn't always possible to describe a relationship with a function. Mathematicians sometimes use another type of graph to visualize relationships among objects. In this unit, you will apply this new type of graph to a variety of real-world problems.

MAKING CONNECTIONS

Many real-world problems involve objects that are related or connected in some way. For example, cities are connected by airline flights and highways; homes and businesses are connected by phone lines. People are connected because they are relatives, but also because they work for the same employer or live in the same neighborhood. Many problems that are suitable for mathematical analysis arise in these situations: a person who is traveling to several cities wants to use the flights that have the lowest total cost; the phone company wants to connect homes and businesses in the most efficient way; a company wants to schedule its employees in a way that makes the best use of employee skills and company facilities.

In this unit you will learn how objects and the connections among them can be represented in a simple but helpful way, and you will develop problem-solving procedures that can be applied to these representations.

LESSON ONE
Connections

KEY CONCEPTS

Graph

Network

Vertex

Edge

Compatibility graph

Conflict graph

Optimization problems

Digraph

Weighted graph

The Image Bank

PREPARATION READING

A Management Problem

*I*magine that you are the manager of a small company. The company does not own all the equipment it needs. It rents equipment that it uses only occasionally. As part of your job, you must arrange to rent this equipment as economically as possible.

Several current contracts require that you rent time on a sophisticated computer. The computer can handle certain kinds of tasks at the same time, but not others.

The table in **Figure 3.1** explains which contracts can share computer time. Each contract requires the use of the computer for about an hour, whether or not it shares time with another contract. Of course, you could rent the computer for five hours, but that may not be necessary. If you are doing your job well, you will assign more than one task at a time to the computer and rent it for fewer than five hours.

Contract	Can share time with
A	B, C, E
B	A, C, D
C	A, B, D
D	B, C
E	A

Figure 3.1.
A table of five contracts.

This situation is typical of many that occur in businesses and other organizations around the world every day. Precious resources (computer time in this case) must be managed as efficiently as possible. Manage poorly, and both the organization and its employees suffer. Manage well, and both prosper. In this lesson, you will consider a visual way to represent situations like this one.

ACTIVITY

A GRAPH MODEL

1

The situation described in the preparation reading can be represented with a mathematical model called a **graph**, also sometimes called a network. Do not confuse this type of graph with the function graphs your graphing calculator produces for linear, quadratic, and exponential functions. Also, do not confuse it with statistical graphs like bar graphs, scatter plots, and pie charts.

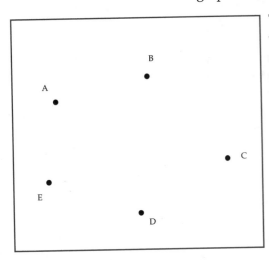

Figure 3.2.
The first step in drawing a graph.

This type of graph is made up of points called **vertices** (the plural of **vertex**) and connecting lines called **edges**. (The connecting lines do not have to be straight.) When these graphs are used to model real-world situations, the vertices usually represent objects and the edges usually represent relationships or connections among the objects.

One way to make a graph to represent the situation described in the preparation reading is to begin by drawing a vertex to represent each of the contracts as shown in **Figure 3.2**. So that you can keep track of which vertex represents which contract, label each vertex with a contract's name.

The next step is to draw edges between vertices according to some criterion. For example, you could draw an edge if two contracts can share computer time. A graph in which edges are drawn when the vertices are compatible is called a **compatibility graph**. But you could also select a different criterion. You could draw an edge if two contracts cannot share computer time. A graph in which edges are drawn when the vertices are not compatible is called a **conflict graph**.

This is an important point: there isn't a single correct way to draw a graph to represent a situation. A good graph is one that clearly represents the information needed to solve some problem. Skill at constructing good graphs comes with practice.

1. Draw a compatibility graph in which edges mean that contracts can share computer time. (Note: It is permissible to draw edges that cross each other because sometimes it is

A GRAPH MODEL

impossible to avoid crossing edges. However, if you draw an edge across another edge, the point at which they cross is not a vertex of the graph. It is permissible to curve the edges to avoid crossings when possible.)

2. Draw a conflict graph in which the edges mean that contracts cannot share computer time.

3. Compare the graphs you made in Items 1 and 2. How are they related? Which do you prefer? Why?

4. Of course, representing the situation with a graph doesn't determine the number of hours of computer rental needed. It just gives you another way to think about the problem. Try solving the problem. Try solving it by using the table and by using each of the graphs you drew. With which representation is the problem easiest to solve? Why?

5. The situation you just modeled with a graph is simple because, as you know, it is best to start simply. Most real-world problems that are similar to this one involve more contracts. Suppose the table in the preparation reading has ten contracts instead of five, as shown in **Figure 3.3**. Consider the table and the two kinds of graphs you made in Items 1 and 2. Which of the three representations do you prefer to use to solve this more complicated problem? (You do not have to make the graphs or solve the problem unless you feel you need to do so to answer the question.)

Contract	Can share time with
A	B, C, E, G, I, J
B	A, C, D, F, G, J
C	A, B, D, F, H, I
D	B, C, F, H, I, J
E	A, F, H
F	B, C, D, E, G, H, J
G	A, B, F, I, J
H	C, D, E, F, I
I	A, C, D, G, H
J	A, B, D, F, G

Figure 3.3.
A table of ten contracts.

Situations in which a graph is a useful tool often are similar to this one in that someone is trying to find the best or optimal way to do something. Problems concerned with the best way to do something are called optimization problems. For example, the company manager in this activity is trying to find the optimal way to schedule expensive computer time. This lesson and this unit are concerned with representations and procedures that can help solve certain kinds of optimization problems. You will recall optimization problems that occur in other units of this program, such as finding an optimal location for a fire station in Unit 1, *Gridville* and choosing an optimal strategy in Unit 2, *Strategies*.

INDIVIDUAL WORK 1

Modeling with Graphs

Choosing the right representation is an important step in solving a problem. The purpose of this individual work is to give you practice creating graphs to represent situations. The situations have been kept simple so that the complexity of the problem does not distract you from making good decisions about how to create the graph.

In Course 1, Unit 1, *Pick a Winner: Decision Making in a Democracy*, a form of graph called a runoff diagram is used to show runoff elections between pairs of candidates. The runoff diagram helps determine the best way to choose a winner in an election using certain criteria. For example, the preference diagrams in **Figure 3.4** indicate the preferences of voters for candidates A, B, and C. (Recall that a preference diagram shows how voters rank the candidates. For example, 24.3% of the voters prefer A to either B or C.)

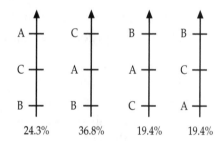

Figure 3.4.
Voter preferences.

24.3%   36.8%   19.4%   19.4%

If an election is held between only two candidates, then there are three different elections possible: A v. B, B v. C, and A v. C. Suppose an election is held between A and B. If you cover C in Figure 3.4 you can see that 24.3% + 36.8% = 61.1% of the voters prefer A to B. Therefore, A beats B in an election between A and B. To use a graph to show that A beats B, draw vertices to represent A and B, draw an edge between them to indicate the runoff, and draw an arrow pointing from A to B to indicate that A beats B (**Figure 3.5**). The arrow is a useful modification because it is important to know who won the election between A and B.

Figure 3.5.
A graph representing a runoff between candidates A and B.

A graph that uses arrows to indicate direction is called a **digraph**, which is short for directed graph.

1. Draw a digraph that shows all three runoffs. Explain how the digraph can be used to determine the Condorcet winner (the candidate who can beat all the others in two-way races.)

Adding an arrow to a graph can be a useful modification in a situation in which the direction of a relationship is important. Another way to modify a graph is by writing a number by each edge. For example, a graph that indicates flights between cities is often more useful if the costs of the flights are shown in the graph. Suppose that the cost of a flight between cities A and B is $267, the cost of a flight between cities A and C is $438, and the cost of a flight between cities B and C is $193. To use a graph to show a flight between cities A and B and the cost, draw two vertices to represent cities A and B, connect them with an edge, and write the cost along the edge (**Figure 3.6**). Writing the cost alongside the edge is a useful modification if the graph is used to plan the least expensive trip.

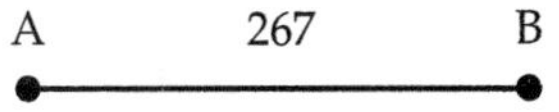

A   267   B

Figure 3.6.
A graph representing a flight between cities A and B.

A graph in which a number is attached to each edge is called a **weighted graph**. Note that in a weighted graph, it is not necessary to draw the edges so that their lengths are proportional to their weights.

2. Draw a weighted graph that shows all three flights.

Digraphs and weighted graphs are the most common variations of basic graphs. Whenever you represent a situation with a graph, it is important to think about whether it is useful to attach arrows or weights to the edges. You should use arrows or weights if the information they convey is essential to solving the optimization problem at hand.

In Items 3-10, show at least one way to represent the situation with a graph. Be sure it is clear what the vertices and edges of your graph represent. Pay particular attention to the optimization problem to be solved and ask yourself whether the direction arrows or weights of the edges are needed to solve the problem. If your graph is a digraph or a weighted graph, explain why you think the modification is important.

3. A large company's headquarters is a complex of five buildings that the company needs to connect with fiber-optic cable. Because of different distances and obstacles between some of the buildings, the

costs of connecting buildings vary as shown in **Figure 3.7**. The company is interested in finding the least expensive way to connect the buildings.

Building	A	B	C	D	E
A		$2,500	$1,700	$3,400	$1,900
B			$2,300	$1,800	$3,100
C				$2,400	$1,600
D					$2,900
E					

Figure 3.7.
Fiber-optic costs.

Note: Every entry in the Figure 3.7 table could be included twice. It is not necessary to do so since, for example, the cost of connecting A to B is the same as the cost of connecting B to A. For readability, tables in this unit usually do not display figures twice.

Leonard Euler (1707–1783) is considered one of the greatest mathematicians of all time. He produced half of his mathematical work after becoming completely blind. Some historians believe that his work on the Königsberg bridge problem is the first use of a graph to solve a problem.

4. The Swiss Mathematician Leonard Euler (pronounced oiler) once used a graph to solve a famous problem. The city of Königsberg in what was then Prussia (later Kaliningrad in Russia) is divided into several parts by the Pregel river. The parts of the city were at that time connected by seven bridges (**Figure 3.8**). Residents of Königsberg tried to walk through the city, cross each bridge exactly once, and return to their starting location. (Hint: In this problem, it is permissible to have more than one edge connecting a given pair of vertices.)

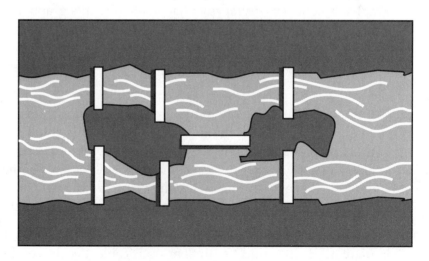

Figure 3.8.
Konigsberg and its seven bridges.

5. Five teams are involved in a round-robin tournament, which means that each team plays every other team. When the tournament is over, team A has defeated B and D; team B has defeated C, D, and E; team C has defeated A; team D has defeated C; and team E has defeated A, C, and D. Tournament officials must decide which team will receive the first-place trophy.

6. Circuit boards used in electronic equipment like stereo systems and computers are soldered by robots. A robot arm travels from point to point on the board, soldering as it goes, then returns to its starting point to await the next board. **Figure 3.9** shows distances (in centimeters) between pairs of points on a circuit board. The robot must be programmed to solder the five points as quickly as possible.

Points	A	B	C	D	E
A		2.3	2.1	3.4	2.6
B			4.0	3.1	2.0
C				3.3	3.7
D					3.5
E					

Figure 3.9.
Circuit board distances.

7. A zoo is updating its enclosures from cages that each hold a few members of a single species to larger habitats holding several species. A habitat cannot hold two species if one preys on the other. The zoo is trying to determine the most economical way to assign species to habitats.

Species A preys on B, C, and E.

Species B preys on C and D.

Species C preys on E and F.

Species D preys on F.

Species E and F do not prey on any of the others.

8. A developer is building a ski resort complex composed of four buildings. Roads must be constructed to connect the buildings, but it isn't necessary to build a road from every building to every other. The developer wants to build the roads as economically as possible. **Figure 3.10** shows the estimated costs the developer obtained from a road contractor.

Building	A	B	C	D
A		$3,500	$3,800	$4,200
B			$2,100	$2,400
C				$2,800
D				

Figure 3.10.
Estimated costs.

9. The manager of a rock band is planning a series of four gigs. The manager has obtained costs of plane fights between each pair of cities in the tour and wants to buy the least expensive set of tickets. Of course, the tour must start and end at the group's home, which is near Chicago. **Figure 3.11** is a table of plane ticket costs between various pairs of cities.

City	Chicago	St. Louis	Denver	Houston	Phoenix
Chicago		$118	$229	$312	$285
St. Louis			$198	$231	$271
Denver				$239	$164
Houston					$182
Phoenix					

Figure 3.11.
Plane ticket costs.

10. In most communities there are services that travel nearly every street: mail delivery, garbage collection, street cleaning, and milk and bottled water delivery. The people who manage these services want to cover the streets as efficiently as possible. Routes must be designed to avoid "deadheading," which is the term used to describe driving back over a street that has already been serviced. Draw a graph to represent the neighborhood in which you live. Explain why you think your graph is a useful model for someone planning a delivery route in your neighborhood.

11. Two people drew the graphs in **Figure 3.12** for the same situation. Could they both be right? Explain.

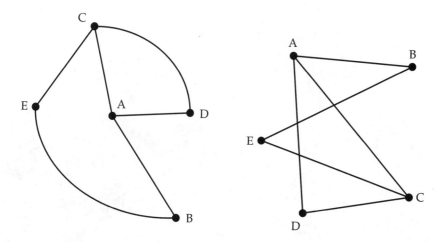

Figure 3.12.
Graphs drawn by two people.

12. In this lesson, you have used graphs to model a variety of situations. Although the situations are all different, the graphs you drew may have led you to believe that some of the situations are quite similar. Go back through the situations and group together those that you think are similar in some way. Explain your groupings.

LESSON TWO
Procedures

KEY CONCEPTS

Connected graph

Algorithm

David Barber

PREPARATION READING

Following Directions

Have you ever bought something that needed assembly? If so, you know that it's helpful to lay out all the pieces in an organized way. Making a graph to represent a problem is a lot like laying out the pieces in a meaningful way.

Once you are familiar with the pieces and have some idea of how they are related, you can begin the task of assembling the object. However, you don't put it together in whatever order comes to mind. You follow the directions. The directions that come with your purchase are a procedure that, when properly followed, takes you from a collection of parts to a finished product.

Similarly, a graph that represents a problem doesn't usually solve the problem. In some very simple cases, the graph and a little trial and error may be sufficient. That's true of some things you buy, too. If the object has only a few parts, and the relationships among the parts are clear, you may be able to assemble the object without reading the directions. However, trying to assemble a complicated object without reading the directions is risky business.

Using trial and error to solve a problem represented by a graph can be risky business, too, especially if the graph has quite a few vertices and edges. You may get an answer, but it may not be the best answer. Similarly, if you don't follow directions when you assemble something, the finished product might work, but it might not work as well as it should.

Solving problems with graphs requires procedures. Just as assembly directions must be written in clear steps that most people can follow, procedures for solving problems with graphs must be written clearly so they can be followed by most people.

How do you develop procedures for solving problems and write them in a clear way so that they can be followed by others?

ACTIVITY

STEPS TO A SOLUTION

2

In Item 3 of Individual Work 1, you made a graph to represent a situation in which a company is looking for the optimal way to connect its buildings with fiber-optic cable. **Figure 3.13** is the table of costs from that item.

Building	A	B	C	D	E
A		$2,500	$1,700	$3,400	$1,900
B			$2,300	$1,800	$3,100
C				$2,400	$1,600
D					$2,900
E					

Figure 3.13.
Fiber-optic costs.

1. Perhaps the problem of finding the cheapest way to connect the buildings can be solved without a graph.

a) Connecting five buildings requires only four cables. Explain why.

b) Since only four cables are needed, the problem can be solved with very simple directions: Construct the four cables that are cheapest. Use a graph to explain why these directions do not produce a good solution in this case.

c) Mathematicians use the term connected graph to describe a graph that is "all in one piece." A **connected graph** is one in which it is possible to get from any vertex to any other vertex by traveling along the graph's edges. Is the graph you made in part (b) a connected graph?

d) What is the minimum number of edges a connected graph with *n* vertices can have? (Hint: Mark several points on your paper. What is the minimum number of edges you need to make a connected graph out of the points?)

STEPS TO A SOLUTION

2. a) Explain how to use your graph to find a solution to the problem.

   b) How much will the fiber-optic connections cost the company?

3. Devise a procedure of your own that uses the graph you made in Item 3 of Individual Work 1 to solve the problem. Write the procedure in clear steps that you think others could follow. Number the steps. Explain why you think this procedure will produce optimal solutions in other problems like this one.

4. To show that your procedure works in other situations, apply it to another problem that has similar structure. You can use another problem from Lesson 1 or you can make up one of your own. If you do not think your procedure transfers well to a similar situation, then revise it.

5. Present your procedure to others in your class. Ask them to try your procedure on the fiber-optic problem as you present the procedure. Revise the wording of your procedure if others have trouble following it.

INDIVIDUAL WORK 2

The Right Procedure

Mathematicians use the word **algorithm** to describe a procedure for solving a problem. Algorithms can be written in sentences and paragraphs, but they are usually written in numbered steps so that they are easy to follow. In this individual work, you will examine several algorithms and write some of your own.

Not every algorithm is a good algorithm. Some do not always find the best solution to a problem. Some find a best solution, but are longer than necessary.

A good algorithm is easy to understand and gives a best solution to the problem it is intended to solve. If two different algorithms do both these things, mathematicians usually prefer the shorter one.

The word algorithm stems from a Latin translation of a mathematical work by the Persian al-Khowarizmi, whose name became Algoritmi in translation. The book was written around 825 A.D. and translated into Latin in the twelfth century.

1. The first algorithms in *Mathematics: Modeling Our World* are in Course 1, Unit 1, *Pick a Winner: Decision Making in a Democracy*. They are procedures for finding the winner of an election. Here, for example, is an algorithm intended to find the winner by assigning points to candidates according to the way voters rank them.

   1. Pick one of the candidates.

   2. Count the first-place, second-place, third-place, etc. votes for the candidate.

   3. Multiply the number of first-place votes by the number of candidates, the number of second-place votes by one less than the number of candidates, the number of third-place votes by two less than the number of candidates, etc.

   4. Add the results of the multiplications in step 3 to obtain the candidate's total.

   a) The algorithm is incomplete. What else must it do to determine a winner?

   b) Write the steps needed to complete the algorithm.

2. A computer or calculator program is a special type of algorithm. Often people hire programmers to write a program to their specifications. Those specifications might be a set of steps written in English describing what the program must do. If the steps are clear, the programmer can do the job.

Here is a set of steps describing a program to animate the horizontal motion of a pixel on a graphing calculator screen, which you may recall doing in Course 1, Unit 5, *Animation/Special Effects*.

> 1. Turn on a pixel on the left edge of the screen.
>
> 2. Turn on the pixel to the immediate right of the pixel in step 1.
>
> 3. Repeat step 2 until you reach the right edge of the screen.

Do you think this is a good algorithm for producing horizontal motion on a graphing calculator screen? Explain. If you think it is not a good algorithm, write a better one.

3. Another example of an algorithm that may be familiar to you is a magic number trick like the ones in Course 1, Unit 2, *Secret Codes and the Power of Algebra*. Devise a number trick of your own and write the directions in numbered steps. Explain what the magician must do to perform the trick.

4. In Course 1, Unit 4, *Prediction*, you used residuals to evaluate mathematical models. Here is an algorithm for finding residuals. Explain why it is or isn't a good algorithm.

> 1. Use a calculator to find a regression (linear, exponential, quadratic) model for the data.
>
> 2. Select any value of the explanatory variable from the original data and apply the model to obtain a predicted value.
>
> 3. Subtract the actual value of the response variable paired with the chosen value of the explanatory variable from the predicted value to obtain the residual.

5. Here is an algorithm for finding a certain point from several other points that have been graphed on a sheet of graph paper.

> 1. Find the median (or median interval) of the x-coordinates.
>
> 2. Find the median (or median interval) of the y-coordinates.
>
> 3. Locate a new point (or points) whose x-coordinate is the number found in step 1 and the y-coordinate is the number found in step 2.

For what purpose is this a good algorithm?

6. Here is another algorithm from Course 1, Unit 1, *Pick a Winner: Decision Making in a Democracy*. Its purpose is to determine the winner of an election by the runoff method. One of the steps of the algorithm is missing. Supply the missing step.

    1. Count the first-place votes for each candidate.

    2. Eliminate all the candidates except the two with the highest and second-highest vote totals.

    3.

    4. Total the first-place votes for the two candidates in the runoff.

    5. The winner is the candidate with the highest vote total.

7. Here is an algorithm for approximating the area of an irregular shape. You used a similar algorithm in Course 1, Unit 3, *Landsat*.

    1. Randomly distribute 50 pebbles over a known area containing the irregular shape.

    2. Count the number of pebbles that landed on the irregular shape.

    3. Divide the number of pebbles in the irregular figure by 50.

Is this a good algorithm for determining the area of an irregular shape? Explain your answer.

8. Here is an algorithm for solving a problem like the fiber-optic problem in Activity 2.

    1. Pick the least expensive edge and mark it in some way.

    2. Pick the least expensive of the remaining (unmarked edges) and mark it.

    3. Check to be sure that the edge you just marked is needed. Reject (erase) the edge if it isn't needed.

    4. Repeat steps 2 and 3 until the graph is connected. The marked edges are the ones to use.

a) Apply the algorithm to the fiber optic problem. What do you think is meant by "needed" in step 3?

b) Do you think this is a good algorithm? Explain.

9. Sometimes people have a hard time developing an algorithm because they can't get started. In a problem like the fiber-optic problem, getting started isn't so hard because you are trying to find the cheapest way to connect the buildings, and it makes sense to start by picking the cheapest edge. When the edges do not have weights, getting started isn't as easy. In such situations, algorithms might start by making a random selection.

Figure 3.14 is a graph for the management problem in Activity 1. Edges indicate that contracts cannot share computer time. Each contract requires an hour of computer time. The problem is to determine the minimum number of hours of computer time to rent.

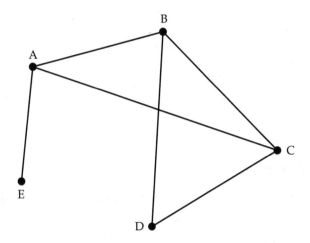

Figure 3.14.

An algorithm starts with this step:

>   1. Randomly pick one of the vertices and label it Hour 1.

a) Suggest a second step. In other words, what might the algorithm do next?

You also can start an algorithm by ranking the vertices in some way. Another algorithm starts with this step:

>   1. Pick the vertex that has the most edges and label it Hour 1.

b) Suggest a second step for this algorithm.

10. Ties can cause problems for algorithms. For example, the graph in **Figure 3.15** represents the round-robin tournament described in Item 5 of Individual Work 1. The vertices are teams, and an arrow means that the team the arrow is pointing away from beat the team it is pointed toward.

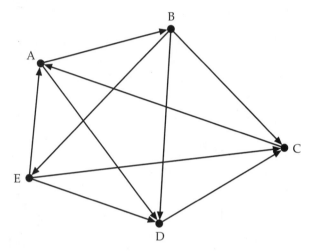

Figure 3.15.
The results of a round-robin tournament.

A simple algorithm for finding the winner of this tournament is:

1. Select a vertex and count the number of arrows pointing away from it.

2. Repeat step 1 for every vertex.

3. The winner is the vertex with the most arrows pointing away from it.

a) What happens if this algorithm is applied to the graph in Figure 3.15?

b) Modify the algorithm so that it is better suited to Figure 3.15.

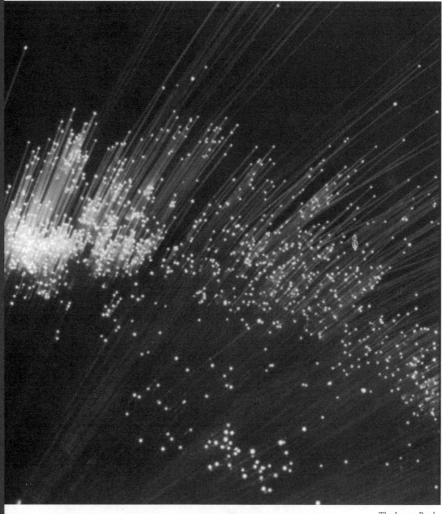

The Image Bank

PREPARATION READING

Trees

There are many kinds of situations that can be modeled with a graph. Some of them are quite similar. For example, the fiber-optic problem in Item 3 of Individual Work 1 is similar to the ski resort problem in Item 8 of Individual Work 1. In each case, the vertices of the graph must be connected with the fewest and cheapest edges possible.

A graph that is connected and has the fewest possible edges is called a **tree**. In a tree you cannot find a group of edges that are closed up in some way. Mathematicians use the word **cycle** to describe a portion of a graph that is closed. A cycle is a path through a graph that begins and ends at the same vertex without repeating any edges or vertices. Thus, a tree is a graph with no cycles.

In the fiber-optic and ski-resort situations, the problem is to find a tree whose edge weights have the smallest possible total. Such a tree is called a **minimum spanning tree**. As you saw in the fiber-optic and ski-resort situations, a minimum spanning tree has one fewer edge than the graph has vertices.

You were challenged to write an algorithm for the fiber-optic problem in Item 3 of Activity 2. You considered algorithms written by others in your class in that same activity. Item 7 of Individual Work 2 asked you to consider another algorithm.

You have had a chance to decide if these algorithms are written clearly. You've also thought about whether they are correct; that is, whether they really do find the optimal solution in the fiber-optic problem. However, in order to be correct, an algorithm must work well in similar situations, including ones that are more complex.

In this lesson, you will consider the correctness of the fiber-optic algorithms from Lesson 2 by applying them to problems that are more complicated (and, therefore, more realistic). If you are dissatisfied with these algorithms, you will revise them.

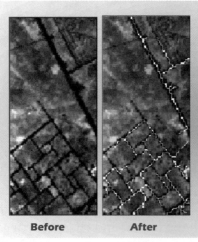

Before   **After**

Researchers in England have found a way to use a minimum spanning tree algorithms to extract features like mountain ranges and rivers from Landsat images. This is a process that was previously done by hand and is quite time consuming.

These photographs show an image before and after application of the technique.

ACTIVITY

GROWING MINIMUM SPANNING TREES

3

The fiber-optic problem is realistic because satisfactory communications can be obtained from a minimal number of connections. If you use more cable than is necessary, you do not improve the communication system. However, the fiber optic problem given in the previous lesson has fewer places to connect than most real applications.

Your task in this activity is to apply an algorithm from Lesson 2 to a more realistic situation. The situation serves as a test of your algorithm. If the algorithm does not handle this new situation, you should revise it. Mathematicians must test their work against reality and revise it when it fails to handle a new situation well.

Figure 3.16 is a table of costs of connecting ten locations with fiber-optic cable. A table of the costs of connections is sometimes called a cost matrix.

	A	B	C	D	E	F	G	H	I	J
A		$5420	$4200	$4780	$4180	$1950	$5020	$449	$5060	$5140
B			$8490	$8750	$9290	$3600	$7880	$2920	$1550	$720
C				$300	$5710	$5230	$8640	$6090	$7440	$8490
D					$6420	$5650	$9330	$6180	$7620	$8810
E						$5960	$4160	$8660	$9170	$8850
F							$5850	$2920	$3220	$3400
G								$8710	$8470	$7230
H									$1590	$3400
I										$1950
J										

Figure 3.16.
Costs of fiber-optic connections.

ACTIVITY

3

GROWING MINIMUM SPANNING TREES

1. Make a graph to represent this situation. (Suggestion: Use a full sheet of paper and make a large graph. You will also find it helpful to arrange the vertices in a circular formation when a graph has many vertices.)

2. Select two algorithms from Lesson 2. You can use the algorithm you wrote for Item 3 of Activity 2, one written by another student, or the algorithm in Item 8 of Individual Work 2. Solve the problem of finding the optimal way to connect the locations by applying both algorithms you selected. Be sure to show which locations are connected and the cost of connecting them.

3. If the algorithms do not handle this situation, explain why. Then modify at least one of the algorithms so that it does handle this situation and solve the problem.

4. Select one of the two algorithms you chose in Item 2 or the modification you made in Item 3 and explain why you think it gives the optimal solution.

CONSIDER:

1. How can you be sure that a minimum spanning tree algorithm will work in all situations?

2. If there is more than one correct minimum spanning tree algorithm, how should you choose which to use?

INDIVIDUAL WORK 3

Optimal Connections

1. A company is building an amusement park with five major attractions connected by a system of canals. **Figure 3.17** shows the distances in meters between each pair of attractions. The cost of building the canals is proportional to the distances, and the company wants to build the canal system in the least expensive way.

	Kiddie Land	Olympic Village	Sea Universe	World Palace	CatMan Coaster
KL		560	750	500	440
OV			1000	800	680
SU				630	300
WP					520
CC					

Figure 3.17.
Amusement park distances.

a) Model this situation with a graph.

b) Apply an algorithm to the graph to find the optimal way to build the canal system.

c) Here is another algorithm for solving minimum spanning tree problems. Apply it to this situation and decide it if it is a good algorithm. Explain your decision.

1. Select any vertex and mark the cheapest edge that is attached to it.

2. Follow the chosen edge to the vertex at the other end.

3. Find the cheapest unmarked edge that is attached to this vertex. If it does not make a cycle, mark it. If it makes a cycle, find the cheapest edge attached to this vertex that does not make a cycle and mark it.

4. Repeat steps 2 and 3 until the number of marked edges is one less than the number of vertices.

2. One reason algorithms are useful is that they can be translated into computer programs. Programs that use algorithms to solve problems about graphs are sometimes used to control the motion of robots. For example, a factory uses robots to move material from one place to another. The robots must be able to get from one location to another along paths that are clearly marked on the factory floor so that employees can keep the paths clear of obstacles. The table in **Figure 3.18** shows the distances in feet between various locations the robots must visit. (Note: Some pairs of locations have obstacles between them, so a robot cannot travel directly from one location to the other.)

	A	B	C	D	E	F
A		35	38		49	30
B				48	61	57
C				32		42
D					52	
E						47
F						

Figure 3.18.
Robot travel distances.

Design a path system that has the smallest possible total length.

3. The company that you connected with fiber-optic cable in Activity 3 has six new locations that it wants to connect. The costs are shown in **Figure 3.19**.

	P	Q	R	S	T	U
P						
Q	$3200					
R	$2500	$4300				
S	$4700	$1950	$6000			
T	$3900	$3000	$2900	$4750		
U	$4400	$5400	$2250	$7100	$2800	

Figure 3.19.
Costs of fiber-optic connections.

a) Design a network of cable to connect these six locations.

b) The company wants to connect this new network to the one you designed in Activity 3. Describe an optimal way to do this.

4. When you make a graph to represent situations like the fiber-optic problem and the canal problem in Item 1, you connect every pair of vertices with an edge. That's because it's possible to connect any pair of locations in the situation the graph represents. A graph in which every pair of vertices is connected is called a **complete graph**.

a) When is it possible to draw a complete graph without crossing any of the edges? Explain.

b) When is it possible to draw a minimum spanning tree without crossing any of the edges?

5. In some situations, like the fiber-optic problem, the graph isn't a complete graph. For example, obstacles may make connecting a pair of buildings impossible. In other situations, it may be unnecessary to attach weights to the edges of the graph. For example, weights are not needed if the cost of connecting each pair of buildings is the same. **Figures 3.20–3.22** represent situations in which neither complete graphs nor weights are necessary. All that's needed is to find a way to connect the vertices with as few edges as possible.

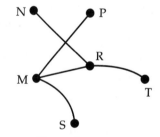

Figure 3.20.

a) In each of Figures 3.20–3.22, find a spanning tree. A **spanning tree** is a tree that includes all the vertices of the graph.

b) Can you draw a graph for which it is impossible to find a spanning tree?

c) Here are the first two steps of an algorithm for finding a spanning tree of a graph. Complete the algorithm.

1. Pick any edge and mark it.

2. Pick an unmarked edge. Check to see if it makes a cycle with any of the marked edges. If it does, reject it. If it does not, mark it.

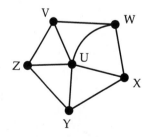

Figure 3.21.

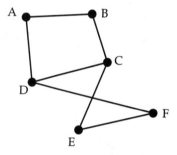

Figure 3.22.

6. The table in **Figure 3.23** is a cost matrix for a situation that can be modeled with a graph.

	A	B	C	D	E
A		118	229	312	285
B			198	231	271
C				239	164
D					182
E					

Figure 3.23.
A cost matrix.

a) Draw a graph to represent this situation.

b) Use an algorithm to find a minimum spanning tree for your graph.

c) A situation that uses this cost matrix is in Item 9 of Individual Work 1. Go back and read that situation. Does the minimum spanning tree you made in part (b) solve the problem? Explain.

7. Devise a situation in which a minimum spanning tree solves the problem. Write your situation on a piece of paper, but do not include an answer. Give the paper to someone else in your class and ask them to make a graph to represent the situation and to use an algorithm to solve the problem. Compare your answer to the other person's. Resolve any differences.

8. Here is an algorithm that mathematicians sometimes use to solve minimum spanning tree problems. It is called Prim's algorithm.

   1. Mark the cheapest edge of the graph and its two vertices.

   2. Find the cheapest unmarked edge that has one marked vertex and one unmarked vertex. Mark this edge and its unmarked vertex.

   3. Repeat step 2 until all the graph's vertices are marked.

   a) Use Prim's algorithm to find the minimum spanning tree for the graph you made in Activity 3.

   b) Compare Prim's algorithm to the algorithm you used for Item 1 of this individual work. Which do you prefer? Explain.

9. a) Complete the table in **Figure 3.24** to show the number of edges in a complete graph, the number of edges in a minimum spanning tree, and the number of edges in a cycle with each number of vertices. Make drawings or use matrices if necessary.

Number of Vertices	Number of edges in a complete graph	Number of edges in a minimum spanning tree	Number of edges in a cycle
2			
3			
4			
5			
6			

Figure 3.24.
Table of vertices and edges.

b) Find mathematical functions that model the relationship between the number of vertices and the number of edges in a complete graph, the relationship between the number of vertices and the number of edges in a minimum spanning tree, and the relationship between the number of vertices and the number of edges in a cycle.

LESSON FOUR

Coloring to Avoid Conflicts

KEY CONCEPTS

Conflict resolution

Graph theory

Vertex coloring

Degree
(of a vertex)

Subgraph

Counterexample

Incidence matrix

The Image Bank

PREPARATION READING

Conflict Resolution

There are many real-world situations that involve conflicts. Conflicts occur whenever it is not possible to group two or more things together. Putting them together may not mean putting them in the same location. It may mean assigning them the same task or the same time slot. Problems about how to group things so that conflicts are avoided are called conflict resolution problems. In conflict resolution problems, the ideal solution is one that avoids all conflicts.

Mathematicians have found graphs helpful in solving conflict resolution problems. But a graph is not a solution, so mathematicians have also devised algorithms to apply to graphs that represent conflict resolution problems.

In this lesson you consider how to use a graph to represent conflict resolution problems and devise algorithms to apply to the graphs.

Conflict resolution problems are different from minimum spanning tree problems in many ways. For one, the graphs that represent the situation are not weighted. Another major difference is that mathematicians have not succeeded in devising an algorithm that always produces the optimal solution in conflict resolution problems. That is very different from minimum spanning tree problems, for which mathematicians know several algorithms that give optimal solutions.

Minimum spanning tree problems and conflict resolution problems are two examples of situations that can be modeled with graphs. The branch of mathematics that studies graphs is called graph theory. You are about to consider an area of graph theory in which mathematicians are still puzzled and in which non-mathematicians have made significant contributions. Perhaps your work in this lesson will lead you to make a contribution of your own someday!

ACTIVITY

HOW MANY TIMES?

4

The director of a community center is planning an evening of "sampler activities" intended to determine which activities are popular enough to schedule regularly.

Figure 3.25 shows the people who have signed up for each activity.

Guitar	Magic	Dance	Comedy	Acrobatics	Singing	Juggling
Alan	Beatriz	Henry	Anna	Alan	Tahira	Brad
Tahira	Brad	Beatriz	Latoya	Tim		Susan
Anna	Latoya	Peter		Susan		
Tim	Henry					
Peter						

Figure 3.25.
Activity sign-up list.

The director wants to schedule the activities in one-hour time slots starting at 7 p.m. and ending no later than 10 p.m. If possible, each person should be able to attend each activity they chose.

Of course, the problem is fairly simple and can therefore be solved by trial and error. Since not all conflict resolution problems are this simple, mathematicians need a procedure for solving the more difficult problems. In this activity, you will consider ways to represent the problem with a graph and how the problem might be solved.

1. Use a graph to represent this problem. In order to do so, you need to consider what the vertices should represent and what criterion to use for drawing edges. Try drawing graphs in different ways. Then discuss which graph might be more useful in solving the problem. (You might consider how you represented similar situations in Lesson 1.)

HOW MANY TIMES?

2. Choose one of your graphs and use it to solve the problem. Since the problem involves finding time slots for each activity, you might write the times 7, 8, and 9 on your graph as you work on the problem. If you decide that the graph you are using isn't very helpful, then reconsider your reasons for choosing the graph and try another of the graphs you made in Item 1.

3. Write an explanation of how you used your graph to solve the problem of scheduling the activities. You can write the description in paragraph form or you can write it as an algorithm with numbered steps. Choose the method that you think makes your explanation clearest to another person.

CONSIDER:

1. In the activity scheduling problem, are the activities in conflict or the people?

2. How can your answer to the previous question be used as a guide to representing a conflict resolution situation with a graph?

3. What guide can you give for drawing the edges in a conflict resolution graph?

INDIVIDUAL WORK 4

Coloring Graphs

1. In Lesson 1, you used graphs to represent a variety of real-world situations. Go back to Lesson 1 and identify the situations that involve conflict resolution.

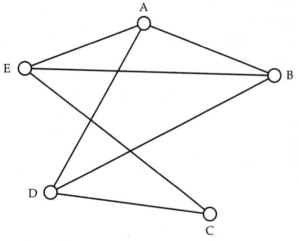

Figure 3.26.
A conflict resolution graph.

2. The graph in **Figure 3.26** represents a conflict resolution situation.

a) Make a table showing the conflicts among the vertices.

b) When you use a graph to solve conflict resolution problems, it is helpful to mark the vertices in some way. Often a second label is added: the label might be times in one situation, zoo habitats in another. Mathematicians frequently use colors as an all-purpose method of labeling vertices and sometimes call conflict resolution problems **vertex coloring** problems. To make the vertices easy to color, they use empty circles for the vertices and shade the circles with colors or write the first letter of a color inside the circle. Find a way to color the vertices of the graph in Figure 3.26 that uses as few colors as possible and has no two connected vertices colored the same.

3. a) Mathematicians call the number of edges that meet at a vertex its **degree**. Find the degree of each vertex in Figure 3.26.

b) What does the degree of the vertex mean in the situation the graph represents?

4. To help determine the minimum number of colors that graphs need, mathematicians study groups of graphs that have something in common. For example, the graphs in **Figure 3.27** are all empty graphs. An empty graph has no edges and represents situations with no conflicts.

a) Explain why every empty graph needs only one color.

Figure 3.27.
Empty graphs with 2, 3, and 4 vertices.

b) **Figure 3.28** shows three different complete graphs. Find the minimum number of colors for each of these graphs. What can you say about the number of colors needed to color complete graphs? Explain.

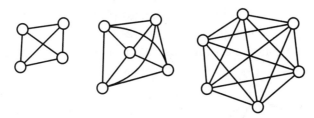

Figure 3.28.
Complete graphs of 4, 5, and 6 vertices.

c) **Figure 3.29** shows four different graphs that are cycles. Find the minimum number of colors for each of them. What can you say about the number of colors needed to color cycles?

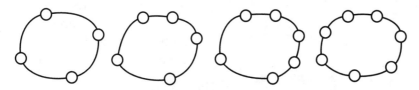

Figure 3.29.
Cycles of 4, 5, 6, and 7 vertices.

d) **Figure 3.30** shows five different graphs that are trees. Find the minimum number of colors for each of them.

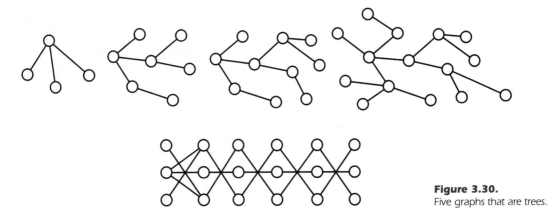

Figure 3.30.
Five graphs that are trees.

What can you say about the number of colors needed to color trees?

INDIVIDUAL WORK 4

5. Mathematicians find it helpful to know something about the number of colors that common graphs such as complete graphs, cycles, and trees require. That's because common graphs are often contained in large, unfamiliar graphs and a large, unfamiliar graph needs at least as many colors as any of its parts. A graph that is part of another graph is called a **subgraph** of the graph that contains it. The vertices and edges of a subgraph are vertices and edges of the original graph. **Figure 3.31** shows a subgraph that is a cycle of three vertices.

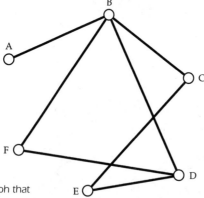

Figure 3.31.
A graph with a subgraph that is a cycle of three vertices.

a) Find another cycle that is a subgraph of Figure 3.31.

b) Use your knowledge of sub-graphs to analyze Figure 3.26 in Item 2 of this individual work. Does what you have learned about cycles prove that you used the minimum number of colors in your answer to part (b) of Item 2?

6. The manager of a chemical storage facility keeps dangerous chemicals in sealed cabinets in case of leakage. However, some chemicals cannot be stored in the same cabinet with others because of the possibility of fire or explosion if the chemicals come in contact with each other. **Figure 3.32** is a table of incompatible chemicals.

a) Represent the chemical incompatibilities in a graph.

b) Use your knowledge of sub-graphs to determine a minimum number of colors for the graph.

c) Try to color the graph with the minimum number of colors you found in part (b).

Chemical	Incompatible with
A	B, D, F, H
B	A, J, K, L
C	I
D	A, L
E	G
F	A, L
G	E, I
H	A, J
I	C, G
J	B, H
K	B
L	B, D, F

Figure 3.32.
Incompatible chemicals.

7. The exhibit director of an aquarium is planning several large tanks containing different species of fish. Some pairs of species cannot be in the same tank because one species has been known to attack the other. The table in **Figure 3.33** shows the incompatibilities.

Chemical	Incompatible with
A	B, C, I, J
B	A, G
C	A, I, J
D	E
E	D
F	G, J, L
G	B, F, I
H	– (H is very compatible!)
I	C, G
J	B, H
K	B
L	B, D, F
M	N
N	M

Figure 3.33.
Incompatible aquarium species.

a) Represent the incompatibilities in a graph.

b) Use your knowledge of subgraphs to determine a minimum number of colors for the graph.

c) Try to color the graph with the minimum number of colors you found in part (b).

DESIGNING A COLORING ALGORITHM

In Individual Work 4, you found a rule that helps determine the number of colors needed to color a conflict resolution graph. That rule says that a graph requires at least as many colors as any subgraph. If, for example, you know that a subgraph requires three colors, then the entire graph requires at least three colors.

Rules like this one are helpful, but only in certain situations. People who regularly solve conflict-resolution problems need an algorithm they can use on any graph—not just certain special types of graphs.

Now your challenge is to write an algorithm that can be used on all graphs to find a vertex coloring that uses the fewest possible colors.

One way to develop an algorithm is to record exactly what you are thinking and doing as you color the vertices of a particular graph. Then use your notes as a rough draft to be refined as you think more carefully about the specific steps. You may, for example, want to make use of the explanation you wrote in Item 3 of Activity 4. Remember that your algorithm should be written clearly enough that other people can follow it correctly on their own graphs without your help.

When you have finished writing your algorithm, try it on some situations such as the one in Activity 4 and the ones in Items 6 and 7 of Individual Work 4.

You will have a chance to enter your algorithm in a contest with other algorithms to see which algorithms color a group of test graphs correctly.

INDIVIDUAL WORK 5

Coloring Our World

No algorithm has been found that colors every graph with the fewest colors possible. To put it differently, for every algorithm that has been proposed, there are graphs that the algorithm does not color with the fewest colors possible. When a graph causes an algorithm to use more than the minimum possible number of colors, it is called a **counterexample** for that algorithm. In this individual work you will consider several vertex coloring algorithms and try to find a counterexample for each of them.

1. The director of a museum is planning a culture day in which twelve different cultural groups will make presentations. The groups require certain facilities for their exhibits. For example, some groups are doing food demonstrations and need cooking or other kitchen facilities. **Figure 3.34** shows groups that need the same facilities and cannot be scheduled at the same time. The director wants to schedule the presentations in the fewest time slots possible.

   a) Represent the situation with a graph.

   b) Use an algorithm from Activity 5 to color the graph.

   c) Did your algorithm use the minimum number of colors possible? Explain.

   d) Is the graph in part (a) a counterexample for your algorithm?

Cultural area	Conflicts with
1	5, 6, 7, 11
2	5, 6, 8, 12
3	6, 10, 11
4	5, 9, 11, 12
5	1, 2, 4, 10
6	1, 2, 3, 9
7	1, 9, 12
8	2, 9, 10, 11
9	4, 6, 7, 8
10	3, 5, 8, 12
11	1, 3, 4, 8
12	2, 4, 7, 10

Figure 3.34.
A table of conflicts.

INDIVIDUAL WORK 5

User	Cannot share with
A	B, F, H
B	A, C, F, H
C	B, D, G
D	C, E, G
E	D
F	A, B, H
G	C, D
H	A, B, F

Figure 3.35.
Mobile radio users.

Variable	Active at same time
A	C, D, G, I, J
B	C, G, J
C	A, B, D, E, F, H
D	A, C, E, G, I
E	C, D, H, I, J
F	C, G, H, J
G	A, B, D, F, J
H	C, E, F, I
I	A, D, E, H, J
J	A, B, E, F, G, I

Figure 3.36.
Computer program variables.

2. Mobile radio communication is quite popular. One challenge presented by this popularity is the assignment of radio frequencies so that one user does not interfere with another. If two users are sufficiently close, they cannot share the same frequency. **Figure 3.35** shows which users cannot share frequencies.

a) Represent this situation with a graph.

b) Use any algorithm to color the graph.

c) Is your graph a counterexample for the algorithm you chose?

3. Computer scientists are interested in finding ways to improve the speed at which computers perform their tasks. One way that computer scientists do this is by assigning variables to hardware registers because variables in registers can be accessed more quickly than variables not in registers. However, a computer program usually has more variables than there are hardware locations, and two variables that are active at the same time cannot share a register. **Figure 3.36** shows which of a program's variables are active at the same time. What is the smallest number of hardware registers the programmer needs?

a) Represent this situation with a graph.

b) Use any algorithm to color the graph.

c) Is your graph a counterexample for the algorithm you chose?

4. Apply the following algorithm to one or more graphs. Try to find a counterexample. When you have found one, draw the graph on your paper and record the order in which you colored the vertices and the colors you used.

    1. Number the vertices in any order from 1 through n (where n is the number of vertices).

    2. Place the colors in a list.

    3. Color the vertices in order. Use the first color that is permissible.

5. The following algorithm is called the Welsh and Powell algorithm. Find a counterexample for this algorithm.

    1. Place the vertices in a list from highest degree to lowest. (Break ties arbitrarily.)

    2. Place the colors in a list.

    3. Color the vertices in order. Use the first color that is permissible.

> Several computer scientists have developed and patented coloring algorithms for register allocation. Of course, none of them produce a minimum coloring for all register allocation problems.

6. The following algorithm is called the color degree algorithm. Find a counterexample for this algorithm.

    1. Place the colors in a list.

    2. Color any vertex with the first color.

    3. Find the uncolored vertex that is connected to the largest number of colored vertices. (Break ties arbitrarily.) Color it with the first color in the list that is permissible.

    4. Repeat step 3 until all the vertices are colored.

7. Graph coloring algorithms are sometimes translated into computer programs because computers can solve large problems much more quickly than people can. Although a graph is a useful way for humans to visualize a conflict resolution situation, that is not the case for computers because computers do not have eyes. Computer scientists use an **incidence matrix** to communicate a graph to a computer. An incidence matrix is a matrix in which the rows and columns are labeled with the vertices of the graph. A 1 is written in the matrix if two vertices share an edge; a 0 is written if they do not.

INDIVIDUAL WORK 5

a) **Figure 3.37** is an incidence matrix for a graph that represents a conflict resolution situation. Draw the graph.

b) A computer can use an incidence matrix to determine certain properties of a graph. For example, if the computer uses a coloring algorithm that colors the vertices in order according to their degrees, the computer can determine the degree of each vertex from the matrix. Explain how the computer does this.

$$\begin{array}{c c} & \begin{array}{c c c c c} A & B & C & D & E \end{array} \\ \begin{array}{c} A \\ B \\ C \\ D \\ E \end{array} & \left[\begin{array}{c c c c c} 0 & 1 & 0 & 1 & 0 \\ 1 & 0 & 1 & 1 & 0 \\ 0 & 1 & 0 & 0 & 1 \\ 1 & 1 & 0 & 0 & 1 \\ 0 & 0 & 1 & 1 & 0 \end{array} \right] \end{array}$$

Figure 3.37.
An incidence matrix.

c) How do you think a matrix could be used to communicate the graph in a minimum spanning tree problem to a computer? Explain.

8. **Figure 3.38** is a map of the lower 48 United States that are west of the Mississippi. Often maps are colored so that no two bordering states have the same color. For centuries mathematicians were unsure of the minimum number of colors needed to color any map drawn on a plane surface. In 1976, mathematicians Kenneth Apel and Wolfgang Haken of the University of Illinois proved that no such map needs more than four colors.

a) Can this map be colored with fewer than four colors? Explain.

b) How is this problem like a vertex coloring problem?

Figure 3.38.
The United States west of the Mississippi.

LESSON FIVE
Traveling Salesperson Problems

KEY CONCEPTS

Traveling salesperson problem (TSP)

Nearest neighbor algorithm

Hamiltonian circuit

The Image Bank

PREPARATION READING

Will I Get Back Home?

*I*n Lesson 4, you read that graph theory is a branch of mathematics in which mathematicians do not yet have all the answers. For example, they lack an algorithm that always produces a minimal coloring in vertex coloring problems.

Because vertex coloring problems have many real-world applications, many people are engaged in the search for better vertex-coloring algorithms. One of the best vertex-coloring algorithms that mathematicians have found is the color-degree algorithm (Item 6 of Individual Work 5).

Another graph theory problem that still puzzles mathematicians and has many real-world applications is called the **traveling salesperson problem** (abbreviated TSP). It is so named because in its original form it involves a salesperson who wants to visit customers in several cities and return home. Of course, the salesperson wants to use the shortest (or fastest or cheapest) routes. The traveling salesperson problem is a part of graph theory because it can easily be represented with a graph. In terms of the graph, the problem is to find the cheapest way to start at a designated vertex, visit each vertex exactly once, and return to the starting vertex.

In this lesson you will consider traveling salesperson problems. Can you find an algorithm that gives reasonably good solutions to these problems?

The Image Bank

ACTIVITY

THE TSP

6

In most companies there is an individual (perhaps an entire office in large companies) who handles travel arrangements. Moreover, in many companies there are employees who take trips to visit branch offices or customers in several cities and return home. In such cases, the order in which the cities are visited often does not matter, and the person who handles travel arrangements is expected to plan the trip as economically as possible.

In this activity you will consider a simple traveling salesperson problem and write an algorithm for solving similar problems.

Figure 3.39 is a table of ticket prices for flying from one city to another. In this case, although it may not be true in all cases, the cost of flying between two cities is the same regardless of direction.

City	A	B	C	D	E
A		$238	$422	$357	$512
B			$258	$489	$333
C				$289	$505
D					$435
E					

Figure 3.39.
Plane ticket costs.

1. Show how this situation can be modeled with a graph.

2. Suppose the traveler's home is in city A. The trip must start and end there, and it must visit each of the other cities. Write an algorithm that you think will result in a reasonably inexpensive trip in this case and in similar situations. When your algorithm is written, apply it to this situation. List the route that it produces and the total cost.

3. Compare your algorithm, the route that it produced, and the total cost that it produced to those of others in your class.

4. What is the cheapest possible route in this case? The only way of knowing for sure is to list all possible costs and select the smallest. Do so. (Note: some people find a tree diagram helpful when listing all possibilities in a situation like this.)

CONSIDER:

1. Is it necessary to examine every possible route to be sure your algorithm produces the lowest cost?

2. Some people confuse minimum spanning tree problems with traveling salesperson problems. How are they alike and how are they different?

INDIVIDUAL WORK 6

No Place Like Home

*I*n this individual work you will consider several issues related to TSPs.

1. An employee of a company in Dallas will visit branch offices in Los Angeles, St. Louis, and Seattle, then return home. The company's travel office has obtained the ticket costs in **Figure 3.40**.

City	Dallas	Los Angeles	St. Louis	Seattle
Dallas		$375	$450	$425
Los Angeles			$525	$475
St. Louis				$800
Seattle				

Figure 3.40.
Plane ticket costs.

a) Model this situation with a graph.

b) Apply your algorithm from Activity 6 to this situation. What route does it produce and what is the total cost?

c) Make a list of all possible costs. Did your algorithm find the cheapest trip?

2. **Figure 3.41** is a list of distances in miles among several popular locations in New England.

	C.T.	F.	L.M.	L.R.G.	M.W.	P.C.P.	S.L.
Cannon Tramway		7	14	17	53	36	46
Flume			7	10	46	29	44
Loon Mountain				8	42	27	37
Lost River Gorge					50	32	45
Mt. Washington						22	47
Polar Caves Park							63
Story Land							

Figure 3.41.
Distances among New England locations.

a) Prepare a graph to represent this information.

b) A tourist plans to start at Polar Caves Park, visit each of the other locations and return to the park. Apply your algorithm from Activity 6 to the problem of finding the shortest route for the tourist. List the route produced and the total distance.

c) Since the list of all possible routes is quite long, it is unreasonable to ask you to make the list. Instead, compare your route to those found by others in your class, or find several routes found by trial and error. Can you find a route that is shorter than the one produced by your algorithm?

3. Return to Lesson 1 and identify the situations that are traveling salesperson problems. Apply your algorithm from Activity 6 to each situation. (If a starting location is not identified, use the first location in the table.) List all possible routes and determine whether your algorithm produces the cheapest route.

Mathematicians at IBM's Tokyo Research Laboratory reduced the time it took a robot arm to drill 1,129 holes in a circuit board by over 75% when they implemented a new TSP algorithm.

Source: IBM Corporation

OLD METHOD   NEW METHOD

4. Here is an algorithm for TSPs. It is called the nearest neighbor algorithm.

> 1. From the starting vertex, travel to the nearest vertex. In other words, travel along the cheapest edge.
>
> 2. From the vertex at which you are now located, travel to the nearest vertex not yet visited.
>
> 3. Repeat step 2 until you have visited every vertex, then travel back home.

a) Apply this algorithm to the problem in Activity 6. What route is produced and what is the total cost?

b) Did the algorithm produce the optimal solution?

5. A traveling salesperson problem is a type of Hamiltonian circuit problem. A **Hamiltonian circuit** is a path in a graph that visits each vertex exactly once and starts and ends at the same vertex. A traveling salesperson problem requires finding a Hamiltonian circuit of minimal total weight.

a) **Figure 3.42** shows three unweighted graphs. In each, find a Hamiltonian circuit that starts at any vertex.

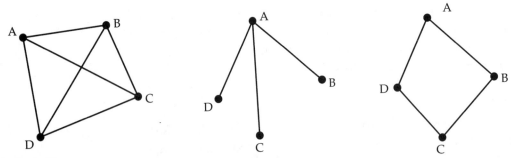

Figure 3.42.
Three graphs.

b) Mathematicians have not found a way to determine if any graph has a Hamiltonian circuit, but they know a lot about certain special kinds of graphs. Experiment with trees, complete graphs, and cycles. When do these special graphs have Hamiltonian circuits?

6. In some problems similar to TSPs, it is not possible to travel between every pair of locations. When this happens, the graph that represents the situation is not a complete graph. **Figure 3.43** represents a situation in which it is not possible to travel between every pair of vertices.

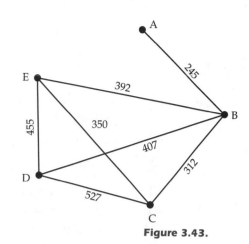

Figure 3.43.

a) In a real situation, why might it not be possible to travel from A to C, D, or E?

b) How can a traveling salesperson algorithm be adapted to this situation?

c) Apply a TSP algorithm to this situation. Find a route and a total cost.

d) Is your solution a Hamiltonian circuit? Explain.

7. Use a map of your neighborhood, town, or state to select five or six locations. Prepare a table showing the distances between each pair of locations. Model the situation with a graph. Use an algorithm to design a minimal Hamiltonian circuit that starts and ends at one of the locations. Try to find a route that is shorter than the one found by your algorithm.

8. The only certain way to find the shortest Hamiltonian circuit in a traveling salesperson problem is to list every possible circuit and pick the shortest one. Is this a reasonable way to solve such problems? To answer that question you need to know more about how many possible circuits must be checked.

a) The simplest TSP involves a complete graph with three vertices. Consider the example in **Figure 3.44**. Explain why there is only one total cost possible in this situation.

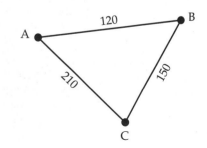

Figure 3.44.
An example of the simplest TSP.

Number of vertices	Number of costs to check
3	1
4	
5	
6	60

Figure 3.45.

b) Complete the table in **Figure 3.45**. Use your earlier work in this lesson whenever possible.

c) Look for a pattern in your table. Predict the number of different costs to check when there are 7 vertices.

d) Use a calculator or a spreadsheet to extend the table to 20 vertices.

e) Computers are very fast. For example, many are capable of performing well over a million operations a second. Assume that an operation is equivalent to listing one route in a TSP. How long would it take a computer that can perform one million operations per second to list all possible routes in a 20-vertex TSP?

San Francisco, Dec. 17, 1996 (Reuters)

Intel Corp. says it has developed the world's fastest supercomputer, capable of performing 1 trillion operations a second, for the Department of Energy.

That speed comes close to tripling the previous record achieved by Hitachi Ltd. in 1995 with a supercomputer capable of doing 368 billion operations a second, Intel and government officials said.

The Intel supercomputer will be used at the government's Sandia National Laboratories to simulate the performance of nuclear weapons, replacing live tests of stockpiled weapons.

LESSON SIX
Matching

KEY CONCEPTS

Bipartite graph

Stable matching

The Image Bank

PREPARATION READING

Is Everybody Happy?

*P*erhaps you can recall a situation from your childhood in which you and someone else wanted the same thing, but there was only one item to be had. Such situations are common, and they often involve more than just two people and more than just one item.

Suppose you are a newspaper editor who must assign reporters to stories about several people in your community. You know that most reporters prefer some stories over others, and that the subjects—the people being interviewed—prefer some reporters over others. Ideally, you'd like to match reporters and subjects so that all of them get their first choice. However, you know that doing so is seldom possible. The solution is to try to assign the stories so that everyone gets a fairly high choice.

What is the optimal way to assign stories to reporters? What does optimal mean in this situation?

This lesson explores problems in which members of one group (reporters) must be matched with members of another (subjects) in the best way possible.

SEARCHING FOR STABILITY

In this activity, you will develop criteria for deciding if a matching is a reasonable one.

As editor, you must assign four of your reporters to stories that center around four important people in your community. The first names of your reporters are Bill, Jenny, Samantha, and Simon. The stories involve the mayor, a coach, the chief executive officer of a corporation, and a bishop.

Figure 3.46 shows the preferences of reporters and subjects.

Subject Preferences	1st choice	2nd choice	3rd choice	4th choice
Mayor	Samantha	Simon	Bill	Jenny
Coach	Samantha	Jenny	Simon	Bill
CEO	Bill	Samantha	Jenny	Simon
Bishop	Simon	Jenny	Bill	Samantha

Reporter Preferences	1st choice	2nd choice	3rd choice	4th choice
Bill	Mayor	CEO	Coach	Bishop
Jenny	Bishop	CEO	Mayor	Coach
Simon	Mayor	CEO	Coach	Bishop
Samantha	Coach	Bishop	CEO	Mayor

Figure 3.46.
Preferences of reporters and subjects.

ACTIVITY

SEARCHING FOR STABILITY

7

There are a variety of ways to represent the information in these tables. One way is to make a graph showing the preferences of all the individuals. **Figure 3.47** is the first stage of such a graph.

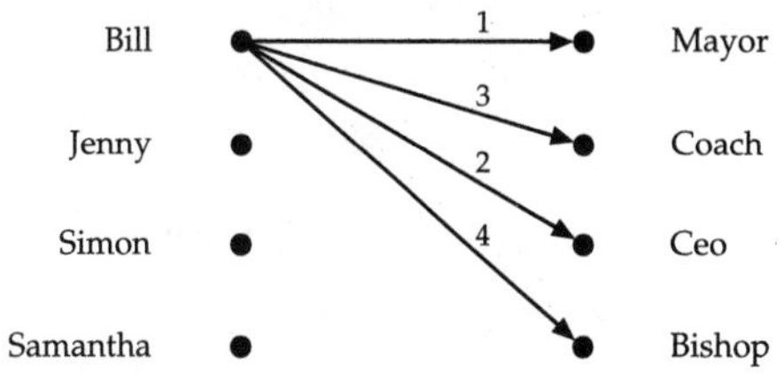

Figure 3.47.
The first step of a graph showing preferences of reporters and subjects.

1. The graph in Figure 3.47 is called a **bipartite graph**. In a bipartite graph the vertices fall into two separate groups, each of which has no edges among its vertices. If the bipartite graph in Figure 3.47 were finished, do you think it would be a more helpful representation than the table in Figure 3.46?

2. The information can also be represented in two tables that are somewhat different from those in Figure 3.46. **Figure 3.49** is one of the two tables. It shows the reporters' preferences.

	Mayor	Coach	CEO	Bishop
Bill	1	3	2	4
Jenny	3	4	2	1
Simon	1	3	2	4
Samantha	4	1	3	2

Figure 3.48.
Reporter preferences.

Make a similar table for the preferences of the subjects. Do you think this type of table is more helpful than the tables in Figure 3.46 or the graph in Figure 3.47?

3. Is it possible to assign reporters so that both reporters and subjects get their first choice? Explain.

4. Find a matching that you think is reasonable. When you have finished, compare your matching to those of others in your class. When you make a comparison, try to decide if one matching is better than the other in some way.

5. What criteria might you use to decide whether a matching is acceptable?

INDIVIDUAL WORK 7

A Good Match

S ince it is usually not possible to match first choices with first choices, a good matching is usually based on the idea of stability. A matching is stable if you cannot find two people who want to be paired with each other rather than the people to whom they are assigned. To put it differently, a **stable matching** is one in which no two unmatched people prefer each other to their assigned partners.

1. **Figure 3.49** shows a different set of preferences for reporters and subjects.

Reporter preferences	Mayor	Coach	CEO	Bishop
Bill	1	3	2	4
Jenny	3	4	2	1
Simon	2	3	1	4
Samantha	4	1	3	2

Subject preferences	Bill	Jenny	Simon	Samantha
Mayor	2	3	4	1
Coach	2	4	3	1
CEO	1	3	4	2
Bishop	1	4	3	2

Figure 3.49.
Tables of reporter and subject preferences.

a) Suppose Bill is assigned to the CEO, Samantha to the mayor, Simon to the bishop, and Jenny to the coach. This matching is not stable because Simon and the coach would rather be paired with each other. Explain.

b) Suppose Jenny is assigned to the CEO, Bill to the mayor, Simon to the coach, and Samantha to the bishop. This matching is also unstable. Find an unmatched pair who prefer each other to their partners.

c) Suppose Jenny is assigned to the bishop, Bill to the mayor, Simon to the CEO, and Samantha to the coach. Is this matching stable?

2. If you have trouble deciding whether a matching is stable, a bipartite graph may help. For example, suppose reporter and subject preferences are those shown in Figure 3.49. Also suppose that you match Samantha with the coach, Jenny with the CEO, Bill with the bishop, and Simon with the mayor.

Make a bipartite graph in which the matched reporters and subjects are directly across from each other. Draw an arrow from each reporter to the subjects the reporter prefers to the one that is assigned. For example, in **Figure 3.50**, the matched reporters and subjects are across from each other, and the edges indicate that Bill prefers all of the other three to his assigned subject, the bishop.

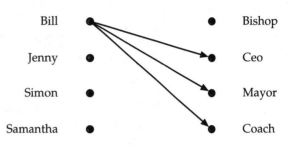

Figure 3.50.
A bipartite graph.

a) Complete the graph by adding edges for the other reporters and subjects.

b) Is this matching stable? Explain.

c) How can a graph of this type be used to determine whether a matching is stable?

3. **Figure 3.51** is another set of reporter and subject preferences.

Reporter preferences	Mayor	Coach	CEO	Bishop
Bill	4	1	3	2
Jenny	2	3	1	4
Simon	2	3	1	4
Samantha	4	1	3	2

Subject preferences	Bill	Jenny	Simon	Samantha
Mayor	1	3	4	2
Coach	1	3	4	2
CEO	1	3	4	2
Bishop	1	3	4	2

Figure 3.51.
Reporter and subject preferences.

a) If Jenny is assigned to the mayor, Simon to the CEO, Bill to the coach, and Samantha to the bishop, is the matching stable? Explain.

b) If Jenny is assigned to the CEO, Simon to the mayor, Bill to the coach, and Samantha to the bishop, is the matching stable? Explain.

c) If Jenny is assigned to the CEO, Simon to the mayor, Bill to the bishop, and Samantha to the coach, is the matching stable? Explain.

4. Can the definition of stable matching at the beginning of this individual work be applied to the task of matching socks after they have been laundered? Explain.

Items 5–7 ask you to find a stable matching. As you do these items, think about the process you are using. Item 8 asks you to write your process as an algorithm.

5. Three medical school graduates have ranked hospitals at which they hope to become residents, and the hospitals have ranked the graduates. The results are shown in **Figure 3.52**.

Graduate preferences	Deaconess	General	Trinity
Kim	3	2	1
Lee	3	1	2
May	3	1	2

Hospital preferences	Kim	Lee	May
Deaconess	1	2	3
General	1	2	3
Trinity	2	1	3

Figure 3.52.
Graduate and hospital preferences.

Since the 1950s, the National Residency Matching Program has matched medical school graduates with hospital residency positions. Graduates and hospitals submit preference lists to the program, which uses an algorithm to match the graduates with the residencies. In a recent year, about 33,000 candidates competed for about 23,000 residencies.

Source: Association of American Medical Colleges

Is there a stable matching for the graduates and the hospitals?

6. The local community college has a nursing program in which second-year students are assigned as advisors to the new students. Both the new students and the second-year students rank each other, and the director of the program then tries to find a stable matching of the two groups. Is there a stable matching for the preference matrices in **Figure 3.53**?

New students' preferences

	V	W	X	Y	Z
A	3	1	2	5	4
B	2	3	4	5	1
C	1	3	2	5	4
D	3	1	2	5	4
E	3	2	1	5	4

Second-year students' preferences

	A	B	C	D	E
V	3	5	2	1	4
W	5	3	2	1	4
X	4	1	3	5	2
Y	2	1	3	4	5
Z	1	5	2	4	3

Figure 3.53.
New and second-year students' preferences.

7. The director of a school play has a problem. There are four lead parts in the play (a rewritten Aesop fable) and four student actors that badly want those parts. The director has strong preferences for which parts should be played by which students, based on their auditions. The actors have strong preferences for which parts they play. The preference matrices in **Figure 3.54** represent the actors' preferences and the director's preferences. Everybody agrees to accept any matching that is stable. Find at least two different stable matchings for this problem.

Actors' preferences

	Eel	Fox	Gnat	Hare
Ann	1	4	2	3
Bill	2	1	3	4
Cal	2	3	4	1
Deb	2	3	1	4

Directors' preferences

	Ann	Bill	Cal	Deb
Eel	4	3	2	1
Fox	1	4	3	2
Gnat	3	2	1	4
Hare	3	1	4	2

Figure 3.54.
Actors' and directors' preferences.

8. Describe the procedure you used to find matchings in Items 5–7. Write it as an algorithm that can be applied to similar situations.

Wrapping Up Unit Three

1. Seven teams A, B, C, D, E, F, and G compete in a tournament in which each team played each of the others once. These are the results: A beat B and F; B beat C and D; C beat A, D, E, and F; D beat A and F; E beat A, B, and D; F beat B and E; G beat A, B, C, D, E, and F.

   a) Represent this situation with a graph.

   b) Tournament officials must decide a final ranking of the teams. Suggest a way.

2. **Figure 3.55** shows five different graphs that are wheels. A **wheel** is a cycle with a central hub that is connected to each of the other vertices. Find the minimum number of colors for each of them. What can you say about the number of colors needed to color wheels?

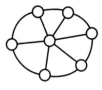

Figure 3.55.
Five graphs that are wheels.

3. Each morning the driver of a delivery truck starts at the company's warehouse (W), delivers goods to four customers (A–D), and returns to the warehouse. The distances (in miles) between locations are shown in **Figure 3.56**. What is the optimal way for the driver to complete the deliveries?

	W	A	B	C	D
W		12	8	16	4
A			10	16	12
B				18	8
C					14
D					

Figure 3.56.
Delivery route distances.

4. A zoo keeper is stocking aquariums and needs to know how many aquariums are needed to hold all the fish safely. The fish that are not compatible with each other are shown in **Figure 3.57**. Represent this situation with a graph, coloring its vertices to indicate the minimum number of aquariums the zoo keeper needs. Explain how you obtained your answer and why you are sure you have used the minimum number of colors.

Species	Incompatible with
A	B, E, G
B	A, C, D
C	B, E
D	B, E, F, G
E	A, C, D
F	D, G
G	A, D, F

Figure 3.57.
Aquarium species.

5. To prepare for the high school prom, five committees are needed: Food, Decoration, Favors, Music, and Cleanup. Paul, Stacey, Lynn, Don, and Megan have volunteered to head a committee. They have preferences, of course. So they leave it up to the class president to make the committee assignments. She decides to list each of the five people in the order of their abilities for the various jobs, as she sees them. Use **Figure 3.58** to help her make a stable assignment.

	Food	Decor	Favors	Music	Cleanup
Paul	2	3	4	1	5
Stacey	2	3	4	1	5
Lynn	1	4	3	2	5
Don	2	3	4	1	5
Megan	3	2	1	4	5

	Paul	Stacey	Lynn	Don	Megan
Food	5	4	1	3	2
Decor	4	2	3	5	2
Favors	3	1	5	4	1
Music	2	3	1	4	5
Clean	1	5	2	3	4

Figure 3.58.
Student preferences and job rankings.

6. Disasters like hurricanes and earthquakes often destroy communications, roads, and other vital links between important locations. After a disaster, those links must be restored as quickly as possible. **Figure 3.59** shows distances among five communities. Develop a plan for reestablishing telephone communications among these communities if a disaster wipes out all communications.

	A	B	C	D	E
A		12	14	18	29
B			17	22	31
C				9	42
D					35
E					

Figure 3.59.

Mathematical Summary

GRAPHS AS REPRESENTATIONS

Many mathematical problems involve objects and relationships among them. A graph is a valuable representational tool in many such problems.

For example, the vertices of the graph in **Figure 3.60** might represent people, cities, or other objects. The edges might indicate that the people are compatible or incompatible in some way or they might represent roads or airline flights among cities.

Graphs are sometimes modified to show additional information that is important to the situation the graph represents. The two most important types of modified graphs are the directed graph and the weighted graph.

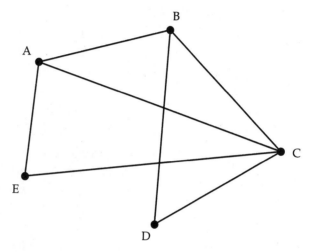

Figure 3.60.
A graph.

In a directed graph, the edges have arrows that indicate direction. Direction, for example, might mean that the street connecting two locations is one-way or that one team beat another. In a weighted graph, a number is written along each edge. Weights, for example, might be the cost of an airline ticket between two cities or the distance between two locations.

OPTIMIZATION PROBLEMS

Problems in which graphs are useful are often concerned with the best or optimal way to do something. For example, the director of a community center may be interested in scheduling activities so that there are as few conflicts as possible; a company may be interested in sending a representative to several clients as economically as possible; a newspaper editor may be interested in assigning interviews to reporters so the reporters and the subjects of the interviews are matched in the best way possible. Regardless of motivation, whether it be efficiency, competition, or the desire to achieve harmony, optimal solutions are important to people in a wide variety of situations.

TYPES OF OPTIMIZATION PROBLEMS

Four common types of optimization problems are prominent in this unit.

One type of problem is called a minimum spanning tree problem. In such problems, there is a need to connect the vertices of a graph with the fewest possible edges that have the smallest possible total weight. For example, a company that wants to connect a group of offices or buildings with fiber-optic cable needs only to build the fewest and cheapest cable lines.

Another type of optimization problem is the vertex coloring problem. In vertex coloring problems, the vertices of a graph must be assigned colors (times, locations) to avoid conflicts. For example, the director of a community center must assign activities to time slots so that no activity conflicts with another. Two activities conflict if they are scheduled at the same time and the same person wants to participate in both of them.

Traveling salesperson problems occur when each vertex of a graph must be visited, the trip must start and end at the same vertex, and the total weight of the edges used must be as small as possible. For example, a robot that solders a circuit board must solder each joint and return to the starting position to await the next board on the assembly line. Keeping the total distance traveled as small as possible means that more boards can be processed in a given amount of time, which helps keep the company competitive.

In the fourth type of optimization problem considered in this unit, optimization is defined in terms of stability. In this problem, individuals from two groups must be matched in a stable way. The matching is considered stable if there are no two people who prefer to switch to each other. That is, there are no two unmatched people who prefer each other to their assigned partners.

PROCEDURES FOR SOLVING OPTIMIZATION PROBLEMS

Optimization problems are solved with procedures called algorithms. The best algorithms are those that always produce an optimal solution and that are easy to use.

There are several algorithms that produce optimal solutions for minimum spanning tree problems, and there are algorithms that produce optimal solutions for matching problems. However, no known algorithm always produces optimal solutions in traveling salesperson problems. The same is true of vertex coloring problems. Because of the lack of perfect algorithms for these two types of problems, both are the subject of current mathematical research.

Glossary

ALGORITHM:
A procedure for solving a problem. Algorithms are usually written in numbered steps.

BIPARTITE GRAPH:
A graph in which the vertices fall into two groups, each of which has no edges among its vertices.

COMPABILITY GRAPH:
A graph in which edges are drawn when the vertices are compatible.

COMPLETE GRAPH:
A graph in which every pair of vertices is connected.

CONFLICT GRAPH:
A graph in which edges are drawn when the vertices are not compatible.

CONNECTED GRAPH:
A graph in which it is possible to get from each vertex to every other vertex by following a path made up of one or more edges.

COUNTEREXAMPLE:
A situation that shows that a conjecture is false. In this unit, counterexample is used to describe a problem in which an algorithm does not produce the optimal solution.

CYCLE:
A path through a graph that begins and ends at the same vertex without repeating any edges or vertices.

DEGREE:
The number of edges that meet at a vertex.

DIGRAPH:
A graph in which the edges have direction. Direction is indicated with an arrow.

EDGE:
A curved or straight line segment connecting two vertices in a graph.

GRAPH:
A representation consisting of points and connecting lines, which may be curved. The points usually represent objects and the lines usually represent relations among the objects.

HAMILTONIAN CIRCUIT:
A path in a graph that visits each vertex exactly once and begins and ends at the same vertex.

INCIDENCE MATRIX:
A matrix that shows whether the vertices of a graph are connected.

MINIMUM SPANNING TREE:
A tree whose edge weights have the smallest possible total, and also includes every vertex.

STABLE MATCHING:
A matching in which no two unmatched people prefer each other to their assigned partners.

SUBGRAPH:
A graph whose edges and vertices are edges and vertices of another graph.

TRAVELING SALESPERSON PROBLEM (TSP):
A graph problem in which the purpose is to find the cheapest way to start at a designated vertex, visit each vertex exactly once, and return to the starting vertex.

TREE:
A graph that has no cycles. A tree is a connected graph that uses the fewest possible edges.

VERTEX (PLURAL: VERTICES):
A point of a graph.

WEIGHTED GRAPH:
A graph in which numbers are attached to the edges. The numbers are called weights.

WHEEL:
A cycle with a central hub that is connected to each of the other vertices.

UNIT

4

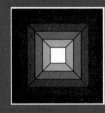

The Right Stuff

Packages are geometric: soft drinks, vegetables, and soups are packaged in metal cylinders; boxes that contain everything from electronic equipment to shoes are rectangular. Sometimes packages that are geometric shapes contain packages that are other geometric shapes. For example, the cylinders that contain soft drinks are often sold in rectangular boxes that contain six or more of the cylinders.

The design of efficient packages requires a knowledge of geometry. Since packages are three-dimensional objects with two-dimensional sides, volume and area play important roles in the geometry of packaging. In this unit you will consider how geometry can be used to create a definition of efficient packaging and how a knowledge of geometry can be used to improve package design.

IT'S A PACKAGE DEAL

It is difficult to imagine life without packaging. You encounter packaging every day of your life. Indeed, packaging seems more a necessity than a convenience. The packages that contain the food you eat keep the food from spoiling and protect it from insects, thereby reducing the risk of certain kinds of diseases. Packaging, however, can create problems. It accounts for about 30% of the material in U. S. landfills, many of which are overburdened. By contributing to the volume of the goods it contains, packaging puts additional demands on space in warehouses, in delivery vehicles, and on store shelves. To minimize the problems created by packaging, packages must be well designed.

LESSON ONE
Packaging Models

KEY CONCEPTS

Mathematical modeling

Modeling criteria

Modeling factors

The Image Bank

PREPARATION READING

The World of Packaging

Nearly everything that is manufactured and sold is packaged in some way. Many products undergo several levels of packaging. Foods, for example, may be packaged in cans, which in turn are packaged in cartons that are stacked on pallets. Some things are packaged several times during their existence. Household goods, for example, are often repackaged when their owners move or take a trip.

Because packaging is pervasive in the modern world, it affects many people. Who are the people most concerned with packaging? What are their concerns? Do the concerns of some people conflict with the concerns of others, or are their concerns compatible? Can their concerns be quantified so that they can be analyzed mathematically?

In this lesson you will consider the impact of packaging and ways in which efficient packaging can be defined. Your definition of efficiency will be used in subsequent lessons of this unit to consider ways in which packaging can be improved.

PACKAGING CONCERNS

In order to use mathematics to improve the efficiency of packaging, you must first consider the concerns of people who are affected by packaging. Those concerns will help you select criteria you can then use to create a definition of efficient packaging.

Optimizing packaging schemes is a very broad problem. Here, as in all modeling efforts, it is best to start by narrowing the focus. Start by considering the secondary packaging of soft drinks. Soft drinks, like many products, have several levels of packaging. The primary packaging is the can or bottle that contains the soft drink. Secondary packaging holds several of the cans or bottles. Secondary packaging is often made of a material called paperboard, which is similar to cardboard, but is not corrugated.

Soft drink cans are all the same size, so analyzing their packaging is easier than analyzing packaging of objects that are not all the same size. Later in this unit, after you have defined efficient soft drink packaging and developed one or more mathematical models that can be used to improve efficiency, you will turn your attention to the design of a single efficient package for several different sizes of honeydew melons.

Consider the issue of secondary packaging of soft drinks. Secondary packaging is commonly used to hold six, 12, or 24 cans. (Unless otherwise indicated, in this unit "soft drink packaging" refers to secondary packaging. The cans contained in the secondary packaging are the standard-sized 12 oz. can.)

1. Make a list of some people who are affected by the way soft drink cans are packaged.

2. List some concerns of each of the individuals you listed in Item 1. (You may want to interview people in your community.)

3. Use the list of concerns you made in Item 2 to create a list of several criteria that could be used to create a mathematical model. For example, in several units in the *Mathematics: Modeling Our World* program you have used mathematical models to solve optimization problems. Your criteria might include an optimization statement such as, "Good soft drink packaging should minimize (or maximize) _____ ."

INDIVIDUAL WORK 1

Packaging Criteria

1. Whenever you begin the task of applying mathematics to a real-world problem, it is interesting to ask whether the problem is worth solving. According to the National Soft Drink Association (NSDA), in 1995 Americans consumed 62.6 billion cans of soft drinks. One criterion that members of the NSDA have for efficient packaging is that it be economical. Suppose you find a way to save the soft drink industry one-tenth of a cent ($.001) on the packaging of each six-pack sold.

   a) Estimate the total annual savings to the soft drink industry.

   b) If you receive royalties worth about 10% of the savings to the industry, estimate your annual income for the use of your innovation in the United States.

   c) Can you think of anyone who might be opposed to a decrease in the cost of packaging?

   d) What are some factors that could affect the cost of soft drink packaging?

2. One criterion that could be used to judge the efficiency of soft drink packaging is that the packaging should minimize the amount of space needed to store the soft drinks.

   a) In what units might storage space be measured?

   b) Who would approve of a reduction in storage space?

   c) Who might disapprove of such a reduction?

   d) What are some factors that could affect the space needed to store packages of soft drinks?

3. One criterion that could be used to judge soft drink packaging efficiency is that the packaging maximize sales.

   a) How would you measure the ability of packaging to increase sales?

   b) Who are the people likely to benefit from a sales increase?

   c) Who might suffer adverse effects from a sales increase?

   d) What are some factors that might affect the ability of packaging to increase sales?

4. A criterion that could be used to judge soft drink packaging efficiency is that the packaging minimize the amount of packaging material.

   a) How might the amount of packaging material be measured?

   b) Who are some people who would approve of a reduction in the amount of packaging material?

   c) Who might suffer adverse effects from a reduction in the amount of packaging material?

   d) What are some factors that might affect the amount of packaging material?

5. In Item 3 of Activity 1, you listed several criteria that might be used to create a mathematical model for packaging efficiency. Select one or more of those criteria.

   a) Discuss how you would measure efficiency.

   b) Discuss people who would benefit and people who might object if efficiency improved.

   c) List one or more factors that might affect efficiency.

6. Designing efficient secondary packaging for soft drinks is very similar to designing an efficient container to hold several honeydew melons. Explain.

7. How is the problem of designing efficient secondary packaging for soft drinks similar to the problem of designing cartons to fit in a moving van? How are the two problems different?

ACTIVITY

2

PACKAGING EFFICIENCY

In this activity, you play the role of a manager of a soft drink company. Your company pays bonuses to employees who find ways to save the company money. The bonus policy means that you occasionally have to consider employee suggestions and decide whether to enact them. Thus, you have to establish decision-making criteria for new suggestions.

Today you are considering a suggestion for the redesign of your most basic secondary packaging, the six-pack. One of your employees has suggested that the six-pack be based on a "staggered" configuration as shown in **Figure 4.1**. Previously, your company has used a conventional arrangement in which the cans are "stacked" against each other as shown in **Figure 4.2**. The employee thinks the new design is more efficient and will save the company money.

Figure 4.1.
A staggered six-pack.

Figure 4.2.
A conventional stacked six-pack.

Your task in this activity is to select a criterion for deciding whether the new six-pack design is more efficient than the old. You do not actually have to make the decision. When you have settled on a criterion, you can send the design and your criterion to the company's mathematicians for evaluation.

PACKAGING EFFICIENCY

The following are two possible criteria.

- Packaging should minimize the total amount of package space not used by the cans.

- Packaging should maximize the percentage of package space used by the cans.

1. Note that the first criterion above involves the total amount of space. List at least two other criteria that involve the total of some quantity.

2. Note that the second criterion above involves a percentage, which in this case is really the ratio of space used by the cans to space used by the package. List at least two other criteria that involve a ratio of two quantities.

3. Discuss the pros and cons of each criterion. A good criterion must be clearly stated. It should be based on a quantitative definition of efficiency that can be explored mathematically.

4. Write a memo to the company's mathematicians describing the new design and the criterion you have selected for judging it.

CONSIDER:

1. In general, do you prefer criteria that involve total amounts of some quantity or criteria that involve the ratio of two quantities? Explain.

2. If a package design makes efficient use of space in a six-pack, will the six-packs also make efficient use of space on a store shelf?

3. If six-packs make efficient use of space on a store shelf, will they also make efficient use of space in a delivery truck?

INDIVIDUAL WORK 2

Packaging Factors

1. A criterion used to judge the efficiency of soft drink packaging can be based on the amount of storage space used by the cans or by the package. If space is used to define efficiency, then it is important to understand factors that have an effect on the amount of space used and how to measure their effect.

   a) Consider the shape of the package as a factor. Do you think changing the shape of the package affects how well space is used? For example, predict whether a package shape that uses the staggered arrangement in Activity 2 would use space more or less efficiently than a package shape that uses the conventional arrangement. Explain.

   b) How would you determine whether the staggered arrangement uses space more efficiently than the conventional arrangement? That is, how would you measure the way each arrangement uses space?

   c) Consider the number of cans in the container as a factor. That is, predict whether changing the number of cans in a conventional stacked arrangement might result in more efficient use of space.

   d) How would you determine whether changing the number of cans makes more efficient use of space? For example, is a design in which six cans require 200 units of space more or less efficient than a design in which eight cans use 250 units of space?

2. A criterion used to judge the efficiency of soft drink packaging can be based on the amount of packaging material used.

   a) Consider the shape of the package as a factor. Do you think changing the shape of the package has an effect on how well packaging material is used? For example, predict whether the staggered arrangement in Activity 2 uses packaging material more or less efficiently than a conventional arrangement. Explain.

   b) How would you determine whether the staggered arrangement uses packaging material more efficiently than the conventional arrangement? That is, how would you measure the way in which each arrangement uses packaging material?

c) Consider the number of cans in the package as a factor. That is, predict whether changing the number of cans in a conventional stacked arrangement might result in more efficient use of packaging material.

d) How would you determine whether changing the number of cans makes more efficient use of packaging material? That is, how would you compare the use of packaging material by a stacked arrangement of six cans with the use of packaging material by a stacked arrangement of, say, eight cans?

3. Simplifying a problem can help clarify your thinking about it. In this item, consider a simplified version of the soft drink packaging problem. The problem is simplified in two ways. First, the cans are two-dimensional. Second, as shown in **Figure 4.3**, the cans are shaped like squares.

2 cm

Figure 4.3.
A 2 cm square "can."

a) **Figure 4.4** is a 4 x 5 cm package that is intended to hold several of the cans. How many does it hold? Make a drawing to explain your answer.

b) **Figure 4.5** is a 6 x 7 cm package that is intended to hold several of the cans. How many does it hold? Make a drawing to explain your answer.

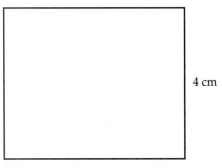

4 cm

5 cm

Figure 4.4.
A 4 x 5 cm package.

c) In which of the two packages do you think the cans use space more efficiently? Justify your answer.

d) A display rack in a store is 12 cm wide and 35 cm long. Compare the number of Figure 4.4 packages and the number of Figure 4.5 packages that fit on the shelf. If necessary, make a drawing to support your answer. Which package permits the store to have more soda on display?

e) How efficiently do the cans use space in a 6 x 6 cm pack? How well do the packs use space on the display rack?

6 cm

7 cm

Figure 4.5.
A 6 x 7 cm package.

4. This item is a simplification of the packaging problem in which the criterion is based on the efficient use of packaging material. Imagine that the packages in Figures 4.4 and 4.5 are made of wire.

a) How much packaging material do the packages in Figures 4.4 and 4.5 use?

b) In which of the two packages do you think packaging material is used more efficiently? Justify your answer.

5. A standard moving van is 8 feet wide, 40 feet long, and 8 feet high.

a) How efficiently do cartons that are 16 inches wide, 16 inches long, and 8 inches high cover the floor of the van?

b) How efficiently do the 16 x 16 x 8 inch cartons use space in the van?

c) Do the cartons 9 inches high use space in the van as well as those in (b)?

d) How efficiently does a single layer of cartons that are 18 inches wide, 18 inches long, and 8 inches high cover the floor of the van?

e) How efficiently do the 18 inch x 18 inch x 8 inch cartons use space in the van?

f) Is the height of the carton a factor when considering how well the cartons cover the floor of the van? Is it a factor when considering how well the cartons fill the van?

The Image Bank

PREPARATION READING

What Makes Packaging Efficient?

*T*he efficiency of a particular soft drink packaging design depends on the criterion chosen to define efficiency. Moreover, not everyone has the same criterion uppermost in their minds. The manufacturer and the seller of soft drinks are concerned with conserving precious space in manufacturing plants, delivery vehicles, and stores. Public officials and conservationists are concerned with minimizing the amount of packaging material that winds up in municipal landfills.

Although there are a variety of ways to define efficient packaging, many of them involve geometric concepts of area and volume. You have studied geometric area and volume formulas in previous courses. How many of them do you recall? How can area and volume be used to define efficiency? How can the formulas you have learned in previous courses help you determine the efficiency of a soft drink package and create new packaging designs?

THE POPZI CHALLENGE

3

In Lesson 1, you considered how the packaging of soft drinks affects people such as the manufacturer, the retailer, and the consumer. By examining their concerns, you were able to formulate criteria by which the efficiency of soft drink packaging can be judged. You also saw one new proposal for redesigning six-packs. Now you will develop one or more designs of your own to evaluate, so you must be able to measure efficiency by the criterion you have chosen.

1. Design a package to hold several soft-drink cans.

   To simplify your first attempt at designing a package, work with a two-dimensional version of the problem. Handout H4.3 has several circles that are the same size as the base of a standard soda can. Cut these out and arrange them according to your design. Draw lines around the design to represent the package.

   The standard number of cans is six, and the standard shape is a rectangle in which the cans are arranged in a stacked configuration. You are free to vary the number of cans, the shape of the container, and the configuration of the cans within the container. The only restriction is that your design be something other than the standard design.

2. Depending on the criterion that you use to evaluate a design, you will need to make one or more measurements and/or calculations in order to find the design's efficiency. When you have finished your design, determine each of the following:

   a) The total area of the package.

   b) The total area of the cans.

   c) The difference between the package area and the area of the cans.

   d) The ratio of the package area to the area of the cans.

THE POPZI CHALLENGE

e) The ratio of the area of the cans to the area of the package.

f) Any other measurements or calculations you feel are necessary to determine the efficiency based on the criterion you selected in Item 4 of Activity 2.

To make area calculations, you need one or more of several area formulas you have used in previous courses. **Figures 4.6–4.9** show several common geometric figures and their area formulas. To calculate areas in your design, measure the appropriate lengths with a ruler, then apply the appropriate formula(s).

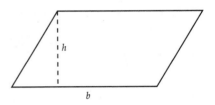

Figure 4.6.
A rectangle. The area is *bh*.

Figure 4.7.
A parallelogram. The area is *bh*.

CONSIDER:

1. Discuss the merits of the following measures of efficiency for investigating two-dimensional soft drink packages:

   a) The total area of the package.

   b) The difference between the package area and the area of the cans.

   c) The ratio of the package area to the area of the cans.

   d) The ratio of the area of the cans to the area of the package.

   e) The ratio of the package area to the number of cans in the package.

2. The two-dimensional soft drink package is sometimes called a **cross section** of the three-dimensional package. What do you think this means?

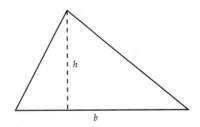

Figure 4.8.
A triangle. The area is 0.5*bh* or $\frac{bh}{2}$.

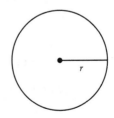

Figure 4.9.
A circle. The area is πr^2.

INDIVIDUAL WORK 3

The Standard Six-Pack

How does the efficiency of your design compare with the efficiency of the standard rectangular six-pack? In this individual work, you will find out. You will also practice making calculations that are useful in evaluating the efficiency of soft drink package designs.

1. **Figure 4.10** is a two-dimensional version of the standard six-pack. It has been reduced in size in order to conserve space on the page of this book. The radius of each can is half the radius of the actual can; the length and width of the six-pack are half the length and width of the actual six-pack.

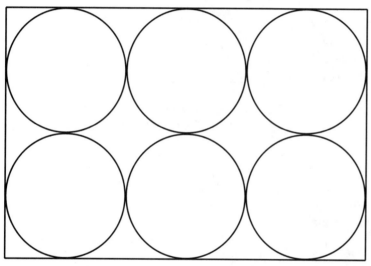

a) Use a ruler to take appropriate measurements, then calculate the area of the six-pack and the area of a single can.

b) Find the efficiency of the design if the criterion is maximization of the percentage of package area used by the cans (the ratio of can area to package area). Also, find the efficiency if the criterion is minimization of the amount of package area per can.

Figure 4.10.
The standard six-pack.

c) In part (a) you found areas for Figure 4.10, which is a reduced version of the actual six-pack. Find the corresponding areas for the actual six-pack.

d) How do the efficiencies for the actual six-pack compare with those you found in part (b)?

e) How do the efficiencies of the standard six-pack compare with your design in Activity 3?

f) Use the method of generalization to show the effect on the percentage of package area used by the cans if the package and can dimensions are all multiplied by a factor of *k*.

2. The mathematical modeling process almost always involves simplification. One reason simplification is important is that it makes problems easier to solve; often you can learn from solving the simpler

problem. One simplification you have used in this unit
is to consider two-dimensional versions of three dimensional problems. In this item you consider the effect of
that simplification.

Just as area is a measure of the space occupied by two-
dimensional objects, **volume** is a measure of the space
occupied by three-dimensional objects. Whereas area is
measured in squares one unit on a side, volume is measured in cubes one unit on a side.

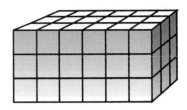

Figure 4.11.
The volume of a rectangular solid is the
area of the solid's rectangular base times
the solid's height.

The package that is the standard six-pack is the three-dimensional
counterpart of a rectangle. It is usually called a **rectangular solid**. If
you multiply the area
of the two-dimensional six-pack by the height of its three-dimensional counterpart, you have its volume. (**Figure 4.11.**)

The can in a standard six-pack is a three-dimensional shape called a
cylinder. A **cylinder** is a three-dimensional solid with a circular top
and bottom and a side surface that is a rectangle when laid flat.
Each cross section of the cylinder is a circle. To find the volume of the
can, multiply the area of the circular cross section by the height of
the can.

a) A standard six-pack is about 12 cm high. Use this height to find
   the volume of the package and the volume of one of the cans.

b) What percentage of the volume of the package is used by the
   cans?

c) What is the package volume per can?

d) What effect does simplifying the problem to two dimensions have
   on the efficiency criterion: percentage of package space used by
   the cans? What effect does it have on the criterion: amount of
   package space used per can?

3. Another criterion that can be used to judge soft drink package design
   involves minimization of the amount of packaging material.

   a) Suppose the two-dimensional six-pack in Figure 4.10 is made of
      wire. The total length of the sides of the rectangle is its **perimeter**.
      Find the perimeter and use it to determine the amount of wire per
      can.

   b) Recall that Figure 4.10 is smaller than the actual six-pack. What
      is the amount of wire per can if Figure 4.10 were the size of the
      actual six-pack?

INDIVIDUAL WORK 3

c) Find the amount of wire per can for your design in Activity 3 and compare it to the amount of wire for the standard six-pack. Which is more efficient by the package material per can criterion?

4. Package material for standard six-packs is sometimes made of paperboard. The entire package is composed of six rectangles (**Figure 4.12**). Handout H4.4 can be cut out to make a scale model of the standard six-pack. Each side of each rectangle in Handout H4.4 is approximately 1/3 the length of the corresponding size of an actual six-pack.

a) Find the areas of each of the six rectangles that make up the actual six-pack. Record them on your paper as well as on your model of the six-pack.

b) What is the amount of package material used per can in the actual six-pack?

c) If you haven't already made a three-dimensional model of your design in Activity 3, do so. Find the area of each face and determine the total amount of packaging material per can. Compare it to the total packaging per can in the standard six-pack. Is your design better by the "package material per can" criterion?

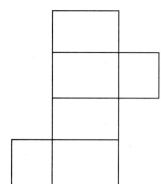

Figure 4.12.
A collapsed view of the standard six-pack.

5. Designing a soft drink package is similar to designing other types of enclosures. For example, consider a company that ships furniture in boxes with bases of three sizes: 3 x 5 ft., 4 x 6 ft., and 8 x 10 ft. Can the company design a storage facility that efficiently holds each of the three kinds of boxes? (Boxes of different types are not stored at the same time.) Explain your answer.

6. Geometric area formulas have applications to many designs. For example, the flags shown in **Figures 4.13(a), (b), and (c)** are based on geometric shapes. Suppose that each is a large 6 x 10 ft. display flag. Find the area used by each of the flag's colors. Explain your work and list any assumptions you made about the flag in order to calculate the areas. (Hint: It is often helpful to subdivide the flag into rectangles and/or triangles.)

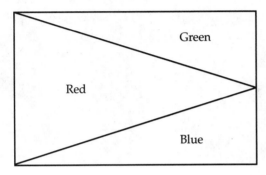

Figure 4.13(a).
Geometric pattern of flag of Eritrea.

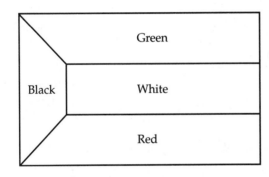

Figure 4.13(b).
Flag of Kuwait.

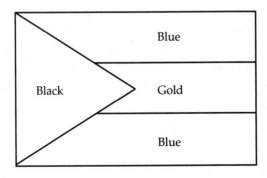

Figure 4.13(c).
Flag of the Bahamas.

LESSON THREE
Techno-logical Solutions

KEY CONCEPTS

Geometric drawing utility

Tangent

Line symmetry

Rotational symmetry

The Image Bank

PREPARATION READING

Gaining Confidence

You have made your first attempt at designing a soft drink package and determining its efficiency by several criteria. These include maximization of the percentage of package space used by the cans and minimization of the amount of packaging material used per can. You have also proposed at least one criterion of your own.

As your knowledge of geometry grows, you will continue to examine package designs by more than one criterion. Thus, it is important to keep in mind that a design that is fairly efficient by one criterion may not be efficient by another.

Regardless of the criterion you have chosen, you may not be satisfied with the values you have obtained from rough drawings and imprecise measurements. Remember that even small errors can make large differences when an industry produces billions of cans of soft drinks annually!

How can you become more confident that the efficiency you calculate for a package design is precise? How can you be sure that your design is more or less efficient than the standard six-pack or another person's design?

Mathematicians have more than one answer for these questions. In this lesson, you will see how technology can help you become more confident in the efficiencies you calculate for your designs.

THE GEOMETRIC DRAWING UTILITY

Until personal computers and sophisticated calculators were developed, people used compasses, rulers, and protractors to make accurate drawings. Even if designs are drawn with the best equipment, however, the precision of measurements made on these drawings does not approach the precision of measurements taken from designs done with electronic drawing equipment.

In order to use technology to make precise calculations of efficiency, you need to know how to draw basic geometric objects with a geometric tool called an electronic drawing utility. In this activity, you will make several basic drawings that you are likely to need when you make an electronic drawing of your package design. Your task is to learn to use the equipment. You do not have to make a model of your package design now. That will come later.

Procedures vary somewhat from one drawing utility to another, so the directions here are general. If you are stumped, consult the manual for your utility or talk to a person who is more experienced with the utility than you are.

Here are 15 basic procedures you should be able to do using your utility. Try them all.

1. Draw a circle.

2. Draw a line.

3. Draw a point.

4. Construct extra points on a line or circle you have already drawn.

> **Note:** Construct is not the same as draw. If you construct a point on an object, the constructed properties can not change as you "drag" the figure. Thus, you cannot move the point off the object, only along it.

ACTIVITY

THE GEOMETRIC DRAWING UTILITY

4

5. Copy a circle so that you have one or more others just like it.

6. Change the size of a circle you have drawn.

7. Construct a perpendicular to a line through a point that is on the line. Also construct a perpendicular through a point that is not on the line. (Again, remember that construct is not the same as draw. You can draw two lines that look perpendicular; you can even measure the angle they form to be sure it measures 90°. But if you disturb the drawing by dragging part of it, the lines may no longer be perpendicular. When you construct a perpendicular, the lines remain perpendicular when you try to move any part of the construction.)

8. Construct a parallel to a line through a point not on the line.

9. Construct points of intersection for two circles. Do the same for a line and a circle.

10. Construct a line or segment through two points.

11. Hide objects.

12. Move objects.

13. Measure the length of a segment.

14. Measure the area of a circle.

15. Hide, show, and customize labels of objects.

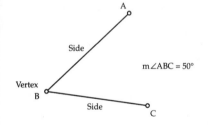

Figure 4.14.
An angle drawn and measured with a drawing utility.

Most utilities use special symbols that you see when you measure parts of your drawings. Among those that your utility might use are "m" for "measure," ∠ for "angle," and Δ for "triangle." If you draw and measure an angle, you may notice that your utility uses three letters to name an angle. For example ∠ABC means "angle ABC." When three letters are used to name an angle, the middle letter must be the vertex of the angle and each of the other two letters must be a point on one of the angle's sides, but not both on the same side (**Figure 4.14**).

In everyday usage, the name of an angle is often not distinguished from its measure. Thus, instead of saying "the measure of angle ABC is 50°," people often abbreviate to "angle ABC is 50°."

INDIVIDUAL WORK 4

Symmetry

With the aid of a geometric drawing utility, you can make excellent drawings of soft drink package designs. Even with a geometric drawing utility, however, it can be difficult to get some aspects of a drawing right. For example, it can be hard to get two circles to touch in exactly one point.

Because soft drink packages contain cans (circles) that are all the same size, the packages are symmetric. Roughly speaking, that means that the packages look the same from different viewpoints. If you understand the symmetries of these packages, you can learn certain "tricks" that will enable you to make very accurate electronic drawings of your package design.

In this individual work, you will examine the symmetries of some basic soft drink can configurations. Your knowledge of these symmetries and the drawing utility skills you developed in Activity 3 will enable you to make accurate drawings and precise efficiency calculations, which you will do in the next activity.

There are two basic configurations for soft drink cans: stacked and staggered. Both are based on two circles that touch in a single point as shown in **Figure 4.15**. In a stacked configuration, the centers of additional circles are aligned either vertically or horizontally with the centers of these two circles. In a staggered configuration, centers are offset as shown in **Figure 4.16**.

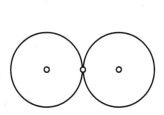

Figure 4.15.

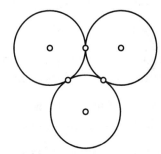

Figure 4.16.

Before you begin the first item of this individual work, there are a few terms you should know.

INDIVIDUAL WORK 4

When two circles intersect (touch) in a single point, they are **tangent** to each other. Similarly, when a circle and line intersect (touch) in a single point, the line is tangent to the circle (in fact, the line is often called a tangent to the circle).

A figure has **line symmetry** if you can fold the figure along a line so that the two parts coincide. If the two parts coincide, then they have the same size and shape. In geometry, when two figures have the same size and shape, they are called **congruent**. (Mathematicians use the symbol ≅ for congruent. Thus, the sentence ΔABC ≅ ΔDEF says that triangle ABC is congruent to triangle DEF.)

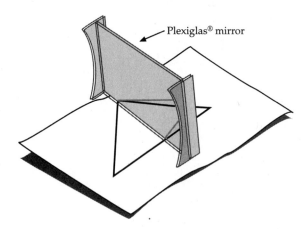

Plexiglas® mirror

Mathematicians sometimes use the words **bisect** and **perpendicular** when discussing lines of symmetry. A line bisects something when it divides it into two equal parts. A line is perpendicular to another line when the two lines form right (90°) angles.

Folding a figure along a line is one way to check for a line of symmetry. Another method, shown at left, is to place a Plexiglas® mirror on the suspected line and check to see if the reflection of one side falls directly on the other side. Some Plexiglas mirrors that are sold for this purpose have a stand to keep them upright. If your mirror does not, be sure to hold it so that it is perpendicular to the paper.

1. a) Figure 4.15 has two lines of symmetry. Draw the lines of symmetry on a copy of the figure or on Handout H4.5.

   b) How is the vertical line of symmetry related to the two circles?

   c) How is the vertical line of symmetry related to a line segment that connects the centers of the two circles?

   d) How are the two lines of symmetry related to each other? Explain. (Hint: What is the size of each of the angles they form? Why?)

2. a) How many lines of symmetry does Figure 4.16 have? Draw the lines of symmetry on a copy of the figure or on Handout H4.5.

   b) Are any of the lines of symmetry tangent to any of the circles?

   c) How are the lines of symmetry related to each other?

3. On a copy of Figure 4.16, connect the centers of the three circles.

a) What can you say about the figure that is formed?

b) Does the figure that is formed have any lines of symmetry?

4. A figure has **rotational symmetry** if there is a central point about which you can rotate the figure so that the figure coincides with itself. One way to test for rotational symmetry is to trace a copy of the figure, hold the copy steady with your pencil placed on the point you think is the central point, then slowly turn the copy until the copy coincides with the original. You can usually determine the size of the rotation by noting the fraction of one full rotation (360°) needed to make the copy coincide with the original figure.

a) Figure 4.15 has rotational symmetry. Identify the central point and the size of the rotation.

b) Does Figure 4.16 have rotational symmetry? If so, identify the central point and determine the size of the rotation.

5. Sometimes it is possible to find points through which part of a figure can be rotated to coincide with another part. Knowing that such points exist can be useful when using a geometric drawing utility to design soft drink packaging. **Figure 4.17** is the same as Figure 4.16, but it has labels so that the points and circles can be distinguished.

a) Can you rotate circle 1 about point B so that it coincides with circle 2? If so, what is the size of the rotation?

b) Is there a point about which you can rotate circle 1 to obtain circle 3? If so, name the point and the size of the rotation.

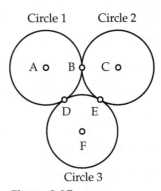

Figure 4.17.

6. The natural world is replete with symmetry. Describe the symmetry in each of these photos of living things.

Figure 4.18.

Figure 4.19.

Figure 4.20.

Figure 4.21.

7. Geometric shapes often have lines of symmetry or rotational symmetry. Describe the symmetry of each of the following.

   a) A circle.

   b) An isosceles triangle. (An **isosceles triangle** has two sides equal.)

   c) A right triangle. (A **right triangle** has one 90° angle.)

   d) A square.

   e) A rectangle.

   f) A parallelogram.

   g) A regular pentagon (**Figure 4.22**). A regular pentagon has five equal sides and five equal angles.

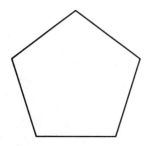

Figure 4.22.
A regular pentagon.

ACTIVITY

THE ELECTRONIC DESIGN

5

In Lesson 1 you established criteria for judging the efficiency of soft drink package designs. In Lesson 2 you designed a soft drink package and used geometric formulas to calculate the efficiency of your design based on criteria established in Lesson 1. However, the measurements and calculations you made in Lesson 2 lack precision. In this lesson you have learned to do basic geometric constructions with an electronic drawing utility and analyzed the symmetries of basic soft drink can configurations. It's time to put the utility to work to improve the precision of your efficiency calculations.

1. Use a drawing utility to produce an accurate drawing of the soft drink package design you made in Activity 3.

2. Have the utility measure or calculate the necessary areas and find efficiency in terms of the percentage of package area used by the cans.

3. When your drawing is finished, try resizing it. You can do this by dragging the center of the circle with which you began the drawing. If necessary, reconstruct any parts that do not automatically resize. What happens to the efficiency when you resize your drawing?

4. Compare the efficiency of your design to the efficiencies of other designs in your class.

5. If time permits, have the utility calculate the efficiency of your design by other criteria. (Or, you may wish to experiment with an entirely new design.)

SUGGESTIONS:

Make the design a convenient size for the display on which you are working. It can be resized to the actual size of the standard soft drink can later, or you can do the scaling using what you know about similar figures. The standard soft drink can has a radius of 3.3 cm.

ACTIVITY

THE ELECTRONIC DESIGN

5

Start by drawing the first can (circle), then use rotations to make the remaining circles. Before drawing the "package," think back to your work in Individual Work 4 to determine appropriate centers and angles of rotation.

Geometric drawing utilities use positive and negative angles to distinguish between counterclockwise and clockwise rotations. Counterclockwise is usually positive.

If you make a mistake, most utilities have a menu item that undoes your last step. Learn to use this feature.

When you have all your circles in place, you need to enclose them in a "package" made of line segments. Although you want only segments, you will probably find it best to construct tangent lines first, even though these lines are longer than necessary. (You can draw the lines, but it is better to construct them.) Individual Work 4 should give you some ideas about how to construct lines tangent to a circle. After you have constructed all the tangent lines, you can construct their points of intersection, hide the lines, and connect the intersection points with line segments.

Your utility can measure the area of any of your circles, but it's really only necessary to measure one because they are all the same. Your utility can also find the area of the package, which is probably a type of polygon. A **polygon** is a closed figure with three or more sides that are line segments. If all of the sides are the same length and all the angles are the same size, the polygon is called a **regular polygon**. Check your utility's manual to learn how to find the area of a polygon.

Your utility can make calculations, but you can also use measurements made by the utility and do the calculations on your calculator.

3 Sides
Triangle

4 Sides
Quadrilateral

5 Sides
Pentagon

6 Sides
Hexagon

7 Sides
Heptagon

8 Sides
Octagon

9 Sides
Nonagon

10 Sides
Decagon

Names of common polygons.

ACTIVITY

THE ELECTRONIC DESIGN

5

SAMPLE SOLUTION:

Figure 4.23 is a sample finished electronic drawing of a mini-package containing two cans. Several parts used in the drawing have been hidden to avoid cluttering the drawing. The efficiency has been calculated as a decimal. Note that there is a discrepancy between the area you would calculate (4.80 x 2.40 = 11.52) and the area measured by the utility. Since the displayed zero in the measurements of the base and height of the rectangle indicate they are measured to three digits, the area calculation should be rounded to three digits. (Most drawing utilities allow you to control the precision of measurements and calculations.)

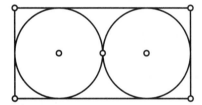

Can radius = 1.20 cm
Package base = 4.80 cm
Package height = 2.40 cm
Area of package = 11.51 square cm
Area of one can = 4.52 square cm
Efficiency = 0.79

Figure 4.23.
A soft drink two-pack.

A CAREER NOTE

If you enjoy solving packaging dilemmas, you might want to consider becoming part of an exciting, growing industry as a packaging professional. With the demand for packaging expertise high, several colleges are providing related courses. With such training, you could be involved in finding the safest, most economical, and environmentally responsible way to package anything from peanuts to computers! You might want to contact the Institute of Packaging Professionals in Herndon, VA, or explore their website at http://packinfo-world.org/wpo/PEF.html where you will find a list of colleges offering packaging courses and other information related to packaging careers, including salaries.

INDIVIDUAL WORK 5

Some Packages

Y ou now have the ability to calculate efficiencies quite precisely. One reason precision is important is that even small differences can add up to a lot in an industry as large as the soft drink industry. Are the designs that you and others in your class have made the most efficient possible? In this individual work, you examine a variety of soft drink package designs. As you do, compare their efficiencies to those of the designs developed in your class.

1. **Figure 4.24** is an electronic drawing of a four-pack.

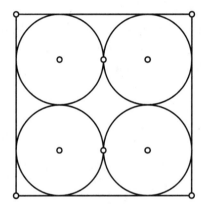

Radius of one circle = 1.20 cm

Figure 4.24.
A four-pack.

a) Find the efficiency in terms of the percentage of package area used by the cans.

b) What is the efficiency if the figure is resized so that the circles have the same radius (3.3 cm) as a standard soda can?

c) What is the efficiency in terms of package material used per can? To answer this question, you need to find the area of the two squares and four rectangles that make up the three-dimensional four-pack. The height of a standard can is 12.2 cm. Use the standard can radius of 3.3 cm.

d) How do the efficiencies in terms of percentage of package space used by the cans and in terms of package material used per can for the two-pack in Activity 5 compare with those of the four-pack? (Be sure to resize the two-pack so that each can has a radius of 3.3 cm.)

2. In Individual Work 3, you made approximate measurements of a scale model of the standard six-pack. Now that you know more precise measurements of the radius of a standard can (3.3 cm) and the height of a standard can (12.2 cm), recalculate the efficiency in terms of percentage of package space used by the cans and package material per can.

3. **Figure 4.25** is a nine-pack. How efficient is it in terms of percentage of package space used by the cans? In terms of package material used per can?

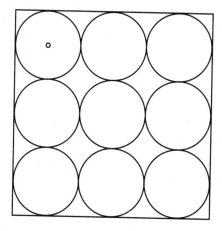

Figure 4.25.
A nine-pack.

4. **Figure 4.26** is a triangular three-pack.

   a) Find its efficiency in terms of percentage of package space used by the cans.

   b) The package surface of the three-dimensional package consists of two triangles and three rectangles. Find the efficiency of the package in terms of package material used per can.

   c) If the triangular design were expanded to six cans, do you think either of its efficiencies would change?

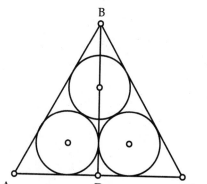

Radius of one circle = 3.3 cm

AC = 18.02 cm

BD = 15.61 cm

Figure 4.26.
A triangular three-pack.

5. **Figure 4.27** is an interesting design. You might call it the "cans in a can" design.

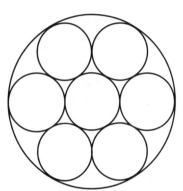

Figure 4.27.
A cylindrical seven-pack.

a) How well do the cans use space in this package?

b) The surface of the cylindrical package is composed of a circular top and a cylindrical side surface. The cylindrical side surface can be made into a rectangle by cutting it from top to bottom and laying it flat (**Figure 4.28**). The base of the rectangle is the same as the **circumference** of the circle, and the height of the rectangle is the same as the height of the package. The formula for circumference (the distance around a circle) is $2\pi r$. Find the efficiency in terms of package material used per can.

Figure 4.28.
The side surface of a cylinder as a rectangle.

6. **Figure 4.29** is an electronic drawing of a hexagonal seven-pack. Only three measurements were made. Is it possible to determine the efficiency in terms of package space used by the cans or in terms of package material used per can from these three measures alone?

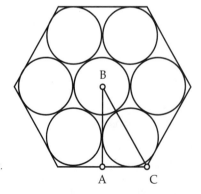

Figure 4.29.
A hexagonal seven-pack.

Radius of one circle = 3.3 cm

AB = 9.00 cm

BC = 10.4 cm

7. In Activity 5, you found the percentage of space used by the cans in your package design. Now find the efficiency in terms of package material used per can. Compare your design's efficiencies to the efficiencies of other designs in your class and to those you have seen in this individual work.

8. You have seen several soft drink package designs besides your own. What does the method of generalization tell you about efficiencies of packages that have each of the following shapes? (Note: To apply the method of generalization, use variables for the radius and, if necessary, the height of the can. Then calculate the efficiency.)

   a) The standard six-pack.

   b) The four-pack in Item 1 of this individual work.

   c) The nine-pack in Item 3 of this individual work.

   d) The triangular three-pack in Item 4 of this individual work.

   e) The cylindrical seven-pack in Item 5 of this individual work.

   f) The hexagonal seven-pack in Item 6 of this individual work.

9. A pizza maker offers a large rectangular pepperoni pizza that is 12 x 24 in. The pepperoni slices are circular with a 3/4 in. diameter. If the pepperoni slices are arranged in a stacked configuration to cover as much of the pizza surface as possible, what percentage of the pizza surface is covered with pepperoni?

10. Write a summary of what you have learned about soft drink package design. Here are several questions you might address in your summary:

   If a design is more efficient than another by one measure of efficiency, is it also more efficient by another measure?

   How does the shape of the package affect efficiency?

   How does the number of cans in the package affect efficiency?

11. A common form of soda can packaging uses a ring-like plastic device to hold cans in groups of six. Discuss how efficiency criteria you have used in this unit might apply to this type of packaging and factors that might affect the criteria.

LESSON FOUR
Getting the Facts

KEY CONCEPTS

Deductive reasoning

Inductive reasoning

Proof

Parallel lines

Angles

Right triangles

The Image Bank

PREPARATION READING

Certainty in Geometry

Computers and calculators are tremendous aids to numerical calculation. They are also very useful for making geometric measurements. However, electronic drawing utilities give only more precise approximate answers than earlier methods of calculation. Although these "precise approximate" answers are sufficient in most modeling situations, such answers are not exact. If you know the proper geometric facts, you can obtain exact answers by applying logical reasoning to geometric figures even if the figures are not accurately drawn. An important advantage of exact answers is that they often can be generalized so that you never have to solve the problem again.

You already know some of the facts you need to obtain exact answers. For example, you know that a rectangle with a base of 6 cm and a height of 4 cm has an area of 24 cm². It seems foolish to start a computer or calculator drawing utility, construct the rectangle to scale, then have the utility measure its area. Because you know the proper geometric fact, you can determine the area of the rectangle from a very rough drawing or perhaps without the aid of a drawing.

Moreover, some things aren't easy to do with a drawing utility. For example, it isn't easy to use a drawing utility to construct a three-dimensional figure, measure the area of its surfaces, and calculate the amount of package material per can.

Drawing utilities have at least one additional problem because they can lead you to make incorrect generalizations. You might, for example, conclude that all triangles have a certain property just because a few that you drew on the utility have that property.

Geometric facts are established by a process called **deductive reasoning**. Deductive reasoning is the process of drawing logical conclusions from facts that you already know. The method of generalization is a form of deductive reasoning with which you are already familiar. However, you probably were not able to apply the method of generalization to some types of package designs such as triangles. If so, it was because you do not know enough geometric facts about triangles.

In this lesson, you will see how deductive reasoning can establish geometric facts that you can use to calculate the efficiency of package designs exactly. The knowledge you gain from a deductive investigation of soft drink package designs will help you evaluate designs of cartons to hold melons of differing sizes in Lesson 5.

A professor of mathematics at the famous University of Alexandria at the time of its founding around 300 B.C., Euclid used deductive reasoning to establish most of geometry on the basis of a very few assumptions. His work, titled simply *Elements*, is considered by many historians to be, with the exception of the Bible, the most widely studied work in the history of humanity. It is still in print today.

ACTIVITY

6

DRAWING CONCLUSIONS

Figure 4.30 is an example of a soft drink package design for which it is difficult to calculate the efficiency exactly.

You know that the radius of each can, one of which is drawn in the figure, has a length of 3.3 cm. It is not hard to calculate the area of the six cans, but it is not clear how the area of the triangular package can be found from the length of the radius alone.

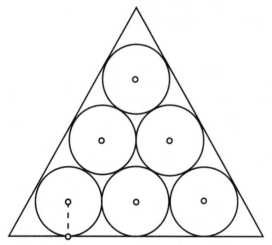

Figure 4.30.
A triangular six-pack.

One approach to finding the area of the triangular package is to divide it up into smaller pieces and find the area of each of the pieces. For example, the segments drawn in **Figure 4.31** appear to form a right triangle and a rectangle. Drawing additional segments appears to form more of the same (**Figure 4.32**). Perhaps there are only two areas that need to be determined. However, you must first be able to conclude that these pieces are right triangles or rectangles, and you must then be able to find (without measuring) the lengths needed to determine the areas.

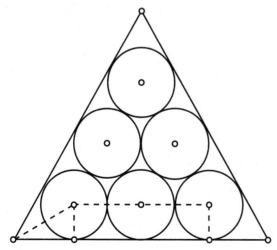

Figure 4.31.

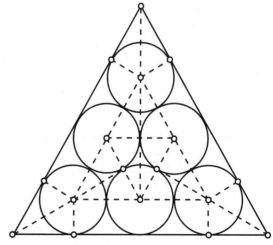

Figure 4.32.

ACTIVITY

DRAWING CONCLUSIONS

6

It is a guess based on appearances that the pieces in Figure 4.32 are right triangles or rectangles. To conclude that a figure is a right triangle or a rectangle requires knowing that one or more angles measure 90°. Therefore, drawing conclusions about efficiencies for some types of soft drink packages appears to require knowing something about the angles of triangles. In this activity you will examine ways to draw logical conclusions about the measures of angles of triangles by considering what is perhaps the most basic of geometric facts about angles of triangles.

Scientists often use a process called **inductive reasoning**. Inductive reasoning is drawing conclusions based on observations: if you see something happen enough times, you believe that it will always happen.

For example, **Figure 4.33** shows the results of an investigation done with a drawing utility. (This utility uses a lower-case "m" to mean "the measure of" and the symbol ∠ to mean "angle." It also uses the three-letter convention for naming angles.)

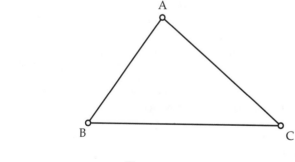

m∠BAC = 85°

m∠ABC = 55°

m∠ACB = 40°

m∠BAC + m∠ABC + m∠ACB = 180°

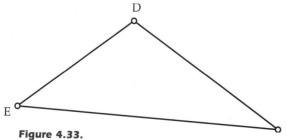

m∠EDF = 109°

m∠DEF = 40°

m∠DFE = 31°

m∠EDF + m∠DEF + m∠DFE = 180°

Figure 4.33.
An investigation with a drawing utility.

ACTIVITY

DRAWING CONCLUSIONS

6

1. Based only on these results, is it wise to conclude that the sum of the measures of the angles of a triangle is always 180°? Explain.

A deductive argument that establishes a geometric fact is called a **proof**. Once a fact has been proved, mathematicians often call the fact a **theorem**. You may already know that the sum of the measures of the angles of a triangle is 180°. In the next part of this activity, you will consider a proof of this fact.

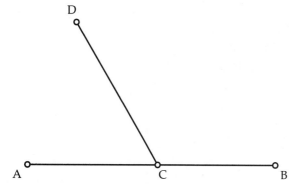

Figure 4.34.
∠ACD and ∠DCB form a straight line and m∠ACD + m∠DCB = 180°.

A proof is based on previously known facts. They may be known because they themselves were proved or because they are so simple that everyone agrees on them.

An example of a geometric fact that is simple enough to be agreed on by most people is that when angles form a straight line, the sum of their measures must be 180° (**Figure 4.34**).

An example of a fact that can be proved is that when two parallel lines are crossed by a third line, each pair of **alternate interior angles** are equal. (A line that crosses two parallel lines is called a **transversal**.) For the purposes of this activity, use this fact as if it were already proved. You will consider how to prove it in Individual Work 6.

When two parallel lines are crossed by a transversal, two pairs of alternate interior angles are formed. In **Figure 4.35**, the angles have been numbered for easy reference. ∠1 and ∠4 are a pair of alternate interior angles; ∠2 and ∠3 are a pair of alternate interior angles. Note that as the name suggests, alternate interior angles are on opposite (alternate) sides of the transversal, and on the inside (interior) of the two parallel lines.

DRAWING CONCLUSIONS

Now consider a triangle that represents any triangle (**Figure 4.36**). Its angles have been numbered for easy reference. Since it is intended to represent *any* triangle, properties that hold *only* for this particular triangle, such as the measures of its individual sides and angles, should not be used in a proof.

A geometric proof is sometimes started by adding one or more lines or segments to the figure. **Figure 4.37** is the same as Figure 4.36, but a line has been drawn through the top vertex of the triangle parallel to one of the triangle's sides.

2. On a copy of Figure 4.37, label any line segments that are transversals. Number all the new angles that are formed by the parallel line and the transversals.

3. List all pairs of alternate interior angles in your figure. What do you know about the angles in each pair?

4. What do you know about the measures of the three angles at the top of the figure?

5. You can combine your answers to Items 4 and 5 to finish a logical argument (in other words, a proof) that the angles of a triangle equal 180°. Explain.

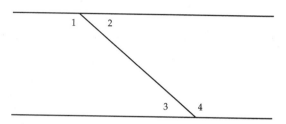

Figure 4.35.
Alternate interior angles

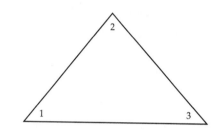

Figure 4.36.
A triangle.

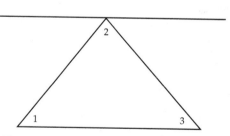

Figure 4.37.

INDIVIDUAL WORK 6

Parallels, Perpendiculars, and Angles

*I*n this individual work, you use the knowledge you gained about angles in Activity 6 to draw conclusions about angles in a variety of figures and to develop methods that can help find exact values for the efficiency of soft drink package designs. In addition, you get a chance to prove that alternate interior angles really *are* equal.

1. One consequence of the fact that alternate interior angles are equal when two parallel lines are crossed by a transversal is that it is possible to use deductive reasoning to find the measures of all angles if the measure of only one angle is known. **Figure 4.38** shows two parallel lines crossed by a transversal. The measure of one angle is shown. Make a copy of the figure on your paper and find the measures of all the other angles without measuring them.

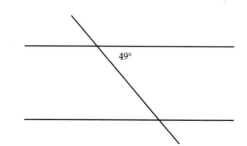

Figure 4.38.

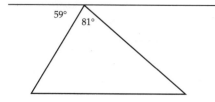

Figure 4.39.

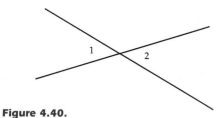

Figure 4.40.
Two intersecting lines.

2. Suppose that a transversal is perpendicular to one of two parallel lines. What can you conclude about the measures of the angles that are formed by the transversal and the parallel lines?

3. **Figure 4.39** is a triangle with a line parallel to one side. Find the measures of all remaining angles.

4. The proof that the sum of the measures of the angles of a triangle is 180° is based on the fact that alternate interior angles formed by two parallel lines and a transversal are equal, which can itself be proved. The proof that alternate interior angles are equal is based on two other facts: (1) vertical angles are equal; and, (2) corresponding angles formed by two parallel lines and a transversal are equal.

   a) **Vertical angles** are a pair of non-adjacent angles formed by two intersecting lines. For example, in **Figure 4.40**, angles 1 and 2 are vertical angles and must be equal.

   There is a second pair of vertical angles in Figure 4.40. Make a copy of the figure and label the second pair.

b) **Corresponding angles** are a pair of angles that occupy identical positions on each of two parallel lines and a transversal. For example, in **Figure 4.41**, angles 1 and 2 are a pair of corresponding angles and must be equal.

Figure 4.41 has three other pairs of corresponding angles. Make a copy of the figure and label each pair.

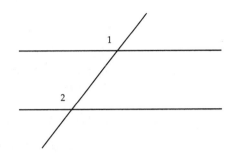

Figure 4.41.
Two parallel lines.

c) Use **Figure 4.42** to explain why facts about corresponding angles and vertical angles prove that alternate interior angles 1 and 2 must be equal.

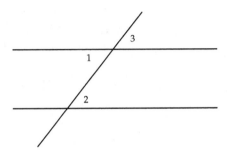

Figure 4.42.
Two parallel lines and a transversal.

5. Once a fact is proved, it can be used to prove new results. To put it differently, whenever you prove a new theorem, your deductive tool kit grows. For example, the fact that the angles of a triangle have a sum of 180° can be used to reach conclusions about the angles of other polygons. Since soft drink package designs are often polygonal in shape, knowing something about the angles of polygons can be useful to the package designer.

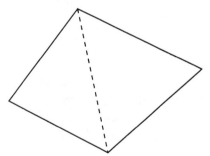

Figure 4.43.
A quadrilateral.

a) **Figure 4.43** is a quadrilateral that has been divided into two triangles by drawing a diagonal between two vertices. (A **diagonal** is a segment that connects two non-adjacent vertices of a polygon.)

Explain what this figure tells you about the measures of the angles of the quadrilateral.

INDIVIDUAL WORK 6

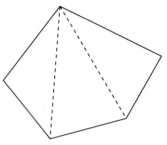

Figure 4.44.
A pentagon.

b) **Figure 4.44** is a pentagon in which two diagonals have been drawn from one vertex. What can you conclude about the measures of the angles of the pentagon?

c) Conduct a similar investigation for a hexagon.

d) Generalize the conclusion. That is, if a polygon has n sides, write a symbolic expression for the sum of the measures of the angles in terms of n.

6. Recall that a regular polygon is one in which all sides are equal and all angles are equal. What does your work in Item 5 tell you about the measure of each angle of:

   a) a regular quadrilateral,

   b) a regular pentagon,

   c) a regular hexagon,

   d) a regular polygon with n sides?

7. When the three sides of a triangle are the same length, the angles must all have the same measure. Similarly, when the angles have the same measure, the sides must have the same length. Is this true of other polygons?

8. **Figure 4.45** is an equilateral triangle with one of its three lines of symmetry. Equilateral triangles are important in soft drink package design because it is often possible to divide a package into pieces in such a way that some of the pieces are equilateral triangles.

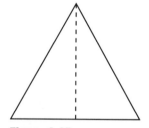

Figure 4.45.
An equilateral triangle with a line of symmetry.

   a) Use deductive reasoning to find the measures of all angles in the figure. Explain your reasoning. Write the measures on a copy of the figure.

   b) **Figure 4.46** is an isosceles triangle with its single line of symmetry. One angle in the figure has been measured. Use deductive reasoning to find the measures of as many other angles as possible. Explain your reasoning.

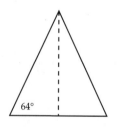

64°

Figure 4.46.
An isosceles triangle with its line of symmetry.

9. **Figure 4.47** shows a triangle and the same triangle with one of its sides extended.

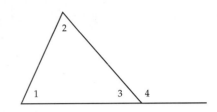

Figure 4.47.
A triangle and the same triangle with one side extended.

   a) How is ∠4, the angle formed by extending the side, related to ∠3?

   b) How is ∠4 related to the other two angles of the triangle (angles 1 and 2)? Explain.

   c) If the triangle in Figure 4.47 is isosceles with ∠1 = ∠2, what can you conclude about the relationship between these two angles and ∠4? Explain.

10. In Course 1, Unit 3, *Landsat*, similar figures are used to estimate measurements based on satellite photos. Recall that similar figures have the same shape, but are usually of different sizes. Pairs of sides of similar triangles (and other polygons) have the same ratio (often called the scale factor). In the case of triangles (and other polygons), similar figures also have pairs of corresponding angles equal.

Figure 4.48 shows two similar triangles. Since pairs of angles are equal, it is possible to place a copy of one triangle over the other so that two angles coincide (**Figure 4.49**).

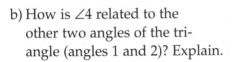

Figure 4.48.
Two similar triangles.

   a) Whenever two lines form equal corresponding or equal alternate interior angles, the lines are parallel. Explain why side AB must be parallel to side DE in Figure 4.49.

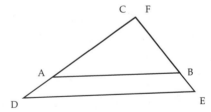

Figure 4.49.
A copy of triangle ABC placed on triangle DEF.

   b) The ratio of the sides of triangle ABC to those of triangle DEF is 3/4. Find the sides of triangle DEF if AB = 4.4 cm, AC = 3.4 cm, and BC = 2.4 cm.

11. Write a summary of what you have learned about angles and their roles in parallel lines and in triangles and other polygons.

ACTIVITY

THE RIGHT GEOMETRIC STUFF

7

Much of the geometry used in package design involves right angles. At first glance, some novel designs such as the triangular package in

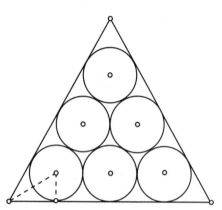

Figure 4.50.

Figure 4.50 appear to involve no right angles. However, it is not obvious how to find the package's area because the only known length in the figure is the radius of each can. When you draw a radius and another segment to form a triangle, the triangle appears to be a right triangle. To find the area of this small triangle you must be certain that it is a right triangle. Once you know that it is a right triangle, you must find the length of one more side before you can calculate its area.

Because right angles are so important in package design, much of the geometry used in package design could be called "the right stuff." In this activity you will consider an important conclusion about right angles and an important conclusion about the sides of right triangles.

To analyze the design in Figure 4.50, you must be able to conclude that the small triangle is a right triangle. This conclusion involves the relationship between a tangent to a circle and a radius of the circle (**Figure 4.51**).

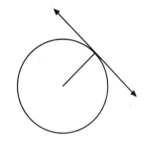

Figure 4.51.
A circle, a tangent, and a radius.

Answer the following questions to prove that the radius and tangent form a right angle.

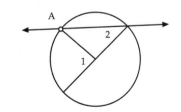

Figure 4.52.

1. In **Figure 4.52**, the radius in Figure 4.51 has been extended across the circle. The tangent has been tilted so that it intersects the circle in a second point A and a radius has been drawn to A. (Now the tangent is no longer a tangent.) What kind of triangle is formed?

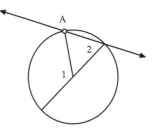

ACTIVITY

THE RIGHT GEOMETRIC STUFF

7

2. How is ∠2 related to ∠1? (Hint: review Item 9 of Individual Work 6.)

3. Suppose point A is closer to the original point of tangency (**Figure 4.53**). Do either of the answers you gave to Items 1 or 2 change?

4. What happens to ∠1 as point A moves closer to the point of tangency? What can you conclude about ∠2?

Figure 4.53.

Now that you have established that the angle formed by a radius and tangent is a right angle, you know that the small triangle in Figure 4.50 is indeed a right triangle. Based on your work in Activity 6 and Individual Work 6, you can conclude that the sum of the measures of the other two angles of the triangle is 90°. You also know the length of one of the triangle's sides, but all these things are not enough to determine the triangle's area. You need to understand how the sides of a right triangle are related, which you will do in this activity. You also need to understand how the relationship applies in this particular situation; Individual Work 7 explains that.

Among all the facts in geometry, perhaps in all of mathematics, none are as important as the Pythagorean formula (also called the Pythagorean Theorem), which says that in a right triangle, the square of the hypotenuse is equal to the sum of the squares of the other two sides. (Note: the **hypotenuse** is the longest side, which is also across from the right, or 90°, angle.)

It is easier to state the formula when referring to a figure. The triangle in **Figure 4.54** is a right triangle. (Note that in a right triangle a small square is sometimes used to mark the right angle.) The lower-case letters a, b, and c represent the measures of its sides. The Pythagorean formula says that as long as the triangle is a right triangle, $c^2 = a^2 + b^2$.

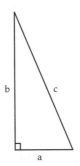

Figure 4.54.
A right triangle.

The Pythagorean formula is named for Pythagoras, a Greek mathematician born on the island of Samos around 540 B.C. Pythagoras settled in Crotona, an Italian seaport town, where he established the Pythagorean school. Members of the school studied mathematics and applied it to science, philosophy, music, and other topics.

ACTIVITY

THE RIGHT GEOMETRIC STUFF

7

Figure 4.55 is an investigation with a drawing utility. Note that the utility uses a bar over two letters to mean "segment." Thus, m$\overline{AB}$ means "the measure of the segment connecting A and B."

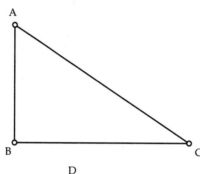

Figure 4.55.
An investigation with a drawing utility.

m $\overline{CA}$ = 6.4 cm

m $\overline{BA}$ = 3.5 cm

m $\overline{BC}$ = 5.3 cm

(m $\overline{CA}$)^2 = 40.3 cm^2

(m $\overline{BA}$)^2 + (m $\overline{BC}$)^2 = 40.3 cm^2

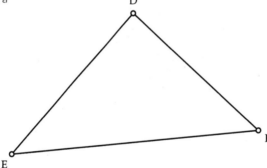

m $\overline{EF}$ = 7.6 cm

m $\overline{FD}$ = 5.1 cm

m $\overline{DE}$ = 5.6 cm

(m $\overline{EF}$)^2 = 57.5 cm^2

(m $\overline{FD}$)^2 + (m $\overline{DE}$)^2 = 57.5 cm^2

5. Explain why this investigation is not a proof of the Pythagorean formula.

In **Figure 4.56**, four copies of the right triangle in Figure 4.54 have been positioned to form a large square with a smaller square in the center.

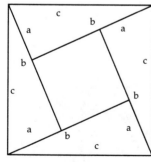

Figure 4.56.
Four right triangles arranged to form a square.

6. How are the areas of the large square, the triangles, and the small square related?

7. Use the variable lengths *a*, *b*, and *c* to write symbolic expressions for the areas of the large square, one of the triangles, and the small square.

8. Combine your answers to Items 6 and 7 to create a symbolic equation describing the relationship among the areas. Then use symbolic procedures to simplify the equation.

The Pack is Back

*I*n this individual work, you use deductive reasoning to find efficiencies for several soft drink package designs. Before doing so, however, you will find a little practice with right triangle calculations beneficial.

The Pythagorean formula is often used to find the measure of one side of a right triangle when the measures of the other two sides are known. For example, consider the right triangle in **Figure 4.57**.

The Pythagorean formula says $4^2 = 2^2 + b^2$ or $16 = 4 + b^2$.

Subtracting 4 from 16, you have $b^2 = 12$ or $b = \sqrt{12}$. Thus, an exact value for b is $\sqrt{12}$; an approximate value to two decimal places is 3.46.

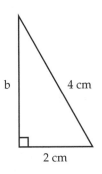

Figure 4.57.
A right triangle.

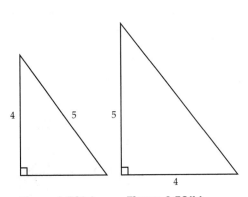

Figure 4.58(a).   **Figure 4.58(b).**

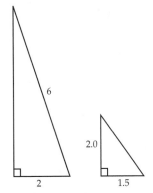

Figure 4.58(c).   **Figure 4.58(d).**

1. a) Use the Pythagorean formula to find the missing side in each of the right triangles in **Figures 4.58(a)–(d)**.

   b) Are there any similar triangles in Figures 4.58(a)–(d)? Explain.

   c) For triangles in general, is it true that the triangles are similar if they have two pairs of corresponding sides with the same ratio? If it is true, explain why. If it is not, give an example.

INDIVIDUAL WORK 7

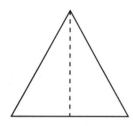

Figure 4.59.
An equilateral triangle and its line of symmetry.

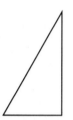

Figure 4.60.
Half of an equilateral triangle.

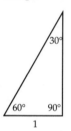

Figure 4.61.

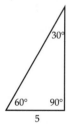

Figure 4.62.

2. **Figure 4.59** is an equilateral triangle with a line of symmetry.

   a) If each side of the equilateral triangle is 6 cm, use the Pythagorean formula to find the triangle's height.

   b) What is the area of the triangle?

3. **Figure 4.60** is half of an equilateral triangle.

   a) On a copy of the figure write the measures of each angle.

   b) If the shortest side of the triangle is 4 cm, find the other two sides.

   c) Apply the method of generalization to Figure 4.60. That is, if x is the measure of the shortest side, find symbolic expressions for the other two sides.

   d) Explain how your general result in part (c) can be used to find the other two sides of the triangle quickly if you know the smallest side.

   e) Can your general result be used to find the shortest side from the side that isn't the hypotenuse? If your answer is yes, give an example.

4. **Figure 4.61** is a right triangle with angles of 30° and 60°. Its shortest side has a length of 1.

   a) Use the short cut you devised in Item 3 to find exact values for the other two sides.

   b) **Figure 4.62** is another right triangle. How is it related to the right triangle in Figure 4.61?

   c) Use the relationship you described in part (b) to find values for the sides of the triangle in Figure 4.62.

   d) Find the sides of the triangle in Figure 4.62 by using the short cut you devised in Item 3.

   e) Use a calculator to compare decimal approximations for the answers you gave in parts (c) and (d)

   f) Use the relationship you described in part (b) to find values for the sides of the triangle in Figure 4.62 if the side labeled 5 has length x instead.

   g) Find the sides of the triangle in part (f) by using the short cut you devised in Item 3.

   h) Use the results you have obtained in parts (e)–(g) to state in words a general rule for the lengths of the sides of a 30°-60° right triangle.

5. You now have enough knowledge of "the right stuff" to tackle some package designs that probably stumped you before. **Figure 4.63** is a triangular six-pack. Recall that the radius of a standard soft drink can is 3.3 cm.

   a) What are the measures of the angles of the small triangle ABC at the lower left of the figure? Explain your reasoning.

   b) Since AC is a radius, AC = 3.3 cm. Find the lengths of AB and BC. (Note: Since these lengths will be used in other calculations, round answers to three decimal places.)

   c) Knowing the length of BC enables you to find the base of the large triangle. Find the base and explain how you found it.

   d) Knowing the base of the large triangle enables you to find its height. Find the height and explain how you found it. (Hint: There are two ways to find it. You can use your knowledge of right triangles or your knowledge of similar triangles.)

   e) Find the efficiency of the triangular six-pack in terms of percentage of package space used by the cans.

6. **Figure 4.64** is an interesting modification of the triangular six-pack in Figure 4.54. Find its efficiency in terms of percentage of package space used by the cans.

7. **Figure 4.65** is a hexagonal seven-pack. Based on a can radius of 3.3 cm, find the efficiency in terms of percentage of package space used by the cans. The broken lines in the figure should help. Explain how you arrived at your answer.

8. Determine the efficiencies of the packages in Items 5–7 in terms of package materials used per can. Compare the most efficient with other designs (for example, designs developed in your class and designs in Individual Work 5).

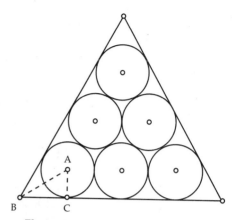

Figure 4.63.
A triangular six-pack.

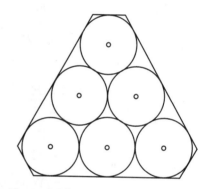

Figure 4.64.
A modified triangular six-pack.

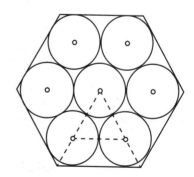

Figure 4.65.
A hexagonal seven-pack.

9. Soft drinks are often sold in larger packages of 12 or more cans. What is the best way to make a "double pack" from one of the configurations you have seen?

10. There are soft drink coolers shaped like a long tube. Is the "cool pack" a good design for a soft drink package?

11. Prepare a report recommending a soft drink package design. You can use a design you developed earlier, one of the designs in this book, or a new design that you develop for this purpose. Your report should include calculations of the design's efficiency in terms of package space used by the cans and in terms of package material used per can. You may also want to consider the efficiencies that result from basing larger packages on your design and the way in which the package uses space on the store floor or shelf. Consider making paper or cardboard models to give others a sense of your design's appeal. To show that your design is really a good one, you may want to compare it to others that you rejected.

12. In Unit 1, *Gridville,* you used both firetruck and helicopter distance. Suppose a fire station is three miles east of house A and four miles south of house B.

   a) Make a sketch showing the fire station and the houses.

   b) What is the helicopter distance from house A to house B?

   c) How much greater is the firetruck distance between the houses than the helicopter distance?

13. Since ancient times, similar right triangles have been used to determine heights that cannot be measured directly. One method uses shadows: you measure the length of the shadow cast by the object whose height you are trying to find, and you measure the length of the shadow of an object of known height such as a stake driven into the ground. For example, suppose a 5.0 ft. stake casts a 3.5 ft. shadow. An unknown object casts a 38 ft. shadow.

   a) Make a sketch that shows both objects and their shadows.

   b) Find the unknown height. Explain how you obtained your answer.

14. The distance you can see on the surface of the earth depends on your height above the earth's surface.

   a) Make a sketch showing the earth as a circle and a point slightly above the earth's surface outside the circle. Draw a triangle by connecting the center of the earth, the point above the surface, and the farthest point on the surface that can be seen. What kind of triangle is it? Explain.

   b) Determine how far you can see if you are 125 meters above the earth's surface. (Use 6,500,000 meters for the earth's radius.)

   c) Use the method of generalization to determine a symbolic equation for finding the distance you can see when you are h units above the surface of a planet with radius r.

   d) Note that your answer to part (b) is about 3,600 times the square root of 125. Does the rule of thumb, "3,600 times the square root of the distance above the earth's surface" give a good approximation for the distance you can see? Explain.

LESSON FIVE
Packaging Spheres

KEY CONCEPTS

Spheres

Aspect ratio

Quadratic models

PREPARATION READING

Unusual Sizes

The first four lessons of this unit are concerned with designing an efficient package for soft drink cans. In those lessons, efficiency is determined either by the percentage of package space used by the cans or by the amount of packaging material used per can.

Soft drink cans have circular cross sections. So do many kinds of fruit that are approximately spherical in shape: oranges, cantaloupes, peaches, and honeydew melons, for example. Unlike a cylinder, the circular cross section of a sphere differs in size according to where the "slice" is made.

Many kinds of fruit are shipped in cardboard or wooden cartons. The fruits are sized according to the number needed to fill the carton. For example, a size 100 orange is called size 100 because approximately 100 such oranges fill a standard carton. Since only 80 size 80 oranges are needed to fill the same carton, a size 80 orange is bigger than a size 100 orange.

The problem of designing a carton for oranges is complicated because a great number of oranges can fit into a carton and because the oranges can be stacked in several layers. Neither of these problems occurs with larger fruit such as honeydew melons. The number of melons packed in a carton is usually less than 10, and sometimes as small as four.

In this lesson, you will consider a problem related to the soft drink package design problem—the problem of designing an efficient carton for melons. Unlike soft drink cans, melons come in different sizes. What, then, is an efficient carton for holding several spherical objects of one size and for holding several similar objects of a different size?

ACTIVITY

8

SEARCH FOUR MELONS

To begin considering the problem of designing an efficient carton for honeydew melons, first consider a simpler problem. For a given carton, what is the largest melon for which exactly four of those melons fit in the carton?

As with the soda package design problem, work with a two-dimensional version in which a rectangle represents the carton and circles of the same size represent the melons. (Since cross sections of melons differ in size, the circles represent the largest possible cross section.) For the purposes of this exploration, define efficiency as the ratio of the total cross section area for the melons (at their "equators") to the area of the base of the box.

Experiment with several different cartons. For each carton you try, find the radius of the largest melon so that exactly four fit in the carton. Use a geometric drawing utility, a compass and a ruler, or mathematical formulas. A typical honeydew melon has a radius of 3"–4", but you should choose a scale that you think is suitable for the tools you are using.

Length	Width	Aspect ratio	Radius	Efficiency

Figure 4.66.
A table of results.

1. Save the final drawing for each carton and record the results in a table like the one in **Figure 4.66**. This chart includes a column for the **aspect ratio** of the carton. The aspect ratio is the ratio of the carton's longer dimension (the length) to its shorter dimension (the width) and is usually expressed as a decimal. Aspect ratio is a measure of a rectangle's shape. A square has an aspect ratio of 1. A very narrow rectangle has a much larger aspect ratio.

When you have finished, discuss each of the following:

2. What shape of box is most efficient for packaging four melons? Explain.

3. Use data from your exploration in Item 1 to describe the relationship between efficiency and box shape.

INDIVIDUAL WORK 8

A Closer Look

*A*ctivity 8 probably convinced you that a fairly simple carton design is best for packaging four melons. However, keep in mind that the same carton used to ship four melons is also used to ship other numbers (5, 6, and 8) of smaller melons. Is the most efficient four-melon carton also efficient for six?

Before you consider packaging different numbers of melons, take a closer look at some packages that contain four melons. When you determine their efficiencies, try to use mathematical formulas rather than measurements made from constructions.

1. **Figure 4.67** is a drawing utility sketch of a simple four-melon carton design.

   a) How efficient is this design?

   b) What is the aspect ratio of the carton?

   c) What is the radius of the melons?

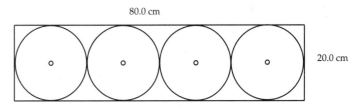

80.0 cm

20.0 cm

Figure 4.67.
A four-melon carton.

   d) Is the design more or less efficient if the aspect ratio increases? Explain.

   e) If both the length and the width of the carton double, what happens to the aspect ratio? What happens to the efficiency?

   f) Does the two-dimensional version of the problem have the same efficiency as the three-dimensional? To find out, you will need to calculate the volume of the carton and the volume of the four melons. The formula for the volume of a sphere is $(4/3)\pi r^3$.

2. **Figure 4.68** is another drawing utility sketch of a simple four-melon carton design.

   a) How efficient is this design?

   b) What is the aspect ratio of the carton?

   c) What is the radius of the melons?

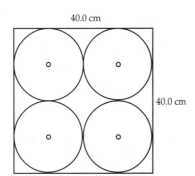

40.0 cm

40.0 cm

Figure 4.68.
A four-melon carton.

INDIVIDUAL WORK 8

d) If both the length and the width of the carton double, what happens to the aspect ratio? What happens to the efficiency?

e) Does the two-dimensional version of the problem have the same efficiency as the three-dimensional?

3. In general, how does the efficiency of a two-dimensional four-pack compare to the efficiency of the three dimensional four-pack? Explain.

4. **Figure 4.69** is a staggered arrangement of melons in a rectangular carton.

   a) If the melons have a 10 cm radius, what are the length and width of the carton? (Hint: Try finding the lengths of the segments shown with broken lines. You may want to draw other segments.) Explain how you found your answers.

   b) What is the carton's aspect ratio?

   c) What is this design's efficiency?

   d) Apply the method of generalization to this design. If r represents the radius of one melon, find symbolic expressions for the length and width of the rectangle, the aspect ratio, and the efficiency, all in terms of r.

5. The sequence of packages in **Figures 4.70(a)–(d)** shows what happens as the aspect ratio of a four-melon carton increases.

As the aspect ratio increases, the distance between the centers of the two upper circles increases and the nature of the right triangle in the figure changes. For some cartons, finding the radius is fairly easy. For others, like those in Figures 4.70(b) and 4.70(d), it is more challenging.

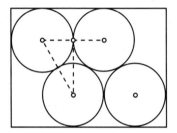

Figure 4.69.
A staggered melon arrangement.

Aspect ratio = 1

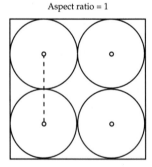

Figure 4.70(a).

Aspect ratio = 1.2

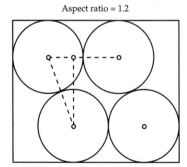

Figure 4.70(b).

Aspect ratio = 1.34

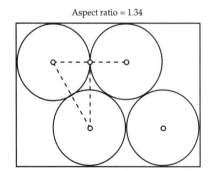

Figure 4.70(c).

Aspect ratio = 1.6

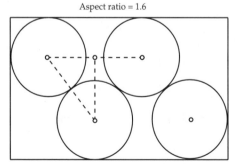

Figure 4.70(d).

Figure 4.71 is a melon carton similar to the one in Figure 4.70(b).

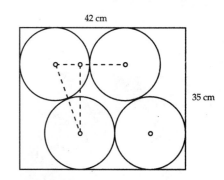

42 cm

35 cm

Figure 4.71.
A melon carton with aspect ratio = 1.2.

a) If *r* represents the radius of one circle, write an expression for the hypotenuse of the right triangle in the figure.

b) An expression for the length of the longer of the other two sides of the triangle is $35 - 2r$. Explain why. (Hint: extend the side so that it reaches the top and bottom of the rectangle.)

c) An expression for the length of the remaining side of the right triangle is $42 - 4r$. Explain why.

d) You can use your answers to parts (a), (b), and (c) and the Pythagorean formula to write a quadratic equation in which *r* is the variable. The equation can be solved by graphing and zooming, by the quadratic formula, or by completing the square. Write the equation and solve it.

6. **Figure 4.72** is a melon box like the one in Figure 4.70(d). Apply an analysis similar to that used in Item 5. That is, use *r* for the radius, find expressions for the three sides of the right triangle, and apply the Pythagorean formula to find *r*.

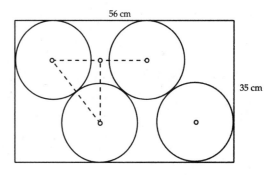

56 cm

35 cm

Figure 4.72.
A melon carton with aspect ratio = 1.6.

7. Use your results from Items 5 and 6 to compute efficiencies and compare them with the results you and your classmates obtained in Activity 8. Describe what you observe.

8. Almost all traditional television screens have an aspect ratio of 4/3. The new high-definition television screens have an aspect ratio of 16/9.

a) How wide is a traditional television screen that is 15 inches tall? How wide is a high-definition screen with the same height? What is the diagonal size of each television?

b) Explain the difference between the two types of television screen.

INDIVIDUAL WORK 8

Tire Width

Sidewall Height

Figure 4.73

9. The sizing of automobile tires often puzzles people. What, for example, is the difference between a P205/60R14 tire and a P205/70R14 tire? The number that precedes R is the tire's aspect ratio, which is obtained by dividing the sidewall height by the tire width, then multiplying by 100 (**Figure 4.73**).

a) Two tires have a sidewall height of 4.5 inches. One is a 60R tire, the other a 70R tire. What is the difference in the width of the tires?

b) A tire with a relatively narrow sidewall has a sporty feel. If you favor such a tire, what kind of aspect ratio is best?

c) How does the definition of aspect ratio for tires differ from the definition used for melon cartons?

10. In this unit you have done a number of calculations involving the Pythagorean formula, the quadratic formula, or both. Often the results of such calculations are square roots of numbers that are not perfect squares—numbers like $\sqrt{32}$. When you obtain such a result, you can give the exact answer $\sqrt{32}$ or an approximate answer rounded to the appropriate number of decimal places. Exact answers such as $\sqrt{32}$ can be expressed in one or more alternate forms if the number inside the square root has a perfect square factor. Since 32 has two perfect square factors, 4 and 16, it can be expressed in two alternate forms, as shown.

$$\sqrt{32} = \sqrt{4} \times \sqrt{8} = 2\sqrt{8}$$
$$\sqrt{32} = \sqrt{16} \times \sqrt{2} = 4\sqrt{2}$$

a) Find decimal approximations for $\sqrt{32}, 2\sqrt{8}$, and $4\sqrt{2}$.

b) Can addition be used to express roots in alternate forms? That is, is it correct to say that $\sqrt{2} + \sqrt{3} = \sqrt{5}$? Explain.

c) Find alternate forms for $\sqrt{75}, \sqrt{48}, \sqrt{200}, \sqrt{63}, \sqrt{27}$.

d) Relate the short cuts you found in Items 3 and 4 of Individual Work 7 to your work in parts (a)–(c) above.

DYNAMIC DUO

9

Your task in this activity is to design a rectangular carton to hold both four melons and six melons efficiently. (It must hold four melons of one size or six melons of another size, but not a mixture of the two sizes.) Prepare a report describing your design and explaining why you think it is an efficient design for holding either four melons or six melons.

Wrapping Up Unit Four

1. **Figure 4.74** is a ten-pack that holds standard soda cans with radius 3.3 cm and height 12.2 cm.

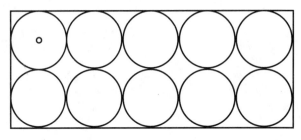

Figure 4.74.
A soda ten-pack.

   a) What is the design's efficiency in terms of percentage of package space used by the cans?

   b) What is the design's efficiency in terms of package material used per can?

2. **Figure 4.75** shows two right triangles.

Figure 4.75.
Two right triangles.

   a) Find the missing side of each triangle.

   b) Are the triangles similar? Explain.

3. One side of an equilateral triangle is 10 cm.

   a) What is the triangle's height?

   b) What is the triangle's area?

4. **Figure 4.76** shows three soda cans in a staggered configuration. If the radius of each can is 3.3 cm, what is the area of the triangle formed by connecting the centers of the circles?

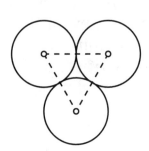

Figure 4.76.

5. Heron of Alexandria, a Greek mathematician of the first century A. D. who is credited with inventing the first jet propulsion engine, used inductive reasoning to conclude that the angle at which a ball strikes a flat surface is equal to the angle at which it rebounds. Observe the path of the ball that is shot at a 45° angle from corner A in each of the four examples in **Figure 4.77**.

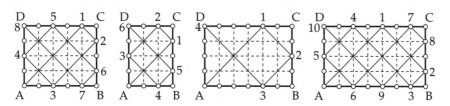

Figure 4.77.

a) What might inductive reasoning lead you to conclude from these examples about the path of a ball on a rectangular table?

b) Is your inductive conclusion correct? Explain.

6. **Figure 4.78** is a parallelopack.

a) The radius of each circle is 3.3 cm. Use your knowledge of the Pythagorean formula or a Pythagorean shortcut to find the base and height of the parallelopack. The triangles shown in the figure may help.

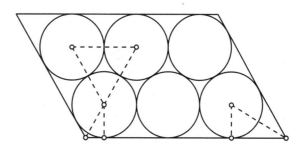

Figure 4.78.
A parallelopack.

b) What is the efficiency of the parallelopack in terms of percentage of package space used by the cans?

7. Is it possible to design a storage facility that can hold either 3 x 8 ft. furniture boxes perfectly or 5 x 9 ft. furniture boxes perfectly? (Either one type or the other, but not a mixture of the two types.) Explain.

8. A quart of water evaporates more rapidly if it is poured into a shallow container with a large surface area than if it is poured into a deep container with a small surface area. Compare the ratio of the surface area of a small cube to its volume with the same ratio for a large cube. Since living things are mostly water, what do the ratios tell you about the ability of living things to withstand dehydration?

9. **Figure 4.79** is a five-melon carton. Each side of the carton is 40 cm.

   a) If r is the radius of each melon, write an expression for the hypotenuse of the right triangle shown in the figure.

   b) An expression for each of the remaining sides of the right triangle is $20 - r$. Explain.

   c) Write a quadratic equation that can be solved to find r. Solve the equation.

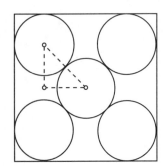

Figure 4.79.
A five-melon carton.

10. There is a convenient shortcut for the Pythagorean formula when the right triangle has angles of 30° and 60°. If you know the short side, double it to get the hypotenuse. Triple its square and find the square root to get the remaining side. There is also a shortcut for a right triangle that has angles of 45° and 45°.

    a) Find the hypotenuse of the right triangle in **Figure 4.80**.

    b) Write an equation to find the other sides if the hypotenuse is 10.

    c) Describe a shortcut for finding the hypotenuse from one of the other sides and a shortcut for finding the other sides from the hypotenuse in any right triangle with a 45° angle.

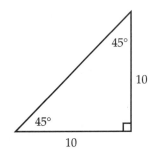

Figure 4.80.
A right triangle.

11. Give a deductive argument that proves that the measures of two angles of any right triangle must have a sum of 90°.

12. A neighborhood committee is reclaiming a vacant lot to use as a playground. They want to cover a 17 x 19 ft. section of the lot with tires like those in **Figure 4.81**. If the tires are arranged in a stacked configuration over a bed of sand, what percentage of the sand is exposed? How many tires are used?

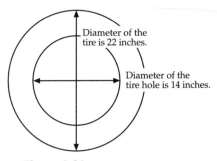

Diameter of the tire is 22 inches.

Diameter of the tire hole is 14 inches.

Figure 4.81.

13. A ramp is constructed for a derby race. The base of the ramp is 24 ft. long. The ramp rises to 10 ft. at its highest point. A 5-ft. support beam for the ramp is placed in the middle of the base, perpendicular to the base of the ramp as in **Figure 4.82.**

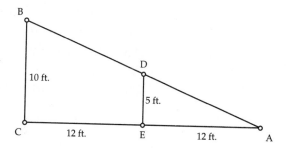

Figure 4.82.

a) What is the length of the ramp?

b) What is the length of the ramp from the middle support beam to the bottom of the ramp (from D to A)?

c) What is the relationship between the answer in part (a) and the answer in part (b)? Why is this not a surprise?

d) Make a list of ratios that are equal.

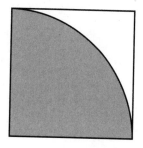

Figure 4.83.

14. **Figure 4.83** is a quarter-circle in a square. What percentage of the space in the square does the quarter-circle use? Explain.

Figure 4.84(a).

15. The aspect ratio of a kite is the ratio of its width to its length. The aspect ratio is the most important factor in determining the kite's flying angle. **Figures 4.84(a) and (b)** show two kites. Which of them has the higher aspect ratio? Explain.

Figure 4.84(b).

16. A sailboat cannot sail directly into the wind. A sailor who wants to sail to a point directly upwind must "tack" at an angle to the wind. A large angle usually results in greater boat speed than a narrow angle; frequently sailors use a 45° angle. **Figure 4.85** shows how a sailor plans to reach a point 10 miles upwind from the boat's current location by sailing at 45° angles to the wind. Determine the total distance the boat sails.

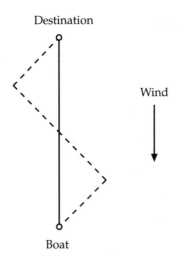

Figure 4.85.

17. Bridges and railroad tracks are built in sections to allow for expansion and contraction that occur because of temperature changes. To understand how important it is to have gaps between these sections, suppose a railroad track 1,000 ft. long is laid without gaps and that expansion causes the length of the track to increase 1 in. If the track is anchored firmly at each end and the track buckles straight up at its center, how far is the center off the ground? Make a sketch and show how you obtained your answer.

Mathematical Summary

*T*he mathematical modeling process requires selection of a criterion on which the model can be based. Two criteria are central in this unit: 1) percentage of package space used by soda cans; and 2) amount of packaging material used per soda can. In the first case the modeling problem is one of maximization; in the second it is one of minimization. Both criteria are measures of packaging efficiency.

Problem simplification is another important part of the modeling process. The problem of maximizing the percentage of space used by cans can be simplified to a two-dimensional problem of determining the percentage of space used by several circles in a polygon. The problem of minimizing the amount of package material per can requires determining the area of several two-dimensional surfaces. Thus, although soda cans and their packages are three-dimensional, much of the mathematics of package design is the geometry of two-dimensional shapes.

Mathematical modeling also requires the identification of one or more factors that affect the chosen criterion. If the criterion is maximization of the percentage of package space used by the circles (cans), the arrangement of the circles and the shape of the package are factors that affect efficiency. However, for a given container shape and can arrangment, changing the number of circles in the container has no effect on the efficiency. If the criterion is minimization of the amount of package material per circle, then the number of circles in the package becomes an important factor.

Most package designs are based on polygons that contain circles arranged in either a stacked or staggered configuration. Since soda cans are a standard size, precisely measuring the efficiency of a design requires calculation of a polygon's area from the radius of one of the circles the polygon contains. "Precise approximate" measurements can be obtained by doing accurate constructions with a computer or calculator drawing utility. Exact results can be obtained by applying geometric area formulas and deductive reasoning.

The deductive reasoning process consists of drawing logical conclusions from established facts. In geometry, established facts are either simple assumptions with which all people agree or facts that have already been established by deductive reasoning.

An example of an assumption with which everyone agrees is that when two or more angles form a straight line, the sum of the measures of their angles is 180°. An example of a fact that can be established by deductive reasoning is that the sum of the measures of the angles of a triangle is 180°. The deductive demonstration, or proof, is based on the fact that when parallel lines are crossed by a transversal, they form equal alternate interior angles. Once the fact that the measures of the angles of a triangle total 180° is proved, it can be used to draw other deductive conclusions about the angles of polygons with more than three sides. In general, if a polygon has n sides, the sum of the measures of its angles is $(n - 2)180°$.

Much of the geometry used in this unit is based on right angles and right triangles because the circles that represent soda cans are tangent to the sides of the polygons that contain them and because a radius drawn to a tangent forms a right angle with it. Deductive reasoning can be used to prove the fact that the radius and tangent are perpendicular.

One of the most important facts about right triangles, indeed one of the most important facts in all of mathematics, is the Pythagorean formula. It says that when the measure of a right triangle's longest side is squared, the result is equal to the sum of the squares of the measures of the other two sides. The Pythagorean formula can be proved deductively in a variety of ways, several of which involve area.

One of the most common right triangles in this unit has angles of 30° and 60°. Two such triangles are formed when an equilateral triangle is folded along its line of symmetry. The Pythagorean formula produces a short cut when it is applied to this type of right triangle. Since a 30°-60° right triangle is half an equilateral triangle, its shortest side is half its hypotenuse. The Pythagorean formula shows that the remaining side is the short side times the square root of three.

The problem of designing an efficient package for several cylinders of the same size is related to the problem of designing an efficient package for several spheres of a given size. Both problems can be simplified to a two-dimensional counterpart involving circles and polygons.

When a rectangle holds several circles of a given size, the percentage of space used by the circles is related to the rectangle's aspect ratio. However, a rectangle that gives a fairly high efficiency for four circles does not give a high efficiency for six smaller circles, a geometric fact that complicates the modeling process when it is applied to the problem of designing an efficient package for spherical fruit like honeydew melons.

Glossary

ALTERNATE INTERIOR ANGLES:
A pair of angles formed by two parallel lines and a transversal. The two angles are inside the two parallel lines and on opposite sides of the transversal.

ASPECT RATIO:
The ratio of the longer side of a rectangle to the shorter side.

BISECT:
A line or point bisects a figure when it divides it into two equal parts.

CIRCUMFERENCE:
The distance around a circle. It is the circular counterpart of a polygon's perimeter.

CONGRUENT:
Geometric figures are congruent if they have the same size and shape.

CORRESPONDING ANGLES:
A pair of angles that occupy identical positions on each of two parallel lines and a transversal.

CROSS SECTION:
A cross section of a three-dimensional solid is the two-dimensional shape that is formed when a cut is made parallel to either the top or the bottom of the solid.

CYLINDER:
A three-dimensional solid with a circular top and bottom and a side surface that is a rectangle when laid flat. A soda can has a cylindrical shape.

DEDUCTIVE REASONING:
The process of drawing logical conclusions from established facts.

DIAGONAL:
A segment connecting two non-adjacent vertices of a polygon.

HYPOTENUSE:
The side of a right triangle that is across from the 90° angle. The hypotenuse is the longest side of a right triangle.

INDUCTIVE REASONING:
The process of drawing conclusions from observations.

ISOSCELES TRIANGLE:
A triangle with two equal sides. An isosceles triangle also has two equal angles.

LINE SYMMETRY:
A figure has line symmetry when it can be folded along a line so that the two halves coincide.

PERIMETER:
The total length of the sides of a polygon.

PERPENDICULAR:
Two lines are perpendicular if they form right angles.

POLYGON:
A closed figure with three or more sides that are line segments. Triangles, rectangles, and hexagons are examples of polygons.

PROOF:
A deductive argument that establishes a fact.

RECTANGULAR SOLID:
The three-dimensional counterpart of a rectangle. A rectangular solid has six rectangular surfaces, or faces.

REGULAR POLYGON:
A polygon with all sides the same length and all angles the same size.

RIGHT TRIANGLE:
A triangle with one 90° angle.

ROTATIONAL SYMMETRY:
A figure has rotational symmetry if there is a central point about which it can be rotated so that it coincides with itself.

TANGENT:
Two circles or a circle and a line are tangent when they intersect in only one point. When a circle and a line are tangent, the line is called a tangent to the circle.

THEOREM:
A fact that has been established by deductive reasoning.

TRANSVERSAL:
A line that intersects two parallel lines.

VERTICAL ANGLES:
A pair of non-adjacent angles formed by two intersecting lines.

VOLUME:
A measure of the amount of three-dimensional space that an object contains.

UNIT

5

Proximity

Many mathematical modeling problems are geometric in nature. Proximity problems are geometric because they involve objects like rocks or locations like fire stations and hospitals that can for many purposes be represented by points. Boundaries that help determine which of several facilities is closest to a given location are often represented by lines. The part of a region that is closest to a given fire station or hospital is often a triangle, quadrilateral, or other type of polygon. In this unit you will apply geometric concepts from Unit 4, *The Right Stuff,* and also develop new geometric tools to help you create mathematical models to solve proximity problems.

The Image Bank

HOW CLOSE?

A wide variety of real-world situations involve questions about proximity. For example, a robot traversing the unfamiliar terrain of an unexplored planet must determine how close it can come to obstacles like rocks and craters. Similarly, the management of a restaurant chain wants to know how close to its existing franchises it can open a new one. Police, fire, ambulance, and other emergency services must decide which available vehicle is closest to the site of a crime, fire, or accident.

In this unit you will develop procedures for answering proximity questions in situations that include estimation of rainfall over the state of Colorado and planning a path for a robot.

LESSON ONE
Colorado Needs Rain !

KEY CONCEPTS

Samples

Mean

Area

Volume

Weighted average

PREPARATION READING

Colorado Needs Rain!

*A*n accurate estimate of rainfall over an area is important because many decisions are based on rainfall estimates. For example, allocation of water resources and disaster relief funds is based, in part, on the amount of rainfall a region receives.

Consider estimation of rainfall for the state of Colorado. The state has 240 stations that measure rainfall. The measurements at these stations are a sample of the rainfall for the entire state. From these samples the state must estimate the total amount of rainfall for all of Colorado. If rainfall is inadequate, the state may need to obtain water from other sources.

The state could ensure an accurate estimate of rainfall by using thousands of rain gauges. Doing so is impractical, however, because of the cost of distributing the gauges and hiring staff to check them.

How should the state use its existing gauges to estimate rainfall for the entire state accurately? Should each gauge be given the same weight in the calculation? How can the existing gauges be used to estimate the rainfall at a location that does not have a gauge?

ACTIVITY

1

A FIRST ESTIMATE

Since the primary goal of this unit is to develop a model for estimating rainfall in Colorado, in this activity you will make your first attempt at developing such a model. The mathematical modeling process often involves several cycles: a model is created, tested against reality, found wanting in some way, and the process is repeated. The knowledge the modeler gains from creating a model, even if the model isn't acceptable, helps perfect the solution.

An important strategy in developing a mathematical model is simplification. Therefore, suppose that the state of Colorado uses only eight rain gauges to estimate rainfall. Colorado is approximately rectangular in shape. The rectangle in **Figure 5.1** has an aspect ratio of 1.36, which makes it similar in shape to Colorado.

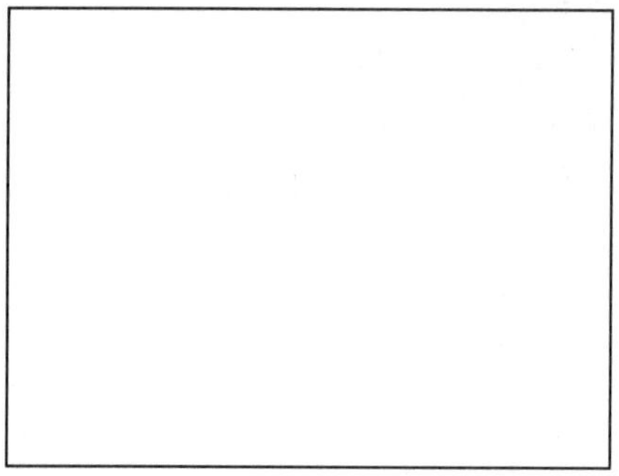

Figure 5.1.
A rectangle with aspect ratio 1.36.

1. Sketch a distribution of eight rain gauges that you think would produce an accurate estimate of rainfall for the entire state. Number the gauges to make it easy to refer to them.

ACTIVITY

A FIRST ESTIMATE

1

2. Rainfall is commonly measured in inches. For example, if an inch of rain falls on a city, then the entire volume of rain would be an inch deep if spread equally across the city. If the rainfall at each of the eight gauges is as shown in **Figure 5.2**, estimate the depth of rainfall for the entire state.

Gauge	1	2	3	4	5	6	7	8
Rainfall	2.11″	2.15″	1.21″	4.45″	2.67″	2.51″	3.11″	2.43″

Figure 5.2.
Rainfall at eight sites in Colorado.

3. The terrain of a region and its population distribution are two factors that can prohibit placing rain gauges in ideal locations. **Figure 5.3** shows locations of eight rain gauges in Colorado. Discuss strengths and weaknesses of the model you used to make your estimate in Item 2. Suggest ways that you could improve your estimate.

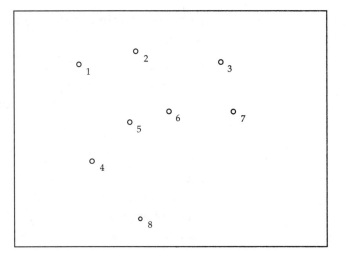

Figure 5.3.
Rain gauges in Colorado.

CONSIDER:

1. Recall that a simple arithmetic mean of a set of data is found by adding the data and dividing by the number of items. Does the mean of the rainfall measured at several gauges give an accurate estimate of the rainfall over a region? Explain.

2. What are some examples of situations in which the mean is useful? What are some examples of situations in which it is not?

INDIVIDUAL WORK 1

Rainfall and Other Averages

Since it is not possible to measure the rainfall at every point in a region, models for estimating rainfall usually employ some type of average. In this individual work you will consider several aspects of rainfall measurement and how averages are used in several different situations. When you finish this individual work, you should have some ideas about how to improve your model for estimating the rainfall in Colorado.

1. Colorado is approximately rectangular, with a length of 375 miles and a height of 275 miles.

    a) What is Colorado's area?

    b) What is Colorado's aspect ratio?

2. Rainfall is sometimes measured in the volume of water that falls over a region.

    a) Convert Colorado's dimensions to feet and determine the number of cubic feet of water that falls on Colorado if the state gets one inch of rain. (1 mile = 5,280 feet.)

    b) What is the change in the total volume of water if the rainfall increases from 1 inch to 1.1 inches?

    c) A flat roof is a relatively inexpensive roof for a large building. However, a flat roof must have adequate drainage so that it does not collapse from the weight of rain and snow. Determine the volume and weight of the water that falls on a 100 ft. x 100 ft. gymnasium roof when one inch of rain falls. (A cubic foot of water weighs approximately 62.5 lbs.)

    d) Rainfall can also be measured in gallons. There are approximately 7.5 gallons of water in a cubic foot. Determine the volume of water in gallons that falls on a 100 ft. x 100 ft. roof during a one-inch rain.

3. a) **Figure 5.4** shows the eight rain gauges from Item 3 of Activity 1 and 14 cities in Colorado. For each rain gauge, list the cities for which you think the gauge provides the best estimate of rainfall.

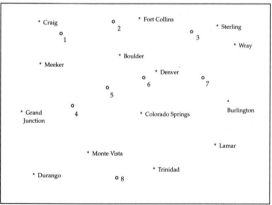

Figure 5.4.
Eight rain gauges and 14 Colorado cities.

b) Give one or more ways to use these gauges to estimate rainfall in Boulder.

c) Should gauge 7 be used to determine Denver's rainfall? Explain.

4. A weather office serves three counties, each of which is approximately rectangular in shape. Each county has a single rain gauge located at its center. **Figure 5.5** shows the dimensions of the counties and the rainfall measured at each gauge during a recent month.

Dimensions	Rainfall
11 mi. x 15 mi.	1.25"
12 mi. x 8 mi.	2.17"
14 mi. x 12 mi.	1.63"

Figure 5.5.
Counties and rainfall.

a) Use a simple arithmetic mean to estimate the rainfall for the tri-county region.

b) Do you think each county's gauge should be given the same weight in estimating the total rainfall for the region? Explain.

c) Estimate the total volume of the rainfall in the region.

d) Show how the volume you found in part (c) can be used to calculate the average depth of rainfall for the region. How else can the depth be calculated?

5. As every student knows, averages are used to determine report card grades. Here is a list of a student's grades in the four categories an instructor uses to determine report card grades:

Tests: 89%

Quizzes: 85%

Homework: 99%

Class participation: 75%

a) Find the mean of these grades.

b) The teacher in this class counts tests as 50% of the final grade, quizzes as 20%, homework as 20%, and class participation as 10%. The teacher calculates the report card grade by computing the **weighted average** of the four category grades. The weighted average is found by multiplying each category by the decimal weight attached to that category and finding a total. Determine the weighted average for these four grades.

    c) What is the sum of the weights used to determine the weighted average in part (b)?

    d) Which category has the greatest impact on the weighted average? Which category has the least impact? In other words, to which category is the report card grade most sensitive, and to which category is it least sensitive?

    e) Could a weighted average be more useful than a simple mean in finding a rainfall estimate for a region? Explain.

6. A family has three children. The parents have budgeted $20 weekly for the children's allowances.

    a) If the allowances are distributed equally among the children, how much does each child receive?

    b) The ages of the children are 5, 7, and 8. If the parents decide to distribute the $20 so that the allowances are proportional to the children's ages, how much does each child receive?

    c) What portion of the $20 does each child receive in part (b)? How are these portions similar to weights?

    d) If the amount budgeted for allowances remains the same, find each child's allowance one year from now. Explain how you obtained your answers.

    e) Examine the long-term implications of this policy.

7. Can weighted averages be used to improve your model for determining rainfall in Colorado? Explain.

ACTIVITY

DIVIDE AND SAVE

2

An accurate estimate of the rainfall in Colorado could be obtained from a very simple model if the rain gauges in the state were distributed uniformly. However, some gauges are closer together than others, which means that some gauges are useful for estimating rainfall in much larger portions of the state than other gauges are. In this activity, you will consider how to determine the part of a region to assign to a given gauge.

Again, it is best to start with a simple situation, so consider a region much smaller than Colorado: a beach. Instead of eight rain gauges, use lifeguard stands. Each lifeguard is assigned the part of the beach that is closest to that lifeguard's stand.

Figures 5.6–5.11 show several different arrangements of two or three lifeguard stands. In each case, draw lines to show the part of the beach assigned to each lifeguard. Estimate the portion of the beach for which each lifeguard is responsible.

1. See Figure 5.6.

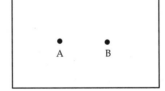

Figure 5.6.

2. See Figure 5.7.

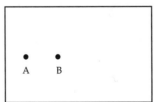

Figure 5.7.

3. See Figure 5.8.

Figure 5.8.

DIVIDE AND SAVE

4. See Figure 5.9.

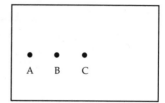

Figure 5.9.

5. See Figure 5.10.

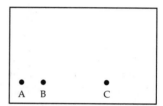

Figure 5.10.

6. See Figure 5.11.

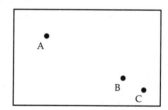

Figure 5.11.

7. In Items 1–6, suppose that the beach is the state of Colorado and the lifeguard stands are rain gauges. The rainfall measured at gauges A, B, and C is 1.25″, 1.80″, and 2.27″, respectively. Show how a weighted average could be used to estimate Colorado's rainfall in each of Items 1–6.

8. Reconsider the map of Colorado in Figure 5.3 or Figure 5.4. Rank the gauges according to which you think should be given the greatest weight in determining a rainfall estimate for the state. Explain your ranking.

INDIVIDUAL WORK 2

Regional Rainfall in Colorado

*A*lthough it is a challenge to determine the best way to combine rainfall readings from eight gauges to produce a single estimate, combining measurements for two or three gauges is simpler. In this individual work, you will make rainfall estimates for portions of Colorado that contain only two or three gauges. In doing so, you apply two models: one that uses a simple average and another that uses a weighted average. The answers you obtain are likely to be imprecise. Ways to improve precision are considered in the lessons that follow. For now, you should concentrate on becoming comfortable with the models.

Figure 5.12 is the table of rainfall measurements at eight gauges in Colorado that you first saw in Activity 1.

Gauge	1	2	3	4	5	6	7	8
Rainfall	2.11″	2.15″	1.21″	4.45″	2.67″	2.51″	3.11″	2.43″

Figure 5.12.
Rainfall at eight sites in Colorado.

1. **Figure 5.13** represents the northwest part of Colorado and the two gauges in that region.

   a) Use a simple average to estimate the rainfall in the region.

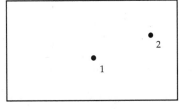

Figure 5.13.
Northwest Colorado.

   b) Copy the drawing onto your paper. Draw a line to divide the region into two parts so that the points in each part are closest to the corresponding gauge. Estimate the portion of the area in each part and use a weighted average to estimate the region's rainfall.

2. **Figure 5.14** represents the northeast part of Colorado and the two gauges in that region.

   a) Use a simple average to estimate the rainfall in the region.

   b) Copy the drawing onto your paper. Draw a line to divide the region into two parts so that the

Figure 5.14.
Northwest Colorado.

points in each part are closest to the corresponding gauge. Estimate the portion of the area in each part and use a weighted average to estimate the region's rainfall.

INDIVIDUAL WORK 2

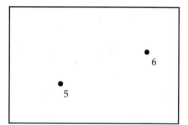

Figure 5.15.
Central Colorado.

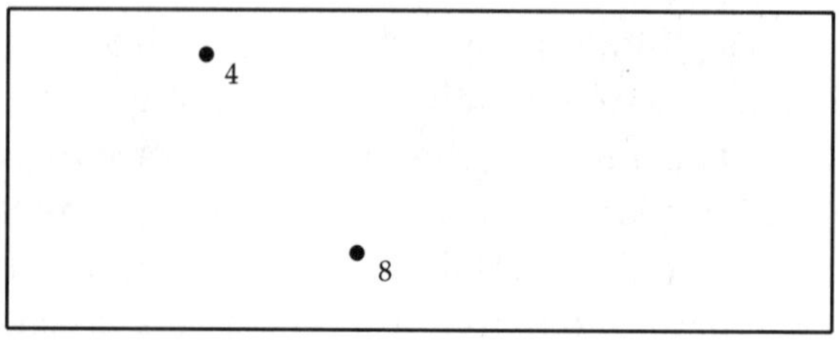

Figure 5.16.
Southern Colorado.

3. **Figure 5.15** represents the central part of Colorado and the two gauges in that region.

   a) Use a simple average to estimate the rainfall in the region.

   b) Copy the drawing onto your paper. Draw a line to divide the region into two parts so that the points in each part are closest to the corresponding gauge. Estimate the portion of the area in each part and use a weighted average to estimate the region's rainfall.

4. **Figure 5.16** represents the southern part of Colorado and the two gauges in that region.

   a) Use a simple average to estimate the rainfall in the region.

   b) Copy the drawing onto your paper. Draw a line to divide the region into two parts so that the points in each part are closest to the corresponding gauge. Estimate the portion of the area in each part and use a weighted average to estimate the region's rainfall.

5. Discuss at least two ways that the estimates you obtained by using weighted averages in Items 1–4 could be used to estimate the rainfall for the entire state of Colorado.

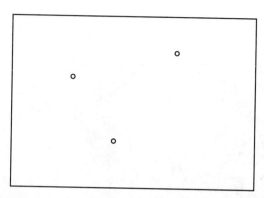

Figure 5.17.

6. A region has three rain gauges as shown in **Figure 5.17**.

   a) Copy the figure onto your paper. Draw approximate boundaries to divide the region into parts so that each gauge is closest to the points in its part.

   b) Estimate the portion of the total area of each part.

   c) Compare the boundaries you drew in part (a) with those you drew in Items 4–6 of Activity 2. How are the results similar or different?

7. Write a brief summary of what you have learned in this lesson about the way rainfall is measured and the way rainfall for an area is estimated. In addition, discuss obstacles that you need to overcome in order to obtain an accurate rainfall estimate for the state of Colorado.

LESSON TWO
Neighborhoods

KEY CONCEPTS

Centers of
influence

Regions of
influence

Voronoi
diagram

The Image Bank

PREPARATION READING

Voronoi Diagrams

Figure 5.18 is an example of a simple rain gauge situation: one with only two gauges. The boundary line separates the rectangle into two regions. All the points in one region are closer to A than to B, and all the points in the other region are closer to B than to A. Points A and B are called **centers of influence** for their respective regions. The region in which each of these points is located is called its **region of influence**. A diagram that is divided into regions so that all points in a region are closer to that region's center of influence than to any other center of influence is called a **Voronoi** (pronounced Vornoy) **diagram**.

Voronoi diagrams are named after the mathematician Georgii Voronoi (1868–1908). For convenience, centers of influence, regions of influence, and boundaries between regions of influence are often called **Voronoi centers**, **Voronoi regions**, and **Voronoi boundaries**, respectively. Voronoi diagrams are an important geometric tool in proximity modeling problems.

Rain gauges and lifeguard stands are just two of the many things that the centers of influence of Voronoi diagrams can represent. If points A and B in Figure 5.18 represent rain gauges, a reasonable estimate of the rainfall for the entire rectangular region can be obtained by calculating the weighted average of the rainfall measured at A and B. Calculating the weighted average requires knowing the portion of the total area that is in each region.

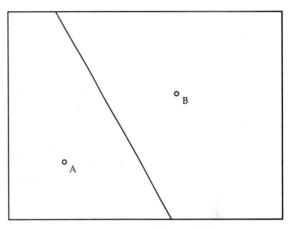

Figure 5.18.
A Voronoi diagram with two centers of influence.

Mathematicians often use symbols to express calculations they do frequently, and that is true of weighted averages. If x represents the rainfall at a gauge and w represents the weight (or decimal portion of the total area) for that gauge's region of influence, then the weighted average is expressed by Σxw. The Greek letter Σ (sigma) is often used in mathematics to indicate that several things are added. Put simply, Σxw says multiply each x by its weight and add the results.

How can the boundaries of regions of influence be accurately drawn? How can precise estimates of the areas of regions of influence be made? These questions must be answered before you can use a weighted average model to obtain an accurate estimate of rainfall for the state of Colorado.

What may be the first recorded use of a Voronoi diagram appeared in the writings of René Descartes (pronounced day–CART) in 1644. Descartes used his drawing in a discussion of the disposition of matter in the solar system.

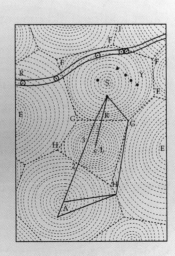

ACTIVITY

ON THE BORDER

3

How much weight should a given rain gauge receive? To apply a weighted-average model to the problem of estimating rainfall in Colorado, you must be able to answer that question. Therefore, the next step in applying a weighted-average model is to develop a method of constructing boundaries for the rain gauges' regions of influence. Your knowledge of geometry from Unit 4, *The Right Stuff,* enables you to develop more than one method. In this activity, you will examine the boundary of a simple Voronoi diagram with several geometric tools.

Handout H5.2 has two copies of the Voronoi diagram in Figure 5.18. You may want to tear the handout in half so that you can work with the copies individually.

Investigate the Voronoi diagram with each of the tools described in Items 1–3.

1. A compass. Mark a point somewhere on the boundary line, but not directly between points A and B. Put the compass point on A and the pencil point on the point you selected. Swing an arc with the compass. Repeat using B instead of A. What do you notice?

2. A Plexiglas mirror. Place the edge of the mirror on the boundary line. What do you notice? Move the mirror. What happens?

3. Paper folding. Fold the paper along the boundary line. What do you notice?

4. Draw a triangle by connecting points A, B, and the point you selected in Item 1. What kind of triangle is it? How is the boundary line related to this triangle? How is it related to the side that connects A to B?

5. Mark two points (centers of influence) on a piece of paper. Show how each of the techniques you explored in Items 1–3 can be used to establish the boundary between your two points.

6. Drawing borders between centers of influence is something you will do often in this unit. Discuss the merits of the three methods. Which one do you prefer? Why?

INDIVIDUAL WORK 3

You Have to Draw the Line Somewhere

Congratulations. You have developed not one, but several methods of establishing a boundary between two centers of influence. Situations with two centers are much simpler than the eight-center Colorado problem, and simplification is an important part of the modeling process. Once you have learned something from the simplification strategy, you need to consider how the results of your work can be extended to more complicated situations. In this individual work, you will begin considering how to find boundaries when there are more than two centers of influence.

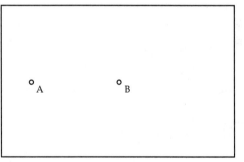
Figure 5.19.

1. a) Trace **Figure 5.19** onto your paper. Use one of the methods you developed in Activity 3 to find the boundary between the two centers of influence. Use a second method to check the accuracy of your boundary. If the two methods do not produce the same line, resolve the discrepancy.

   b) Repeat part (a) for the two points in **Figure 5.20**.

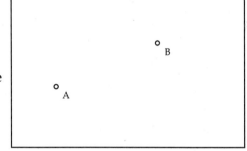
Figure 5.20.

   c) A rain gauge at A measures 1.40", and a rain gauge at B measures 2.30". Estimate the rainfall over the entire rectangular region for Figure 5.19 and Figure 5.20. Explain how you obtained your estimates.

   d) Of which rainfall estimate are you more confident? Explain.

2. **Figure 5.21** represents a portion of the surface of a planet on which scientists want to move a rover to a boulder they have named Scooby Doo. There are craters at A and B that the rover may fall into if it comes too close. Because there is treacherous terrain to the left of A and to the right of B, scientists prefer to have the rover pass between A and B. Suggest a method of getting the robot to the boulder.

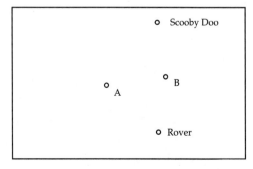
Figure 5.21.

3. Many animals and birds are territorial. For some territorial species, two animals that live in the same region defend the portion of the territory that is closer to their own homes (dens, nests, etc.).

a) **Figure 5.22** shows the dens of two bears. Copy the figure onto your paper and show the portion of the region that each bear defends.

b) The region in Figure 5.22 is a pentagon. How is the procedure for finding the boundary different when the region is not a rectangle?

c) How would the procedure differ if the region were circular?

4. **Figure 5.23** shows three rain gauges in a rectangular state.

a) Construct the Voronoi boundaries between each pair of points.

b) Are the boundaries for each pair of points also boundaries for the entire Voronoi diagram? Explain.

c) Use the boundaries you drew in part (a) to indicate the regions of influence for points A, B, and C. Explain your method.

d) Experiment with other configurations of three points in a rectangular region. Does the procedure you used in part (c) always work for three points?

e) Connect the three points in your drawings to form a triangle. Do you notice any relationship between the angles of the triangle and the boundaries?

5. An ornithologist has attached a band to a bird's leg to help identify the bird in the wild. After the bird is released, the ornithologist sees the bird at locations A, B, and C in **Figure 5.24**.

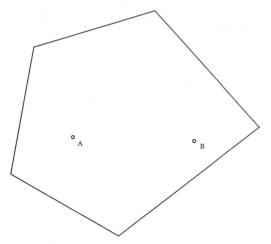

Figure 5.22.

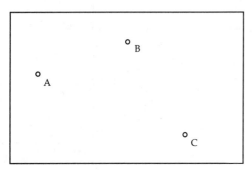

Figure 5.23.

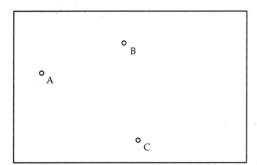

Figure 5.24.

a) Copy the figure onto your paper and suggest a location at which the ornithologist might look for the bird's nest. Explain how you found it.

b) How is the point you have selected related to points A, B, and C? (You may want to use a compass to demonstrate the relationship.) Explain.

c) Suppose the ornithologist sees the bird at the locations shown in **Figure 5.25**. Where do you suggest looking for the nest? Explain.

"I think I've spotted a red-hooded ornithologist."

6. Your experience in Unit 4, *The Right Stuff*, might make you wonder whether an electronic drawing utility is a useful tool for constructing Voronoi boundaries. It is indeed, but getting the drawing into the utility can be an obstacle. Many drawing utilities have a coordinate feature that is helpful.

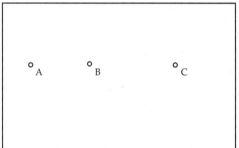

Figure 5.25.

a) **Figure 5.26** shows the rectangular region in Figure 5.19 after it has been transferred to a drawing utility with a coordinate feature. Estimate the coordinates of the rectangle's vertices and the coordinates of points A and B.

b) If you have access to a drawing utility in your classroom, elsewhere in your school, or in your home, use the utility's coordinate features to replicate the drawing in Figure 5.26.

c) When you have the drawing accurately transferred to the utility, use the utility's construction features to establish the boundary. Explain the features you used.

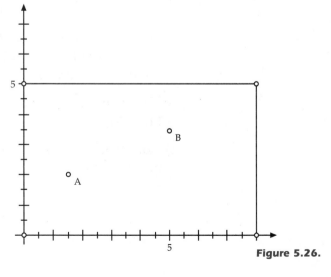

Figure 5.26.

After you have the boundary, to avoid clutter you may want to hide the segment joining A and B and the midpoint of the segment. You should also eliminate the part of the boundary that extends beyond the rectangle by constructing the intersections of the boundary and the sides of the rectangle, constructing a segment between the intersection points, and hiding the original boundary line. You can also hide the axes of the coordinate system if you prefer.

7. **Figure 5.27** represents Figure 5.23 after it has been transferred to a coordinate system.

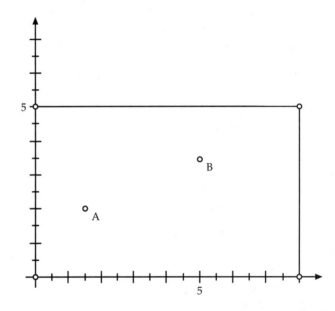

Figure 5.27.

a) Estimate the coordinates of the rectangle's vertices and the coordinates of points A, B, and C. Use your estimates to transfer the figure to an electronic drawing utility.

b) Use the procedure described in Item 6 to construct Voronoi boundaries for centers of influence A, B, and C.

c) Adapt the method you developed in Item 4 to your drawing. In other words, decide which portions of the boundaries to keep.

SEARCHING FOR A METHOD

Your work with Voronoi diagrams in this and the previous lesson represents an important part of the mathematical modeling process: simplification. By developing methods of constructing boundaries when there are two or three centers of influence, you have done the groundwork necessary to estimate rainfall in Colorado from readings taken at eight centers of influence. In this activity, your task is to describe a method that works for three or more centers of influence; you should choose one that you are comfortable using. (Mathematicians have found several ways to construct Voronoi diagrams, and not everyone prefers the same method.)

1. **Figure 5.28** shows a region with three centers of influence. Use the figure to write an algorithm to find the Voronoi boundaries in situations with three centers of influence. Test your algorithm by trying it with other configurations of three centers of influence. Remember that a good algorithm must produce correct results and that it should be as easy to use as possible.

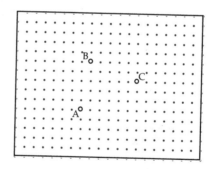

Figure 5.28.

Note: Because you may prefer to use drawing utilities for some of your Voronoi diagram constructions, most figures in the rest of the unit are displayed on a rectangular grid.

ACTIVITY

SEARCHING FOR A METHOD

4

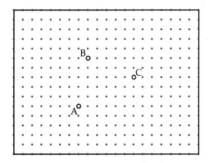

Figure 5.29.

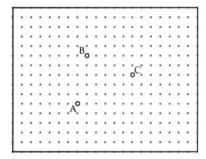

Figure 5.30.

2. In **Figures 5.29 and 5.30**, a fourth point has been added to the three in Figure 5.28. Modify your algorithm for situations with more than three points and show how to use the modified algorithm to find the boundaries in Figures 5.29 and 5.30.

As you work, pay particular attention to the boundaries and to the points at which they intersect. The points at which boundaries intersect are important in deciding which part of a boundary to keep and which to erase. The point at which two boundaries intersect is called a **Voronoi vertex**. To avoid confusing a Voronoi vertex with a Voronoi center, you may want to use two colors.

CONSIDER:

1. From your investigations in Activity 4, what can you conclude about the relationship between the number of Voronoi regions and the number of Voronoi centers in a proximity situation?

2. What can you conclude about the number of Voronoi vertices when there are three centers of influence? When there are four?

3. What can you conclude about the number of Voronoi boundaries when there are three centers of influence? When there are four?

INDIVIDUAL WORK 4

Three and Four Centers

The number of different kinds of Voronoi diagrams that are possible for a given number of centers increases with the number of centers. It usually takes some practice with a variety of situations before people feel confident and comfortable about their ability to apply an algorithm like the one you developed in Activity 4 to new situations. In this individual work, you will construct Voronoi diagrams for a variety of configurations of points and also a variety of contexts. When you have finished, you should have enough skill to plan your approach to the eight-gauge Colorado situation.

Familiarity with common proximity situations helps you deal with unfamiliar ones, and many people find it helpful to have a record of situations they have seen before. As you work with new situations, organize your observations into a table like the one in **Figure 5.31**.

Number of centers	Number of regions	Number of boundaries	Number of vertices
2			
3			
4			
5			

Figure 5.31.

In some columns, such as the third, you can choose the type of information to enter. For example, you might want to enter only the maximum number of boundaries you have encountered for that number of centers, or you may want to list every number you have seen. You might also want to keep a reference to pages and problems in the book that produced the numbers you have recorded.

1. At this point, you should be able to enter information into the first three rows of the table. Do so. When you enter information into the last column, remember that a point is a Voronoi vertex only if it is the intersection of Voronoi boundaries. Do not count a point as a Voronoi vertex if it is at the intersection of a Voronoi boundary and a boundary of the entire region.

INDIVIDUAL WORK 4

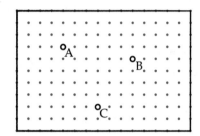

Figure 5.32.

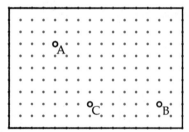

Figure 5.33.

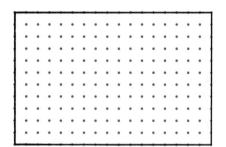

Figure 5.34.
A rectangular domain.

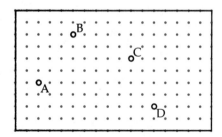

Figure 5.35.

2. Many people find it helpful to visualize a diagram or make a rough sketch before actually constructing the diagram. **Figures 5.32** and **5.33** are regions with three centers of influence.

a) Copy Figure 5.32 and make a rough sketch of the Voronoi diagram without using any of the techniques you have developed. Then use one of the techniques to check the accuracy of your rough sketch.

b) Copy Figure 5.33 and make a rough sketch of the Voronoi diagram without using any of the techniques you have developed. Then use one of the techniques to check the accuracy of your rough sketch.

c) An **acute triangle** is a triangle in which all the angles measure less than 90°, but more than 0°. An **obtuse triangle** is a triangle with one angle that measures more than 90°, but less than 180°. Connect the three centers in your copies of Figures 5.32 and 5.33. Discuss the relationship between the type of triangle and the location of the Voronoi vertex.

d) If the three centers form a right triangle, where do you think the Voronoi vertex is?

3. The region in which the centers of influence are located is called the **domain**. In most of the problems you have considered in this unit, the domain is a rectangle, but it can be any shape. **Figure 5.34** is a rectangular domain.

a) Place three centers in the domain so that the intersection of the Voronoi boundaries is outside the domain. Construct the Voronoi diagram.

b) Complete the Voronoi diagram for your three points.

4. **Figure 5.35** is a domain with four centers of influence.

a) A helpful strategy when constructing a Voronoi diagram in situations with several centers is to first construct a diagram for only some of the centers. Ignore center D and construct a Voronoi diagram for centers A, B, and C.

b) Now construct boundaries for A, C, and D. Explain why you need to construct only two.

c) **Figure 5.36** is another configuration of four centers. Repeat the process you used in parts (a) and (b) to complete the Voronoi diagram.

d) **Figure 5.37** is another configuration of four centers. Repeat the process you used in parts (a) and (b) to complete the Voronoi diagram.

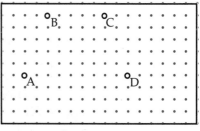

Figure 5.36.

5. Several different kinds of Voronoi diagrams can result when there are four centers of influence.

a) Summarize what you think are the common types of diagrams when there are four centers.

b) If your summary in part (a) did not include situations in which three or four of the centers are collinear, describe the Voronoi diagrams for such cases.

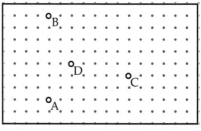

Figure 5.37.

6. In Unit 1, *Gridville*, you used two kinds of distance (firetruck distance and helicopter distance) to study optimal locations for fire stations and other facilities. **Figure 5.38** represents a city in which helicopter ambulances are stored at the five locations labeled A–E. Draw lines to indicate the region of the city that is best served by each ambulance. Explain the method you used.

7. **Figure 5.39** is a hexagonal domain with five centers of influence (labeled A–E).

a) For each center, determine the number of Voronoi boundaries for the center's region.

b) The Voronoi regions in Figure 5.39 represent five communities. When two communities are neighbors, they have potential for sharing expensive services like schools and law enforcement. Which communities do you think have the greatest potential for such cooperation? Explain.

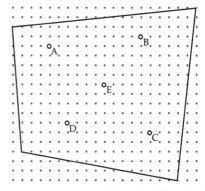

Figure 5.38.

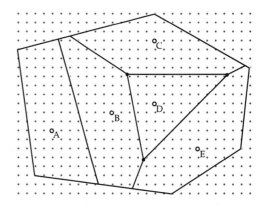

Figure 5.39.

c) The term "neighbor" has fairly obvious meaning in part (b). In general, how would you define neighboring Voronoi regions?

d) The communities centered at B, C, and D decide to pool resources and build a medical center to serve all three communities. Suggest a location for the center. Explain your reasoning.

8. A city has gradually expanded from its original location—point O in **Figure 5.40**—on the bank of a river to its present roughly semi-circular shape. The city's three high schools are at A, B, and C in Figure 5.40. The boundary lines that determine which high school a student attends are shown as broken lines.

a) Why do you think the boundary lines are drawn in this way?

b) The city occasionally receives requests from residents to redraw the boundaries. Suggest an alternative to the current boundaries that you think is reasonable and explain your reasoning.

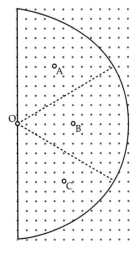

Figure 5.40.

9. A restaurant chain has four franchises in a city. The franchises are shown at A, B, C, and D in **Figure 5.41**. Customers are likely to patronize the restaurant that is closest to them, so the Voronoi boundaries in the figure show the part of the city from which each restaurant draws most of its customers. A competing chain opens a restaurant at location X.

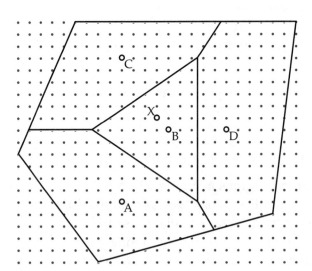

Figure 5.41.
Four restaurant franchises in a city.

a) Determine the region of the city that is most likely to patronize the new restaurant.

b) From which restaurants does the new restaurant take business? Explain.

c) What factors other than proximity could affect a person's choice of restaurant?

d) Why do you think the competing chain chose to locate so close to another restaurant?

10. Now that you have developed and practiced a method for constructing Voronoi diagrams for several centers, you are ready to resume work on the Colorado rainfall-estimation problem. **Figure 5.42** shows the eight Colorado gauges. Write a description of what you think is a reasonable approach to the problem of constructing the Voronoi boundaries. In addition to your description, also discuss any work that must be completed before a finished sketch can be used to estimate the state's rainfall.

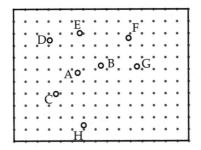

Figure 5.42.
The eight Colorado rain gauges.

LESSON THREE
Rainfall

KEY CONCEPTS

Area of irregular figures

Heron's formula

Pick's formula

Simulation

The Image Bank

PREPARATION READING

Regions and Areas

Voronoi boundaries can be established by a variety of means. They can be constructed with the aid of a compass, a Plexiglas® mirror, or simply by folding the paper. If the coordinates of the centers are known, they can easily be transferred to a drawing utility and the boundaries can be constructed with the aid of the utility's features (that is, connecting two centers with a segment, constructing its midpoint, and constructing a perpendicular at the midpoint).

A Voronoi boundary is a perpendicular bisector of the segment joining two centers. A **perpendicular bisector** is a line that passes through the midpoint of a given line segment and forms right angles with it. Each perpendicular bisector must be inspected carefully, however, since usually only a portion of it is a Voronoi boundary.

As the number of centers grows, extra care must be taken to be sure the right portion of each perpendicular bisector is kept. Constructing a Voronoi diagram when there are several centers is done in a variety of ways. Some people use an algorithm that divides the problem into smaller pieces, then combine diagrams from the smaller pieces. Some people use a more intuitive approach, first inspecting the entire diagram, then concentrating first on certain kinds of points, such as those that are interior to several others.

The skill you have developed with simpler situations should enable you to construct a Voronoi diagram for the eight Colorado rain gauges. In this lesson, you will remove the simplifications you made in earlier lessons and complete the Colorado construction. However, completing the construction does not complete the model; in this lesson you will also consider ways in which the area of each Voronoi region can be found.

ACTIVITY

GETTING THE RAIN RIGHT

5

In this activity, you will determine the Voronoi regions for each of the eight Colorado rain gauges and find the area of each region.

In **Figure 5.43**, Colorado is represented on a grid in which one unit = 25 miles. The gauges are labeled with letters instead of numerals.

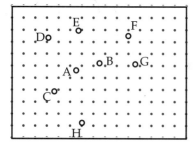

The coordinates of the gauges (in grid units) are

A: (5.5, 5.8)	B: (7.4, 6.3)
C: (3.7, 4.0)	D: (3.1, 8.5)
E: (5.8, 9.1)	F: (9.9, 8.6)
G: (10.5, 6.3)	H: (6.0, 1.3)

Figure 5.43.
Eight rain gauges in Colorado.

1. Use any of the methods you have developed to find the Voronoi regions. (Handout H5.3 is an enlarged copy of Figure 5.43 that you may find convenient.) Write a description of your method.

2. When you have finished constructing the regions, determine approximate values for the area of each and the portion of the state's area in each. Several methods that you might use are described below.

   a. Area formulas from Unit 4, *The Right Stuff*.

   b. A drawing utility.

   c. **Heron's formula**. Heron's formula says that the area of a triangle is $\sqrt{s(s-a)(s-b)(s-c)}$, where s is half the triangle's perimeter and a, b, c are the lengths of its three sides. For example, if the sides of a triangle are 3, 4, and 5 units, s is $0.5(3+4+5) = 6$, and the area is $\sqrt{6(6-3)(6-4)(6-5)} = 6$ square units.

GETTING THE RAIN RIGHT

The formula is named after Heron, a Greek mathematician who lived in the first century A.D. An advantage to this formula is that it requires knowing only the lengths of the triangle's sides and none of its altitudes.

d. **Pick's formula**. Pick's formula says that the area of any polygon whose vertices are points of a rectangular grid is $0.5b + i - 1$, where b is the number of grid points on the polygon's boundary (including the vertices) and i is the number of grid points in the polygon's interior. For example, the polygon in **Figure 5.44** has 12 grid points on its boundary and 10 grid points in its interior. Therefore, its area is $0.5 \times 12 + 10 - 1 = 15$ square units. One advantage of Pick's formula is that it works for any polygon. A disadvantage is that the vertices must be grid points. Pick's formula was discovered and proved by Georg Pick in 1899.

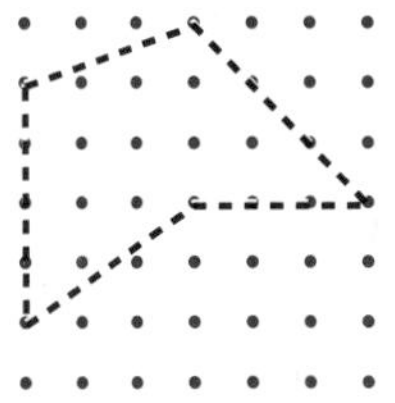

e. A simulation. There are a variety of ways a simulation can be done. One method is to scatter small objects such as grains of rice over a completed diagram and determine the percentage that fall in each region. Keep in mind that a simulation becomes more reliable as the number of trials increases. You might, for example, use 100 grains of rice and repeat several times.

Figure 5.44.
A polygon whose vertices are points of a grid.

3. When you have found the eight areas and percentages, write a description of the method you used. Include in your description a discussion of reasons why your answers may lack precision.

CONSIDER:

1. The estimates of the eight areas in Colorado probably vary somewhat in your class. Which estimate do you think is best? Why?

2. For which region(s) do the area estimates in your class vary most? Why?

INDIVIDUAL WORK 5

Rain and Regions

Figure 5.45.
Colorado rainfall
measurements

Gauge	A	B	C	D	E	F	G	H
Rainfall	2.11″	2.15″	1.21″	4.45″	2.67″	2.51″	3.11″	2.43″

1. **Figure 5.45** shows the rainfall measurements that you first saw in Activity 1.

   a) Use the area estimates of which you are most confident and a weighted average model to estimate the average depth of rainfall in Colorado. Labels on the gauges match those in Figure 5.43.

   b) What is the total volume of rainfall based on your estimate of the average depth? (Remember that Colorado is approximately 375 miles by 275 miles.)

   c) To which rain gauge is the estimate of the state's total rainfall most sensitive? That is, for a given amount of error, say 0.05″, in the rainfall measurement, which gauge's measurement produces the greatest change in the estimate of the state's rainfall?

   d) Your estimate of the area of each Voronoi region in Activity 5 might be off either because your construction is inaccurate or because your measurements lack precision. For example, suppose your estimate of the area of a region is off by one of the grid squares in Figure 5.43. What is the error, in square miles, in the estimate of the region's area?

2. Find the area of the right triangle in **Figure 5.46** by as many methods as possible. Describe each method in writing.

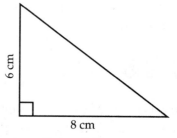

6 cm

8 cm

Figure 5.46.

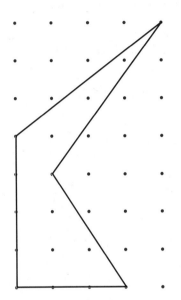

Figure 5.47(a).

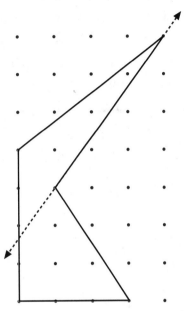

Figure 5.47(b).

3. Most polygons with which you have worked are convex. A **convex polygon** is a polygon in which none of its sides, when extended, intersect other sides. For every pair of points in the interior of a convex polygon, the segment connecting the points is completely in the interior. Sometimes it is necessary to find the area of a polygon that is concave. A **concave polygon** is a polygon in which some of its sides, when extended, intersect other sides. **Figure 5.47(a)** shows a concave polygon. **Figure 5.47(b)** shows an extension of one side intersecting another side.

a) The polygon is drawn on a centimeter grid. Use Pick's formula to find its area.

b) Use another method to find the area.

4. **Figure 5.48(a)** is a graphing calculator screen of the Voronoi boundaries for the eight Colorado rain gauges. **Figure 5.48(b)** is the same eight boundaries after a calculator program has plotted 50 random points in the eight regions.

a) Estimate each region's percentage of the state's area based on this simulation.

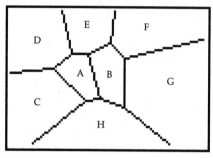 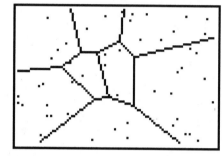

Figure 5.48(a).   **Figure 5.48(b).**

b) How do these simulation results compare with the results you obtained in Activity 5? Explain.

5. Elementary school students sometimes use a device called a geoboard to study geometry. A geoboard has pegs around which you can stretch rubber bands. **Figure 5.49** shows a geoboard shape made by an elementary student. Write a short description that would help the student find the shape's area.

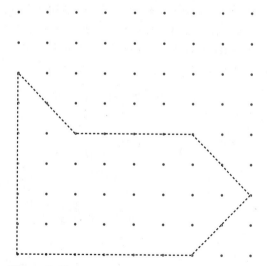

Figure 5.49.

6. Use Heron's formula to find the area of a triangle with sides of 4 cm, 6 cm, and 12 cm. Explain the result.

7. Heron's formula is remarkable in that it does not require knowing the length of any of the triangle's altitudes, only the three sides. No one has ever found a formula for the area of a quadrilateral that uses only the lengths of the four sides. Explain why it is impossible to find such a formula.

8. Sometimes it is necessary to estimate the volume of water in a lake or pond. For example, in response to a number of drowning incidents over the years, residents of Quincy, Massachusetts have proposed draining an old rock quarry now filled with water. Officials estimate that the quarry contains approximately 140,000,000 gallons of water, which presents a problem: how do you remove 140,000,000 gallons of water?

   Figure 5.50 is a lake for which an estimate of the volume of water must be made. As shown in the figure, several depth measurements have been made at locations chosen at random in the lake. Describe a plan for using the readings to obtain an estimate of the amount of water in the lake.

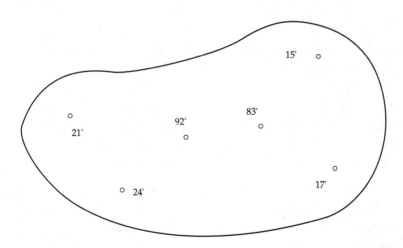

Figure 5.50.
Depth measurements in a lake.

9. Revisit the algorithm you developed in Activity 4. Discuss how well the algorithm works in situations with larger numbers of centers such as the Colorado rain gauge situation. If you do not think the algorithm works well, suggest modifications.

10. Many of the diagrams you have used in this unit have been drawn on a coordinate grid to facilitate transfering them to an electronic drawing utility. In the remainder of this unit, you will see that there are other reasons why coordinate methods are helpful. In this item, you will examine some relationships in Voronoi diagrams that only a coodinate grid can help you see. **Figure 5.51** shows two Voronoi centers and the boundary between them. The centers have been connected with a broken line segment. The origin is at the lower-left corner.

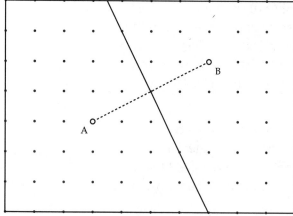

Figure 5.51.

a) Find the coordinates of the point halfway between A and B.

b) Find the distance between the two centers. Explain how you found it.

c) Find the slope of the segment connecting A and B and the slope of the Voronoi boundary. How are they related?

d) Find the equation of the line connecting A and B and the equation of the Voronoi boundary.

11. Use the method of generalization to show that Pick's formula is correct for rectangles with a horizontal/vertical orientation on a grid. That is, if a rectangle whose vertices are points of the grid has length a and width b, write expressions for the number of interior grid points and the number of boundary grid points. Then apply Pick's formula and show that it gives the rectangle's area. (You may want to try one or two specific rectangles first.)

12. Summarize the results of your work on the problem of estimating the rainfall for the state of Colorado. Be sure to discuss any obstacles that you feel still remain to successfully implementing a weighted-average model.

LESSON FOUR

A Method of a Different Color

KEY CONCEPTS

Coordinate geometry

Midpoint

Slope

Perpendicular bisector

Solutions of systems of equations

Distance

The Image Bank

PREPARATION READING

Doing it Descartes' Way

Congratulations! You have created a weighted average model that estimates the rainfall in Colorado.

You can think of your model as having at least two parts. The first part might be called the construction phase, in which you use a compass, a Plexiglas® mirror, paper folding, or a drawing utility to establish the Voronoi boundaries. The second part might be called the measurement and calculation phase, in which you use a drawing utility, measurements, and area formulas, or a simulation to determine the area of each region.

If you have used a drawing utility, you know that it can produce accurate constructions and precise measurements. However, an accurate transfer of a figure to a drawing utility is difficult to achieve unless coordinates of centers are known.

Since coordinates are often established by measuring, the utility is not necessarily free from the measurement imprecision of other methods.

Coordinate geometry provides one of the most important tools available to the mathematician. By using coordinates to describe points and equations to describe lines (and other geometric shapes), it offers a powerful algebraic view of geometry. When the coordinate approach to geometry was invented by René Descartes in the 1600s, it revolutionized mathematics and science and changed civilization forever.

The graphing calculator and the computer are just two of the many technological advances that wouldn't be possible without coordinate geometry. Although you do not always see axes when you look at a graphing calculator screen or a computer screen, both are really coordinate planes. Every pixel has a pair of coordinates that can be used to turn the pixel on or off, or in the case of the computer, to assign the pixel one of a variety of colors. Since every shape (including letters) is really a collection of pixels, the shape is a collection of coordinates.

You have already worked with much of the coordinate geometry you need to do a coordinate analysis of Voronoi diagram problems. You know about equations of lines, slopes, and the Pythagorean formula. In this lesson, you will examine how coordinate geometry can confirm area estimates you have already made and provide you with a new way of thinking about Voronoi diagram problems.

René Descartes (1596–1659) published his method of coordinate geometry in 1637 in an appendix (titled *La geometrie*) to his book *Discours de la méthode*. Note that on this 1937 commemorative French stamp, "sur la" in the title of Descartes' book is incorrect; a stamp with the correct title was issued later.

ACTIVITY

GETTING A LINE ON VORONOI DIAGRAMS

6

The first step in applying coordinate geometry to Voronoi diagrams involves finding equations of boundaries and coordinates of vertices.

Therefore, in this activity, you find equations of Voronoi boundaries and coordinates of Voronoi vertices. As you work, use exact values whenever possible. If you must round, keep several decimal places because the results must be used later to find areas.

The domain and centers in **Figure 5.52** are a simplified version of the Colorado rain gauge situation. The origin of the coordinate system is the lower-left corner. (Assignment of the origin is arbitrary. Placing the origin at the lower-left corner of the domain is a good strategy because it avoids negative coordinates.)

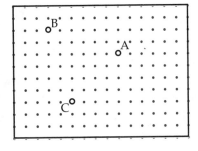

Figure 5.52.

1. **Figure 5.53** shows the first step in finding the Voronoi boundary between points A and B. A and B are connected with a line segment and its **midpoint**, the point that is halfway along the segment, is marked.

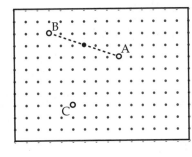

Figure 5.53.

a) Determine the coordinates of the midpoint.

b) Describe a general method of determining the coordinate of a point halfway between two given points. (Experiment with other points on the grid until you think you have a result that works.)

ACTIVITY

GETTING A LINE ON VORONOI DIAGRAMS

6

2. The next step is to construct a line that goes through the midpoint and is perpendicular to the segment connecting A and B (**Figure 5.54**). In coordinate geometry, that means finding the equation of the line. As you know, the equation of a line can be found if you know the line's slope and the coordinates of at least one point on the line. In Item 1 you found a point on the line, so all you need is the slope.

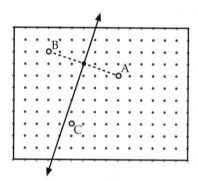

Figure 5.54.

a) Find the slope of the segment connecting A and B, then use it to find the slope of the perpendicular. Then use the slope of the perpendicular to find its equation.

b) Describe a general method for finding the slope of a line perpendicular to a line connecting two points.

3. Repeat the process that leads to the equation of the perpendicular bisector for points A and C.

ACTIVITY

GETTING A LINE ON VORONOI DIAGRAMS

6

4. **Figure 5.55** shows the boundary lines for points A and B and points A and C. Find the coordinates of their point of intersection by solving a system of two equations.

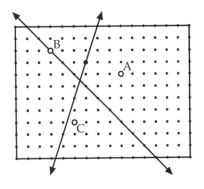

Figure 5.55.

5. **Figure 5.56** shows the completed Voronoi diagram. The boundary between B and C has been added and the unnecessary portions of boundaries erased. Find the equation of the line containing the boundary between B and C. Check that it goes through the point you found in Item 4.

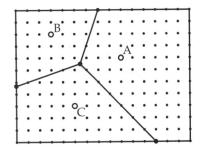

Figure 5.56.

6. Find the coordinates of the points at which the Voronoi boundaries intersect the boundaries of the domain.

CONSIDER:

1. How could you use the coordinates of points of intersection in Voronoi diagrams to estimate areas?

2. A Voronoi diagram often has many boundaries. Finding the equation of each boundary is a repetitive task that can become boring. Boredom leads to mistakes. How might you avoid making such mistakes?

INDIVIDUAL WORK 6

Connect the Dots

Now that you have a method for finding equations of boundaries and coordinates of vertices, you need to use these results to find areas of Voronoi regions.

1. **Figure 5.57(a)** is a completed Voronoi diagram with three centers. This diagram is unusual because all points of intersection are grid points. **Figure 5.57(b)** suggests a way to find the area of each region.

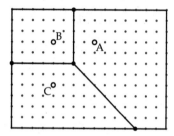

Figure 5.57(a).

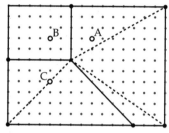

Figure 5.57(b).

a) Find exact values for the lengths of the broken lines in Figure 5.57(b). Explain how you found your answers.

b) Find the area of each Voronoi region by finding the area of each part in Figure 5.57(b). Check your work by using a formula different from the one you used to get your answer.

2. Since lengths of segments are used in area formulas, coordinate methods of solving Voronoi diagram problems require application of the distance formula. **Figure 5.58** shows two points in a coordinate system whose origin is at the lower-left vertex of the rectangle.

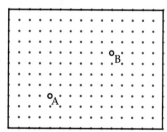

Figure 5.58.

a) Find an exact value for the distance between the two points. Explain how you obtained your answer.

b) Apply the method of generalization to the approach you used in part (a) to find a formula for the distance between two points with coordinates (a, b) and (c, d).

c) Since points are frequently named with capital letters, using lower-case letters for coordinates sometimes results in confusion about whether a letter represents a point or a coordinate. To avoid this confusion, mathematicians use subscripts to distinguish coordinates of one point from those of another. Rewrite the formula you developed in part (b) using (x_1, y_1) and (x_2, y_2) to represent the coordinates of the points.

3. a) Find the equation of the Voronoi boundary between points A and B in Figure 5.58. Explain the steps you used.

b) Apply the method of generalization to the problem of finding the midpoint and the slope of the perpendicular bisector. That is, if (x_1, y_1) and (x_2, y_2) are the two points, find formulas for the coordinates of the midpoint and for the slope of the perpendicular bisector.

c) Use the results of part (b) to find a general equation for the perpendicular bisector.

d) The general results you obtained in parts (b) and (c) are useful in developing a calculator program to find the equation of the perpendicular bisector. Developing a program is worthwhile because finding such equations repeatedly is tedious to do by hand. However, some boundaries have equations so obvious that using a program is silly. Suppose the point $(4, 2)$ is the midpoint of two Voronoi centers. What is the equation of the Voronoi boundary if it is horizontal? What is the equation if the boundary is vertical? What are the slopes of these two lines?

INDIVIDUAL WORK 6

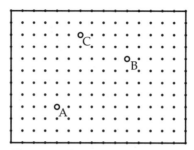

Figure 5.59.

4. **Figure 5.59** is a domain with three centers. Use coordinate geometry to find the coordinates of the Voronoi vertex. (Again, the origin of the coordinate system is at the lower-left vertex of the rectangle.)

5. Find the area of the triangle whose vertices are points A, B, and C in Figure 5.59. Explain your method.

6. Applying the method of generalization often requires some fairly complicated symbol manipulation. However, it is often worth the effort because the results mean that problems of that type never have to be solved again. Because Voronoi problems usually involve quite a few boundaries and vertices, calculator programs that perform routine tasks are worth writing. Write each of the following programs.

a) A program that finds the midpoint and the equation of the perpendicular bisector for two points. The program should ask the user for the coordinates of the two points and output the coordinates of the midpoint and the slope and y-intercept of the perpendicular bisector. You may want to check to see if your calculator has a function to convert decimals to fractions so that the output of this program can be displayed in exact form.

b) A program that finds the area of a triangle. This program is more useful if it can find the area either from the lengths of the sides or from the coordinates of the triangle's vertices. You may want to write the program so that the user is given the choice of side lengths or coordinates for input (an alternative is to write separate programs).

c) A modification of the program in part (b) that finds the area of a polygon from the coordinates of the vertices. The user should be asked for the number of vertices and the coordinates of each vertex. To write this program, you need to think about how to instruct the calculator to divide a polygon into triangles in a systematic way, and how to design a program loop that determines the length of each side of each triangle.

CONFIRMING THE RAINFALL

7

The goal of this activity is to use coordinate methods to confirm the area estimates you made for the eight Colorado rain gauge regions in Activity 5.

Figure 5.60 shows the regions. Remember that each unit of the grid represents 25 miles.

The coordinates of the eight gauges (in grid units) are:

A: (5.5, 5.8)   B: (7.4, 6.3)

C: (3.7, 4.0)   D: (3.1, 8.5)

E: (5.8, 9.1)   F: (9.9, 8.6)

G: (10.5, 6.3)   H: (6.0, 1.3)

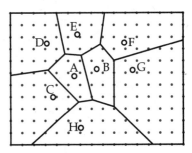

Figure 5.60.

Your task is to find the equations of all boundaries, use the equations to find the coordinates of all vertices, and use the coordinates of the vertices to find the areas of the polygons. Then use the areas of the polygons to find the percentage of Colorado's area in each polygon. Round the percentages to several decimal places and compare them to your previous estimates.

Since the percentages are independent of scale, you can change the coordinates by any convenient factor before beginning your work. For example, you may want to multiply all by 10 to avoid decimals or by 25 to place all units in miles.

Divide the work among several people or groups of people. Use any calculator programs you developed in Individual Work 6. Organize the results into a table showing each gauge, the equations of its region's boundaries, the coordinates of its region's vertices, and the percentage of the state's area in the region.

INDIVIDUAL WORK 7

SOL *the Robot*

*T*his individual work considers a few other applications of coordinate methods.

1. A company is purchasing a robot that it has named SOL to move materials around its factory floor. To avoid accidents, the robot must be programmed to follow paths that are as far away from the eight machines on the factory floor as possible. Your task is to design a network of paths that follow Voronoi boundaries between machines. Programming the robot requires knowing the equations of these paths, the coordinates of their endpoints, and their lengths.

   Figure 5.61 shows the factory floor and the eight machines.

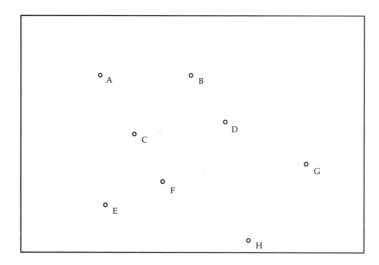

Figure 5.61.
A factory floor and eight machines.

   The coordinates of the eight machines (measured in meters from the lower-left corner) are:

   A: (14, 30)   B: (30, 30)

   C: (20, 20)   D: (36, 22)

   E: (15, 8)   F: (25, 12)

   G: (50, 15)   H: (40, 2)

   Make a diagram of the paths (it can be a rough diagram) and organize the equations of the paths, their vertices and lengths, into a table.

2. On a piece of graph paper, create a rectangular domain with lower-left vertex at (0, 0) and upper-right vertex at (10, 10). Use the points A (1, 3), B (7, 4), C (2, 6), and D (9, 5) as centers. Find the equations of

all Voronoi boundaries and the coordinates of the end-points of all boundaries.

3. Coordinate methods can help solve problems that seem impossible to solve by any other means. For example, here is a version of a problem found in puzzle books. Two ladders are resting against walls 10 feet apart as shown in **Figure 5.62**. One ladder is 15 feet long, the other is 12 feet long. Is there room enough for a child four feet tall to walk under the point where the ladders cross? Use coordinate methods to solve this puzzle.

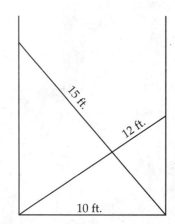

Figure 5.62.

4. Mathematicians sometimes use coordinate geometry as a method of deductive proof. In a sense, their reason for doing so is laziness: some things are a lot easier to do that way.

a) For example, a common technique for finding areas of polygons is to divide the polygon into triangles by drawing several of its diagonals. What properties do diagonals of various polygons have? Consider the square. Since the angles of a square are right angles, a coordinate system can be introduced so that the axes coincide with two of the square's sides (**Figure 5.63**). If s is the length of each side of the square, find coordinates of each of the square's vertices. Then use your knowledge of coordinate geometry to investigate relationships between the diagonals. What can you prove?

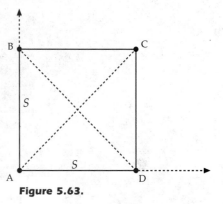

Figure 5.63.

b) Apply a similar coordinate method to a parallelogram. Show how you introduced the coordinate system, the coordinates you used for the vertices, and give a proof of any conclusions you draw about the diagonals.

c) Here is a coordinate proof of a fact you already know. **Figure 5.64** shows an equilateral triangle with a coordinate system introduced so that the triangle's line of symmetry is on the x-axis.

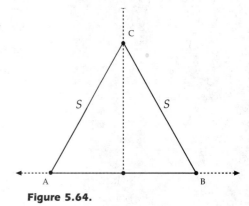

Figure 5.64.

Suppose the length of the altitude is unknown and represent the coordinates of C by $(0, h)$. If s represents the length of each side of the triangle, find coordinates of B. Use the distance formula you devised in Item 2 of Individual Work 6 to write an equation involving the distance from B to C. What can you conclude about h?

Digging for Answers

KEY CONCEPTS

Reflection

Iteration

The Image Bank

PREPARATION READING

Turning the Tables

Phenomena that resemble Voronoi diagrams occur in nature. Dried, cracked mud, for example, forms sections that look like Voronoi regions. Cells viewed through a microscope cluster in Voronoi-like groups. On closer inspection, however, you often find that the centers of the regions are not well-defined.

Are Voronoi diagrams useful in studying phenomena that resemble Voronoi regions? All of the work you have done in this unit involves determining Voronoi boundaries from the centers. Is the opposite possible? That is, if the Voronoi boundaries are known, can centers be found?

In this lesson you will consider modeling problems in which boundaries are known and centers must be found.

UPON REFLECTION

ACTIVITY

8

Archeologists are like detectives: they often must make conjectures based on small pieces of evidence. For example, consider a situation in which an archeologist has gathered clues about the trading habits of ancient people. From the kinds of pottery and other goods that have been unearthed, the archeologist believes there were four distinct trading centers in the region, but has been unable to locate them. **Figure 5.65** is a grid on which the archeologist has separated the region into four sub-regions according to the kinds of goods most abundant in each. (The numbers outside the regions are for easy reference. The scale is 1 square = 10 miles.) To prove the four-trade-center theory, the archeologist must find the four cities that were the trade centers. Your task is to suggest locations at which to begin excavations.

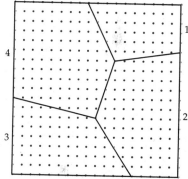

Figure 5.65.
Four archeological regions.

1. Guess a center for one of the regions and mark it on Handout H5.8. How can your knowledge of the relationship between centers of two adjacent regions be used to check whether your point makes sense?

2. Is there more than one way to check your guess?

3. If you have checked the point and found that it isn't correct, make a second guess. Explain how you used the results of your first guess to improve your second guess.

CONSIDER:

1. What modeling assumption(s) is made in order to apply Voronoi diagrams to the archeology problem in Activity 8?

2. How accurately do you think you have located the centers of the regions in Activity 8?

3. How might you improve the results?

INDIVIDUAL WORK 8

What's the Point?

*I*n this individual work, you will apply the procedure you developed in Activity 8 and consider some alternate ways of approaching the center-location problem.

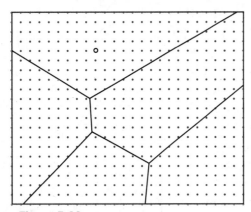

Figure 5.66.

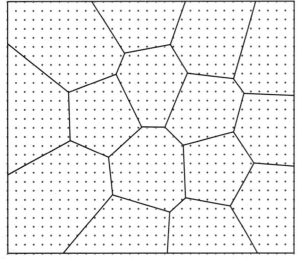

Figure 5.67.

1. **Figure 5.66** shows the results of work done by an archeologist. Evidence indicates that people in this ancient city worshipped at one of five temples, and usually attended the temple closest to them. Approximate boundary lines have been established, and the location of one of the five temples has been discovered. Find approximate locations for the other four temples. Explain your procedure.

2. Biologists who study cells are confronted with the problem of locating cell nuclei when microscopic photographs show cell walls clearly, but do not show interior cell structure clearly. **Figure 5.67** represents cell walls in a cluster of cells.

   a) Determine approximate locations of the nuclei. Explain your method.

   b) What modeling assumption(s) did you make in arriving at your answer?

3. A procedure that repeats the same steps is called an **iterative procedure**. The "guess and check and improve the guess" approach to finding Voronoi centers is an example of an iterative process. The idea is to perform sufficient iterations to achieve the desired degree of precision. It is also possible to examine the center-location problem deductively. For

example, **Figure 5.68** shows four Voronoi centers and their boundaries. Three of the centers have been connected to the Voronoi vertex at which their boundaries meet. The six angles have been numbered for convenient reference.

a) How are angles 1 and 2 related? Why? (You may want to connect two centers to form a triangle.)

b) Does the same relationship hold for any other angles in the figure?

c) How is angle 6 related to angles 2 and 3? Explain. (You may want to think first about how all six angles are related and how that relationship is affected by the relationships you described in parts (a) and (b).)

d) How can your conclusion help locate Voronoi centers from the boundaries? That is, explain how angle measures can help find a Voronoi center in **Figure 5.69**.

4. One way to locate a Voronoi center is with an iterative process: guess a location, reflect it through a boundary into a neighboring region, continue reflecting until each region has a center, and reflect the last point back into the starting region. If the result does not coincide with the guess, then make a new guess by averaging the coordinates of the result and the guess. Repeat the procedure until the guess and the result are as close together as desired. Develop a calculator program to implement this procedure. Inputs of the program are the coordinates of the guess, the number of times the process is repeated (the number of iterations), and lists of the slopes and y-intercepts of the boundaries through which

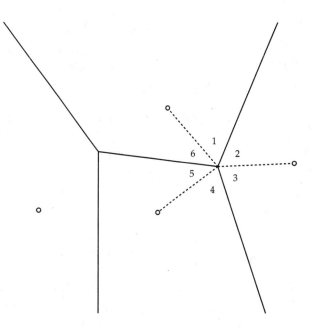

Figure 5.68.

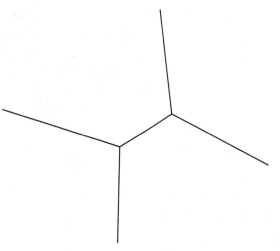

Figure 5.69.

the reflections are done, in the order they are done. Output of the program should be the last coordinate calculated. Optional output is a plot showing the guessed point and the series of approximations to the center of the region in which the guess is made.

To develop your program, you need mathematical formulas for the coordinates of the reflection of a given point through a given line. If the point (c, d) is reflected through the line $y = mx + b$, the coordinates of the reflection are:

$$x = \frac{c(1 - m^2) + 2m(d - b)}{m^2 + 1}$$

$$y = \frac{c - x + dm}{m}$$ (where x is the value given by the previous formula).

These formulas apply only if the reflecting line is neither horizontal nor vertical. For horizontal lines, $x = c$ and $y = 2b - d$. For vertical lines, $x = 2b - c$, $y = d$. Your program must have a way to detect whether a line is horizontal or vertical and make exceptions for those cases.

5. Is it possible to locate a center by making two independent guesses of the location of a region's center? Experiment with two guesses and various reflections of them. If you find a method that appears to work, explain the method and give a deductive argument (proof) that it is correct.

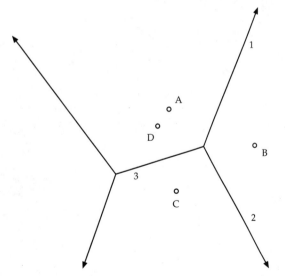

Figure 5.70.

6. Evaluate the following plan for locating the Voronoi centers in **Figure 5.70**. Guess a center in the upper region, labeled A in the figure. Reflect A through line #1 to point B, reflect B through line #2 to C, and reflect C through line #3 to D. Find the point halfway between D and A, and repeat the process several times.

7. How would you locate the centers for a set of Voronoi boundaries that are all parallel lines?

PROXIMITY

Wrapping Up Unit Five

1. A person studying Voronoi diagrams has developed the following algorithm for situations with four centers.

   Step 1.
   Arbitrarily select one of the four centers and label it A. Move around the centers in either clockwise or counterclockwise fashion. As you go, label the remaining points B, C, and D.

   Step 2.
   Construct the perpendicular bisector of each adjacent pair of centers.

   Step 3.
   Determine the portions of each perpendicular bisector to keep.

   A sample implementation of the first two steps of the algorithm is shown in **Figure 5.71**.

   Is this a good algorithm for four-center situations? Explain.

**UNIT
SUMMARY**

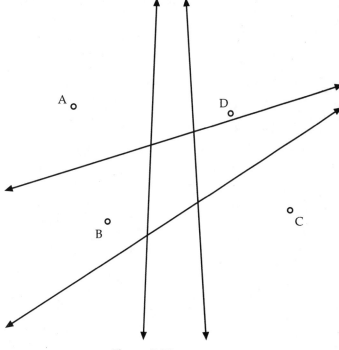

Figure 5.71.

Figure 5.72(a).

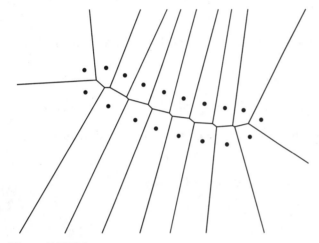

Figure 5.72(b).

Figure 5.72(c).

Figure 5.73.

2. Medical radiology has advanced considerably beyond the x-rays of bone fractures for which it was first used. Today radiologists use sophisticated techniques to help them visualize cross sections of the brain and other organs. However, the images produced by modern equipment are two-dimensional representations of three-dimensional objects. Therefore, the equipment must be able to rotate an image so that it can be seen from various viewpoints. To rotate an image, the equipment's computer must determine an axis about which to rotate the image. One technique used to find the axis of an image involves Voronoi diagrams. First, a collection of points are marked along the boundary (**Figure 5.72(a)**). Then the Voronoi diagram is constructed (**Figure 5.72(b)**). Finally, only the boundaries that are completely within the object's interior are retained (**Figure 5.72(c)**). (In its actual use, this technique involves many more boundary points than this example.)

Predict an "axis" for the set of boundary points in **Figure 5.73**. Then use the technique described above to check your prediction.

3. **Figure 5.74** is a Voronoi diagram with five centers.

   a) This diagram is somewhat unusual for a five-center Voronoi diagram. Explain.

   b) Describe what happens if a sixth point is added interior to the five in Figure 5.74. Include a sketch with your description.

   c) Describe what happens if a sixth point is added somewhere outside the five in Figure 5.74. Include a sketch with your description.

4. a) Find the area of a triangle with sides 10 cm, 15 cm, and 21 cm. Explain how you obtained your answer.

   b) Find the area of the concave polygon in **Figure 5.75**. Explain how you obtained your answer.

   c) **Figure 5.76(a)** shows an irregular shape on a graphing calculator screen. **Figure 5.76(b)** shows the screen after a simulation has distributed 50 random points. If the screen represents a rectangle with an area of 168 mi², estimate the area of the irregular shape.

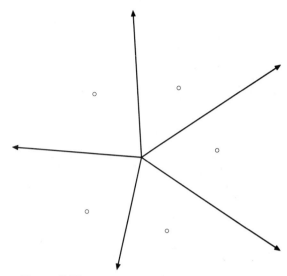

Figure 5.74.

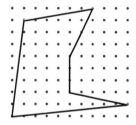

Figure 5.75.

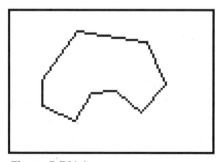

Figure 5.76(a).

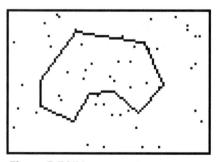

Figure 5.76(b).

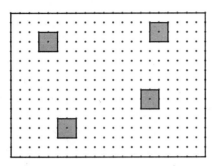

Figure 5.77.

5. Computers are more user-friendly today than in the past. One reason for this is the use of icons to make a computer screen seem more like an actual desktop. Software developers continue to look for ways to make computer desktops more convenient for the user. The U.S. Patent Office has issued a patent to a desktop technique that uses Voronoi diagrams to automatically highlight an icon that the user is most likely to select next so that the user does not have to put the mouse pointer on the icon to activate it. **Figure 5.77** represents a computer screen with four icons.

a) Determine a set of boundaries so that the icon in a region is automatically highlighted if the mouse pointer is in that region. Explain your method.

b) If the user stops the mouse at random, which of the four icons is most likely to be highlighted? Explain.

6. Christopher Gold and Francois Castro of Université Laval in Quebec have developed an "interactive map" that helps boats navigate. The technique uses Voronoi

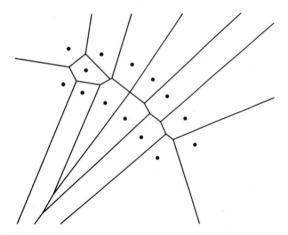

Figure 5.78(a).

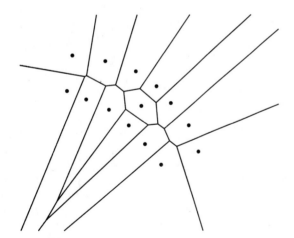

Figure 5.78(c).

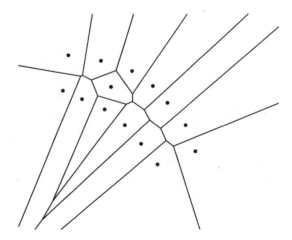

Figure 5.78(b).

diagrams to form a navigational "bubble" around the boat. The boat is steered to stay within the bubble, which is actually a series of bubbles based on the location of the boat and obstacle points that the boat must avoid. For example, **Figures 5.78(a), (b),** and **(c)** represent the bubble as the boat progresses through a narrow passage.

Explain why Voronoi diagrams are useful in protecting the boat from harm.

7. **Figure 5.79** represents a rectangular factory floor with three machines that a robot must be programmed to avoid. The origin is at the lower-left corner.

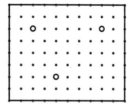

Figure 5.79.

a) Find the equations of three robot paths that stay as far away from the machines as possible. Find the coordinates of their point of intersection.

b) Make a diagram showing the paths.

8. **Figure 5.80** represents four communities. (Grid points are spaced at one-mile intervals.)

a) Locate the Voronoi centers for this domain.

b) Would the centers make good locations for four restaurants that a company plans to open in the area? Explain.

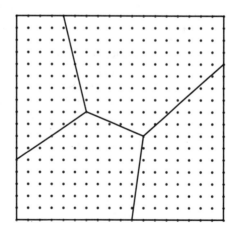

Figure 5.80.

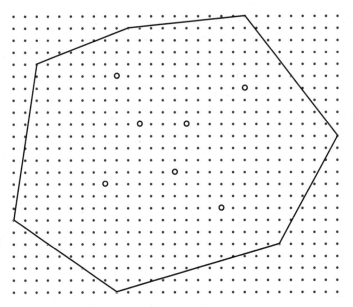

Figure 5.81.

9. A detective investigating a series of robberies has mapped the locations of the robberies as shown in **Figure 5.81**. Based on interviews with witnesses and other evidence, the detective believes the robberies are the work of the same person.

   a) Suggest a location where the detective might search for the robbery suspect.

   b) What assumptions have you made in selecting the point you chose in part (a)?

10. **Figure 5.82** shows a triangle ABC and a coordinate system introduced so that one vertex of the triangle is at the origin and another vertex is on the *x*-axis. The midpoints of two of the triangle's sides are connected with a broken segment.

   a) If the coordinates of B are $(b, 0)$ and the coordinates of C are (c, d), what are the coordinates of the two midpoints?

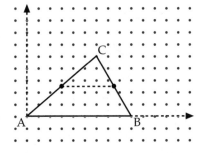

   b) What can you conclude about the relationship between the segment connecting the midpoints and side AB?

Figure 5.82.

Mathematical Summary

*T*he modeling problem from which the mathematics in this unit arises is that of estimating the rainfall for the entire state of Colorado from readings taken at eight rain gauges scattered around the state.

The solution is geometric: divide the state into eight regions so that the points in a region are closer to its gauge than to any other gauge. Weight the rainfall measured at each region's gauge according to the portion of the state's area in that region. The weighted average of the rainfall at the eight gauges estimates the rainfall for the state.

Proximity problems like the Colorado rain gauge problem involve Voronoi diagrams, which are named after the mathematician Georgii Voronoi. To use Voronoi diagrams, you must determine the boundaries of regions from their centers of influence.

The boundaries can be drawn roughly by hand, but answers obtained from rough drawings lack precision. Therefore, the boundaries should be constructed. There are several means of constructing the boundaries. Every Voronoi boundary is the perpendicular bisector of a segment joining two centers. Perpendicular bisectors can be constructed by several methods: 1) by folding a piece of paper and creasing it so that two centers coincide; 2) by placing a Plexiglas® mirror so that the reflection of one center coincides with the other center; 3) by striking intersecting compass arcs from the centers and joining the two points of intersection; 4) or by using the segment, midpoint, and perpendicular construction features of a drawing utility.

Perpendicular bisectors are lines, but Voronoi boundaries are either rays or line segments. Moreover, the perpendicular bisector for some pairs of centers is not a boundary in the Voronoi diagram. Therefore, when perpendicular bisectors are constructed, they must be analyzed carefully to determine which portions to keep. Algorithms for establishing Voronoi boundaries often divide the problem into several smaller problems of, say, three or four vertices, then combine the diagrams that result.

Voronoi regions are usually polygons. (An exception occurs when the boundary of the domain is curved.) Many modeling problems, including the Colorado rain gauge problem, require determination of each regions' area. One way to find a region's area is to divide it into triangles and apply Heron's formula, which finds the area of a triangle from the lengths of its sides. Another method is to apply Pick's formula, which finds the area of any polygon whose vertices are points of a grid from the number of grid points that are on the polygon's border or inside the polygon. Areas can be measured by using a drawing utility's measuring features or estimated by designing and running a simulation.

Another approach to the problem of finding Voronoi boundaries and the areas of Voronoi regions involves coordinate geometry. If coordinates of the centers are known, the equations of boundary lines can be found. First, locate the midpoint of the segment joining a given pair of centers by averaging the x- and y-coordinates. Then determine the slope of the segment joining the two centers and use the opposite reciprocal for the slope of the perpendicular.

After equations of two intersecting perpendicular bisectors are found, the coordinates of the Voronoi vertex at which they intersect are found by solving the system of their two equations. The coordinates of the vertices can be used to determine the area of each region. One way to do this is to divide each region into triangles, use the coordinate geometry distance formula to find the lengths of any unknown sides, then use Heron's formula to find each triangle's area.

Since coordinate geometry develops algebraic formulas for geometric objects from known properties of those objects, the results can be used to develop computer or calculator programs that find equations of boundaries and areas of regions. Since the formulas of coordinate geometry give exact results, the precision of answers obtained from coordinate methods is limited only by the precision of the measurements used in the formulas.

There are some modeling situations in which it is necessary to determine the centers of Voronoi regions from the boundaries. One way to solve such problems is by guessing a center for one of the regions, reflecting it through one of the region's boundaries into an adjacent region, then reflecting the new point into another region, and repeating until a point is located in each region. Then reflect the final point back into the starting region. If this reflection does not coincide with the guess, then an improved guess is made, and the process repeated.

Other methods of locating the center of a region can be devised through a deductive examination of relationships among points, angles, and boundaries in a Voronoi diagram.

Glossary

ACUTE TRIANGLE:
A triangle in which all of the angles measure less than 90°, but more than 0°.

CENTER OF INFLUENCE:
A point used to establish boundaries of regions of influence. All points in a region are closer to that region's center than to any other region's center.

CONCAVE POLYGON:
A polygon in which some of its sides, when extended, intersect other sides.

CONVEX POLYGON:
A polygon in which none of its sides, when extended, intersect other sides. For every pair of points in the interior of a convex polygon, the segment connecting the points is completely in the interior.

DOMAIN:
A region in which centers of influence are located. The domain is the area that is being divided into regions of influence.

HERON'S FORMULA:
The area of a triangle is
$\sqrt{s(s-a)(s-b)(s-c)}$, where a, b, and c are the lengths of the triangle's sides and s is half the triangle's perimeter.

ITERATION (ITERATIVE PROCEDURE):
a procedure that repeats the same sequence of steps over and over. Each cycle is considered one iteration.

MIDPOINT:
A point that is halfway along a segment (equidistant from the segment's two endpoints). In coordinate geometry, the coordinates of a midpoint are found by averaging the coordinates of the two endpoints.

OBTUSE TRIANGLE:
A triangle with one angle that measures more than 90°, but less than 180°.

PERPENDICULAR BISECTOR:
A line that passes through the midpoint of a given line segment and forms right angles with it.

PICK'S FORMULA:
If the vertices of a polygon are points of a grid, then the area of the polygon is $0.5b + i - 1$, where b is the number of grid points on the polygon's border, and i is the number of points in its interior.

REGION OF INFLUENCE:
A region in which each point is closer to the region's center of influence than to any other center of influence.

VORONOI BOUNDARY:
A boundary between two centers of influence.

VORONOI CENTER:
A center of influence.

VORONOI DIAGRAM:
A diagram composed of several centers of influence and their regions of influence.

VORONOI REGION:
A region of influence.

VORONOI VERTEX:
A point at which Voronoi boundaries intersect.

WEIGHTED AVERAGE:
The average found by multiplying each category by the decimal weight attached to that category and finding a total.

UNIT

Growth

In this unit you study functions that model the growth or decay of quantities. You learn to recognize various functions in new, unfamiliar settings by studying the functions and their properties in familiar situations. Knowledge of properties of functions is what permits proposing them in new situations.

Modeling in this unit involves deciding which type of function matches an observed pattern of growth and then choosing constants that produce the most appropriate particular function. In other words, first decide which family of functions to apply to a situation and then decide which member of the family fits best.

The mathematics of arithmetic sequences and series, geometric sequences and series, and mixed sequences form the mathematical core of the unit. Tables and recursive graphs are used to identify patterns for several families of functions, including linear, exponential, and quadratic. Closed-form equations provide tools for making quick predictions. You will add a new function, the logarithm, to your tool kit of functions to assist in your work with exponential growth.

GROWTH PATTERNS

Growth is part of life. Money "grows" in your savings account because you make deposits and earn interest. The population of the United States grows as people immigrate from other countries and as babies are born. The amount of trash in a landfill grows as truckloads of garbage are added to the pile each day. The number of people affected by a flu virus grows as the virus spreads through a community.

Sometimes growth is regulated by laws or contracts. The amount of interest you earn on money you loan to someone else is specified in a contract. Growth that is not regulated often follows a pattern that can be observed and modeled. For example, biologists can model the growth of trout in a lake or rabbits in a habitat because they have studied how trout and rabbits reproduce and survive. Doctors and pharmacists can model the increase or decrease of a prescription drug in the body.

The Image Bank

In many areas of life, people try to predict future growth based on the patterns of the past. Plans are made based on anticipated growth, or decisions are made to regulate growth in order to produce a desired outcome. For example, if you expect the population of your community to increase you can make plans now. You know you will need more schools, parks, and police and fire protection. You plan for these needs by collecting and setting aside money to meet the anticipated need. On the other hand, your community may decide to regulate growth. Voters and community leaders may pass laws to limit the amount or rate of growth. Some cities limit the number of new houses that are built each year.

In this unit you model growth in a variety of situations, including population growth, spread of diseases, investments, use of natural resources, and monitoring of drug levels in the body. You model the growth by determining which family of functions provides the best model and then "fine tuning" the control numbers to find the member of the function family that fits the situation. The goal is to use mathematics to help people make the best decisions possible in situations involving growth.

The Image Bank

LESSON ONE
Growing Concerns

KEY CONCEPTS

Data analysis

Additive growth

Multiplicative growth

Growth factor

Relative rate of growth

The Image Bank

PREPARATION READING

The Right Dose

Jackie is a school nurse at a high school in Oregon. A female student came to Jackie early in the school day and asked for acetaminophen, as she had a severe headache and wanted relief. Jackie asked the student if she had taken other medications. The student's answer prompted Jackie to call for paramedics and have the student rushed to a hospital emergency room. Apparently the student had taken several different medicines and each medicine contained acetaminophen. Jackie concluded the student had taken enough acetaminophen to reach a toxic level. The student escaped harm because the nurse knew the amount of medicine taken might cause serious injury. How does a nurse, a doctor, or a pharmacist determine that a drug dose is too high?

Katie, age 15, had her wisdom teeth removed. The oral surgeon prescribed penicillin to help prevent infection while her gums were healing. How did the surgeon know how much penicillin Katie should take and how often she should take it?

How does the doctor or pharmacist know what dosage to prescribe for *any* medicine? If the doctor prescribes a dosage that is too high, the drug might cause damage to organs in the body. If the doctor prescribes a dosage that is too low, the patient may not experience recovery or relief.

Doctors and pharmacists follow steps similar to those you use in mathematical modeling. They propose a dose based on mathematical formulas and the age and weight of the patient. They monitor a patient's blood serum level and adjust the dose if the level of the particular drug is higher or lower than what is recommended.

In fact, the entire process of diagnosing and treating illness provides a wonderful example of how pharmacists and doctors use modeling principles:

- Exhibited symptoms present a well-defined problem; doctors find the cause of the symptoms and treat the cause and/or provide relief from the symptoms.

- Doctors simplify by formulating a hypothesis based on the most likely causes of the symptoms. (Instead of treating every possible cause, they focus on the most likely cause or the most critical causes.)

- Doctors order medical tests and propose drug doses based on general principles (refer to *Physician's Drug Handbook* or *Handbook of Clinical Drug Data*).

- The model is tested by analyzing blood samples. Then doctors adjust the dose (quantity and/or frequency) based on the results of the test. The process is repeated until the desired equilibrium level is attained, the cause is determined, and/or the patient experiences relief from the symptoms.

In this lesson, you are challenged to make a correct drug dosage adjustment for a patient. Apply what you have learned about functions and modeling as you investigate a new situation involving the growth of a quantity.

BEYOND PRESCRIPTION

Eric had his first seizure one week ago. The doctor prescribed phenobarbital, an anticonvulsant medicine, to control his seizures. He was given an initial dose of 250 mg, and then he took 60 mg of the medicine every 12 hours. The minimum daily concentration of phenobarbital in his blood, measured in micrograms-per-milliliter, is shown in **Figure 6.1**.

The minimum therapeutic level (the amount needed to be effective) for phenobarbital is 15 mcg/ml. Eric's level after one week is 11.3 mcg/ml.

1. Use data analysis methods from earlier units to predict Eric's minimum level after two weeks. Provide evidence and/or give reasons to support your prediction. Assume he continues his present dose of 60 mg two times a day and that all physical processes remain as they have for this first week.

2. The minimum blood level for phenobarbital should be above 15 mcg/ml to be effective, but it must also be below 40 mcg/ml to be considered safe. Phenobarbital levels above 40 mcg/ml are potentially toxic. Levels above 70 mcg/ml may induce a coma. Eric's highest level occurs after each dose and is about 1.5 mcg/ml above the immediately preceding minimum level (which occurs just before each new dose). Predict a dose size (to be taken every 12 hours) that would put Eric's highest level close to 20 mcg/ml after a period of 2 or 3 weeks.

Days	Blood level mcg/ml
1	7.5
2	8.5
3	9.4
4	10.0
5	10.5
6	11.0
7	11.3

Figure 6.1.
Eric's blood levels of phenobarbital.

In this unit you will develop mathematics to model the level of a particular drug in the body. Along the way, you will study the mathematics of growth and decay in many contexts in order to better understand the increase and decrease of a drug in the body. Your goals are to find the best model for determining effective drug doses and to solve real-world problems in a variety of contexts involving the growth of quantities.

ACTIVITY
1

BEYOND PRESCRIPTION

CONSIDER:

1. Suppose you take a single dose of a non-prescription medicine, such as acetaminophen or ibuprofen. The medicine is absorbed into your organs, tissues, muscles, or blood. Eventually it is eliminated from your body. What do you think are some of the factors that affect the amount of the drug in your blood at a given time?

2. Drug dosage may be divided into smaller amounts and introduced into the body at regular intervals, instead of in just one large dose. Describe what you think happens to the amount of the medicine in your body between successive doses. Then give a non-medical example of a similar process.

3. Suppose you saved up money to buy a car—at least enough to make the down payment and begin paying off a loan. As you make monthly payments, the amount of money you owe the bank is affected in much the same way as the amount of medicine in your body as you take regular doses, although perhaps somewhat "backwards." Explain as clearly as you can how these two systems are similar.

4. If you stay with an employer long enough, you expect to get periodic wage raises. Comment on any similarities you see in wages or pay rates and the medicine and car examples above.

ACTIVITY

BEYOND PRESCRIPTION

1

The Consider questions and those in Activity 1 suggest at least two different reasons for investigating the families of functions that describe the behavior of particular processes.

First, it is not uncommon to be confronted with a new and unfamiliar situation. In order to identify its structure, perhaps to make predictions of future behavior, or to deduce the process creating the situation, it is necessary to have a "tool kit" of descriptive functions whose properties are well known. Then you can compare the known functions to data from the unfamiliar situation and look for similarities that might help identify the new situation. This is the kind of problem that Activity 1 presents.

Another reason for studying types of functions is to understand their fundamental properties. By knowing the properties of a function, whenever a situation can be identified using some of those properties, you can automatically know that all other properties common to that function type are "inherited" into the new situation. You don't have to check them all! This may be more like the situations in the car-payment or wage-raise scenarios.

If you completed Course 1, your tool kit already includes linear, exponential, quadratic, and absolute value functions. Linear functions are created by additive processes and exponential functions come from multiplicative processes. Do you still remember all their properties?

INDIVIDUAL WORK 1

Moose Revisited

1. **Figure 6.2** shows a graph representing the concentration of aceta-minophen in a person's bloodstream. The level rises and falls.

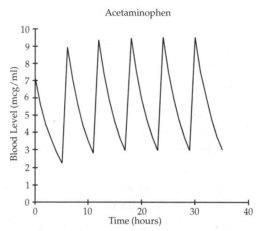

Acetaminophen

Figure 6.2.
Acetaminophen concentration in the blood of an adult.

a) Compare the given graph with your description in Consider question 2. Why does the level increase rapidly every 6 hours? What happens between these "jumps"?

b) The graph in Figure 6.2 is a continuous representation of the concentration of acetaminophen in the blood. Typically, blood concentration is not monitored on a continuous basis. Suppose you graphed the level just before each new dose, when the concentration is at its lowest point. Describe the graph that would result.

c) When a doctor prescribes a particular dose of a medicine for a patient, the doctor wants to achieve a steady-state or equilibrium level that is within a therapeutic range. That is, the doctor wants to know that the concentration of the drug is the same before each new dose, and high enough to help the patient. What do you think is the steady-state minimum concentration for the drug represented in Figure 6.2?

d) You may have studied the growth of a moose population in Course 1, Unit 6, *Wildlife*. In what ways is the study of the growth of a moose population similar to the study of drug levels in the body?

2. **Additive growth** models were introduced in Course 1, Unit 6, *Wildlife*, which you may have studied. An additive process follows the pattern "next = current + constant." Additive growth is defined as a growth process described by recursive equations of the form $p_n = p_{n-1} + k$ where k is a constant.

Suppose park officials determine there are 78 moose in the park. An additive growth model assumes the total number of moose increases by a constant number each year. Assume 6 moose are added to the total each year.

a) Write a recursive equation to represent the total number of moose in the park.

b) Convert your recursive equation to a closed-form equation.

c) Describe the appearance of the graph of total moose population versus time.

d) Describe the appearance of the graph of "change in moose" $(p_n - p_{n-1})$ versus total moose (p_{n-1}).

Home screen iteration is the process of using a calculator's "Answer" command to compute successive values of a sequence directly from a recursive equation. For example, for the moose sequence here, you could type (without the quotes) "78" and press Enter. Then enter the command "Answer + 6" and press Enter again. Now each additional press of the Enter key calculates another term in the list of population values.

e) Use home screen iteration to predict the number of moose after 15 years.

f) Would you use a recursive or a closed-form equation to find the total number of moose in 75 years, assuming the equation accurately models the growth in the number of moose? Find the total number of moose in 75 years.

g) When do you expect the population of moose to reach a total of 500? What equation do you solve? Is your answer realistic? Explain why or why not.

3. **Multiplicative growth** models were also introduced in Course 1, Unit 6, *Wildlife*. Suppose park officials determine there are 78 moose in the park. A multiplicative growth model assumes the total number of moose increases by the same growth factor each year. Suppose you assume a growth factor of 1.08 each year.

a) Write a recursive equation to represent the total number of moose in the park.

b) Convert your recursive equation to a closed-form equation.

c) Describe the appearance of the graph of total moose population versus time.

d) Create and describe the appearance of the graph of "change in moose" $(p_n - p_{n-1})$ versus total moose (p_{n-1}).

e) Use home screen iteration to calculate the number of moose in 15 years.

f) Would you use a recursive or a closed-form equation to find the total number of moose in 75 years, assuming the equation accurately models the growth in the number of moose? Find the total number of moose in 75 years.

g) When do you expect the population of moose to reach a total of 500? What equation do you solve? Is your answer realistic? Explain why or why not.

Different forms of an equation are useful for different tasks. Closed-form equations allow you to evaluate or solve in one or two steps. Recursive equations make it easier to generate the next value, to investigate rates of change, or to classify a particular set of data. Frequently it is useful to convert an equation from one form to another.

4. Convert the recursive equations in parts (a) and (b) to closed form.

a) $a_n = a_{n-1} + 10$; $a_0 = 250$.

b) $a_n = 1.5a_{n-1}$; $a_0 = 20$.

Convert the closed-form equations in parts (c) and (d) to recursive form.

c) $a(n) = 5n + 75$.

d) $a(n) = 3{,}000(1.04)^n$.

Multiplicative growth is described by exponential functions. You may have learned some of the language and mathematics related to this kind of growth in Course 1, Unit 6, *Wildlife*. The equation $p_{next} = b \times p_{current}$ (or $p_n = b \times p_{n-1}$) is the general form of the recursive equation for multiplicative (exponential) growth. That is, p_{next} is proportional to $p_{current}$. The constant of proportionality b is called the **growth factor**. Note that the growth factor describes the amount by which an exponential quantity is multiplied in *one* recursive period. It is sometimes written in the form $(1 + r)$, so that the recursive equation becomes $p_{next} = p_{current} + r \times p_{current}$. In this case, the constant r is called the **relative rate of growth**. The relative rate of growth is the amount of change in a quantity during a given period divided by the amount at the beginning of the period. Note that $b = 1 + r$. For example, if $b = 1.05$, then $r = 0.05$ since $1.05 = 1 + 0.05$.

The corresponding closed-form equation is $p_n = p_0 \times b^n$, an exponential function having base b. This implies that the factor for 2 periods of growth is b^2, for 3 periods is b^3, and so on. Note that p_0 is the coefficient.

Returning to the recursive equation $p_{next} = p_{current} + r \times p_{current}$, subtracting $p_{current}$ from both sides gives an equivalent equation:

$$p_{next} - p_{current} = p_{current} - p_{current} + r \times p_{current}$$

$$p_{next} - p_{current} = r \times p_{current}$$

The expression $(p_{next} - p_{current})$ is the **amount of growth**—the difference between successive values in a growth sequence. This equation reveals that the amount of growth is proportional to $p_{current}$, with the relative rate of growth being the constant of proportionality.

Divide both sides of this last equation by $p_{current}$ and you get the defining expression for r, the relative rate of growth:

$$p_{next} - p_{current} = r \times p_{current}$$

$$\frac{p_{next} - p_{current}}{p_{current}} = \frac{r \times p_{current}}{p_{current}}$$

$$\frac{p_{next} - p_{current}}{p_{current}} = r$$

Therefore, the relative rate of growth r is defined as the change in growth divided by the total population before that growth.

To summarize, for exponential growth (defined by a multiplicative process), the amount of growth per period is proportional to the amount of "stuff" present at the start of that period (proportionality constant r), and the amount of "stuff" at the end of a period is proportional to the amount present at the start of that period (proportionality constant b).

5. Assume a multiplicative recursive growth model for each of the following items:

a) The relative rate of growth is 20%. What is the growth factor?

b) If $p_{current}$ is 640 and p_{next} is 688, what is the growth factor? What is the relative rate of growth?

c) The current population is 13,500. Find the next population (1 year later) if the relative rate of growth is 2.5% per year.

d) If the current population is 8,340 and the relative rate of growth is 4% per year, what is the amount of growth from the current year to the next?

e) Suppose the current population is 5,000 and the population in 4 years is 6,312. What is the growth factor for one year?

f) Describe the graph of relative rate of growth versus current population.

Recall that when the "growth" process results in lower populations, the process is sometimes called **decay** instead of growth. In that case, the relative decay rate is the opposite of the relative growth rate. For example, a growth rate of –5% would be called a decay rate of 5%. The corresponding growth factor would be 0.95 (= 1 + –0.05).

g) What is the relative rate of decay if the population drops from 5,480 to 5,314 in one period? What is the growth factor?

6. Your parents have been giving you $2.00 per week for the last 3 years. In exchange for some chores, and as an incentive to complete your chores, they agree to raise the amount by 5% each week, starting after the first week after the agreement, provided you complete the chores without being asked.

a) Write a recursive equation to represent weekly pay. The increment of time for the recursive equation is 1 week. Let p_n represent the pay for the "nth" week.

b) Write a closed-form equation to represent weekly pay. Let n represent the number of weeks. Let $p(n)$ represent the pay following the nth week.

c) Suppose you complete your chores each week without being asked. What is your weekly pay at the end of 1 year?

d) Suppose your parents offered to raise your weekly pay to a flat rate of $10.00 per week instead of the 5% weekly increase. Which would you choose, the flat rate or the percent increase?

As is the case with other kinds of equations, you may determine values of important constants by substituting known points on the equation's graph and solving the resulting equation(s) for the unknown constants. In general, you will need as many points as there are unknown constants.

7. Suppose that the population of a town was 21,604 in the 1980 census and that the same town had a population of 35,832 in 1990.

   a) Determine the annual growth rate for this town's population over that period.

   b) Write a closed-form equation for the town's population in terms of years since 1980. (So, year 0 is 1980.)

8. A drug is eliminated from the body in such a way that only half remains after 3 days. Assume that the decay is exponential.

   a) Find the hourly "growth" factor.

   b) Write a closed-form equation for the percent remaining in the body in terms of the number of hours since the medicine entered the body.

9. You may have used parametric equations in earlier units to transform graphs of functions. Suppose you inherited $8,000 and immediately opened a savings account that paid an annual rate of 4.8% compounded monthly (0.4% monthly). Then a closed-form equation describing your balance is $y = 8{,}000(1.004)^T$ For a parametric representation of this situation, let T measure time in months since receiving the inheritance. Record X as the number of months and Y as the balance, each in terms of T. Then $X_{1T} = T$ and $Y_{1T} = 8{,}000(1.004)^T$.

   a) Graph this parametric representation of the savings account balance on the window [0, 50] for T, [0, 50] for X, and [8,000, 12,000] for Y (denoted [0, 50] × [0, 50] × [8,000, 12,000], with the T-interval listed first).

   b) Now suppose you had held the money in a non-interest bearing account for 1 year (12 months) before you opened the account described above. Describe in words how the graph of the balance in the interest-bearing account for this new situation would be related to the graph you created in part (a). Again, let T = 0 represent the time at which you received the inheritance.

c) Create a transformation of the previous graph to represent this new situation by graphing $X_{2T} = X_{1T} + 12$, $Y_{2T} = Y_{1T}$. Graph the transformed graph on the same screen as the original graph.

d) What parametric equations would you use to produce a scale transformation from time increments of 1 month to increments of 1 year?

10. The data in **Figure 6.3** show the attendance at a classic automobile race. The total attendance has increased every year from 1983 to 1996.

a) Do you think a single function might describe these data well? How could you decide?

b) How can you determine whether the data are best described by a linear, exponential, quadratic, or some other kind of function you have yet to study?

c) Suppose you decide the attendance pattern is linear. How do you determine the best line to fit the data? How do you know if your line is the best fit?

11. For each of the following scenarios, decide whether multiplicative or additive growth seems to fit the situation. Give reasons for your choice.

a) Situation 1
I started with two fish in my fish tank. A month later there were 12. The next month I counted over 70. By the end of the fourth month there were over 400 fish in my fish tank. The number of fish is growing at a rate of over 50% each week.

b) Situation 2
The amount of water in a reservoir depends on how much water flows into the reservoir each day, how much water is released each day, and how much water evaporates each day. Suppose 235,000 gallons flow into the reservoir each day during the spring months and 172,000 gallons per day are released. Suppose about 0.15% of the water evaporates each day depending on temperature, humidity, and wind.

Year	Attendance (millions)
1980	1.6
1981	1.6
1982	1.5
1983	1.5
1984	2.1
1985	2.1
1986	2.2
1987	2.6
1988	3.0
1989	3.1
1990	3.3
1991	3.4
1992	3.7
1993	4.0
1994	4.9
1995	5.3
1996	5.6

Figure 6.3.
Automobile race attendance.

c) Situation 3

I am reading a book. My goal is to read 25 pages each day. At this rate I expect to finish the book in about 17 days.

12. Chris has an infection. She takes a 250 mg prescription of an antibiotic (tetracycline hydrochloride) every 6 hours. For each dose, only 75% of the medicine reaches the bloodstream. Suppose the minimum effective level is the equivalent of 400 mg in the bloodstream. About 6.70% of the drug is eliminated each hour. Model the minimum levels of tetracycline hydrochloride in Chris's blood every 6 hours for a period of 3 days. Remember to add 75% of 250 mg every 6 hours.

a) Does the prescription maintain a therapeutic level for Chris after 3 days?

b) Describe the effect of giving the same 250 mg dose every 8 hours.

c) Describe the effect of giving a higher dose of 400 mg every 6 hours.

13. Millie opens a savings account that pays 6% annual interest compounded monthly. The beginning balance in her account is the $500 she received as a gift from her grandparents. Millie is working at a part-time job, and she decides to deposit $100 in her savings account every month.

a) Calculate the total amount in her account after 2 years (24 months).

b) How long will it take to accumulate $5,000? Describe or show how you determine the answer.

c) Suppose the interest rate is 5% instead of 6%. How long will it take to accumulate $5,000?

d) Suppose the rate is 6% but Millie decides to deposit $150 each month instead of $100. How long will it take to accumulate $5,000?

e) What other factors affect the total balance in Millie's account?

f) Write a recursive equation to represent the amount of money in Millie's account in relation to the amount from the previous month. Assume her deposits are $150 each month and the interest rate is 6%.

14. Some quantities grow (or decay) because they are regulated by a contractual law or a natural law. The amount of interest added to your savings account is regulated by a contract between you and your bank. Some communities have laws that limit the number of new houses that may be built in a particular year. The rate at which a substance decays or converts to a different form is subject to laws of nature.

 a) Name one quantity that grows because it is regulated by a law.

Some quantities are measured and displayed as data. If the graph of the data appears linear, then a linear equation may be used to represent the data. If the graph seems to curve, then a quadratic equation or an exponential equation may provide a suitable approximation.

 b) Find a table or graph that illustrates a quantity that is growing at a fairly regular rate and can be approximated using a linear or exponential equation.

Rampant Growth: Kudzu

Kudzu, a native plant of Asia, is a weedy flowerting vine that was introduced into the United States at the Philadelphia Centennial Exposition of 1876 in a Japanese garden exhibit. It was widely planted in the U. S. for erosion control in the 1930s and farmers were paid to grow fields of it in the 1940s. Perfectly suited to the climate in the southeastern U. S., Kudzu grows as much as a foot a day during the summer months. Left uncontrolled, it covers trees, power lines. small buildings, and anything else in its path. It has been dubbed "the vine that ate the South" because it now covers millions of acres in that region.

LESSON TWO
Double Trouble

KEY CONCEPTS

Arithmetic sequences

Geometric sequences

Half-life

Recursive graphs

Web diagrams

The Image Bank

PREPARATION READING

Two for One

Population growth is another example of the growth of a quantity. What would life be like if the population in your world doubled?

Imagine twice as many students in your school or twice as many houses in your town. What would traffic be like if there were twice as many cars? Think how it would affect you if you were to double the number of people waiting in lines at the nearest amusement park or waiting in line for a seat at a concert. Imagine feeding 20,000 people with food that would normally be enough for only 10,000.

The Image Bank

If the population of the United States continues to grow at a rate of 1.0% each year, then the current population of 250,000,000 will double to 500,000,000 in less than 70 years. In some countries the population is growing so fast that the total number of people in that country will double within 15 years. Increasing the number of people has an impact on cities, countries, and the world.

If you look at the entire world, when population grows there is a greater need for food to eat, for natural resources to make products, and for fuel to power transportation vehicles or provide heat. If you look at your own city, when population grows there is a greater demand on public services. New schools must be built or new classrooms added to existing schools. Homes are built, roads are widened, police officers are hired, pollution and noise increase. Everyone is affected.

A community has at least two choices when it comes to managing its growth. Some communities allow growth to occur without any intervention. The community lets people build as many buildings as they want wherever they want. If the increase in growth shows a pattern, mathematics can be used to model the growth and help the community plan for the future.

Some communities decide to regulate growth. They may limit the amount of new building each year. Mathematics can be used to design regulations that produce a desired outcome for the community.

The contextual purpose of this lesson is to compare two different growth models, additive and multiplicative, that are based on different assumptions. The mathematical purpose of this lesson is the same as the purpose of the unit; to study a pattern of growth that is familiar so you can recognize patterns in unknown or unfamiliar situations. After you examine the patterns for additive and multiplicative growth, revisit

Activity 1 and apply what you learn from this lesson. Determine whether the level of a drug in the blood follows an additive pattern, a multiplicative pattern, or neither. Based on what you learn, see if you can determine the proper dose for Eric.

The Image Bank

CONSIDER:

1. Can a recursively defined graph be continuous? Give reasons for your answer.

2. You recognize that the equation $y = 5x + 100$ is a linear equation because it matches the form $y = mx + b$ associated with linear equations. Also, the graph of $y = 5x + 100$ is a line. Describe how you recognize the graph and the form of an equation, complete with constants and variables, for each of the following:

   exponential equations

   quadratic equations

   absolute value equations.

3. What assumptions characterize additive growth? That is, how do you know when a situation illustrates additive rather than multiplicative growth? What assumptions characterize multiplicative growth?

ACTIVITY

SLOW DOWN

2

The purpose of this activity is to model and compare two different growth models, one based on additive growth and one based on multiplicative growth. In Activity 3, you will investigate patterns in the tables and graphs associated with these kinds of growth. Study the characteristics of known models, such as the models studied in this activity, so you can recognize whether an unknown situation is modeled best by an additive model, a multiplicative model, some other model, or no model at all.

Suppose your town has 21,000 homes and is growing rapidly. The residents want to regulate the future growth of your town. The planning department is considering two plans for managing growth.

Plan A
Limit growth to 400 new homes each year.

Plan B
Limit growth of new homes to 1.5% each year.

Build a mathematical model for each plan. Prepare a report that explains to the planning department the strengths and weaknesses of each plan. You might want to develop your own Plan C using similar mathematics.

SLOW DOWN

The following items provide ideas to consider as you investigate the plans and write your report.

1. Write recursive and closed-form equations to represent each plan.

2. Use a spreadsheet, graphing calculator, or paper and pencil to prepare a multiple-column table like the one in **Figure 6.4** to represent each plan. Sample values for the first year under Plan A have already been entered.

Time n (years)	Total number of homes p_n	Total homes previous year p_{n-1}	Amount of growth $p_n - p_{n-1}$	Ratio of successive terms p_n/p_{n-1}	Relative rate of growth $(p_n - p_{n-1})/p_{n-1}$
0	21,000				
1	21,400	21,000			
...					
40					

Figure 6.4.
Table for examining plans.

3. Use your results from Item 2 to graph p_n versus n (the time-series graph) for each plan.

Save your work from this activity for use again in Activity 3.

INDIVIDUAL WORK 2

Subscriptions

1. In this item, you practice using subscript notation so you can use it with confidence while working with recursive equations.

   You may have seen subscript notation in Course 1, Unit 5, *Animation*. Subscripts allow you to use one letter to identify a variable and still represent different values for the variable at different times. For example, x_1 may represent the horizontal location of a moving object at time 1 while x_2 represents the horizontal location of the same object at time 2. Or p_{81} might represent the population in 1981 and p_{95} the population in 1995. When you use recursive equations you use expressions such as $p_{initial}$, $p_{previous}$, $p_{current}$, and p_{next} to represent the changes to the variable p from one step to the next. Recall that a "step" is consistently 1 unit for the explanatory variable and often represents 1 unit of time. Subscripts can be words, letters, or numbers.

   a) Find p_5 when $p_0 = 4$; $p_n = 2p_{n-1}$.

   b) $p_{next} = p_{current} + 300$. Find $p_{current}$ when $p_{next} = 5750$.

   c) Find p_{50} when $p_n = p_{n-1} + 20$; $p_0 = 15$.

2. Mario loves listening to music. He now has 175 music compact disks (CD's) in his personal collection. He buys 3 new CD's every week.

   a) Does the growth of Mario's collection represent an additive process or a multiplicative process? Explain your answer.

   b) Write a recursive and a closed-form model for the total number of CD's in Mario's collection. Assume each step in the recursive process is 1 week. Assume Mario has 175 CD's at time 0, the beginning of the first week.

   c) How many CD's will Mario have in 6 months (26 weeks)? Which equation will you use, recursive or closed form?

   d) Mario keeps his CD's in a cabinet that is designed to hold 300 CD's. When will Mario run out of space in the cabinet? What equation do you solve?

3. Gridville has 2,500 homes. The town planning board decides to regulate growth and limit the number of new homes to 2% each year.

   a) Does the growth of Gridville's housing total represent an additive process or a multiplicative process? Explain your answer.

   b) Write both recursive and closed-form models for the total number of homes in Gridville. Assume each step in the recursive process is 1 year. Assume Gridville has 2,500 homes in year 0.

   c) How many houses will Gridville have in 10 years? Which equation will you use, recursive or closed-form?

   d) Gridville's planning board has determined the maximum number of homes will be 5,000. When will Gridville run out of space? What equation do you solve?

4. Suppose you borrow $5,000 from your aunt. Your aunt prepares a repayment schedule as shown in **Figure 6.5.**

   a) Does the growth (decay) of your debt represent an additive process or a multiplicative process? Explain your answer.

   b) Write both recursive and closed-form models for the balance.

Week	Payment	Balance
0	0	$5,000
1	100	$4,900
2	100	$4,800
3	100	$4,700
...	100	
????	100	0

Figure 6.5.
Debt repayment table.

The word **sequence** is often used to describe an ordered list of values. The first six terms of the sequence of balances would be written

5,000, 4,900, 4,800, 4,700, 4,600, 4,500.

A sequence that is created by, or can be modeled by, a recursive additive process is called an **arithmetic sequence**. A sequence that is created by, or can be modeled by, a recursive multiplicative process is called a **geometric sequence**.

The sequence 5, 7, 9, 11, 13, 15, 17, 19, 21 is an example of an arithmetic sequence. It can be represented by the (additive) recursive equation $a_n = a_{n-1} + 2; a_1 = 5$.

The sequence 5,000, 4,900, 4,800, 4,700, 4,600, 4,500 is another example of an arithmetic sequence. It can be represented by the (additive) recursive equation you described in part (b). The amount that is added is a negative number, the equivalent of subtracting the opposite positive value.

The sequence 5, 10, 20, 40, 80, 160, 320 is an example of a geometric sequence. It can be represented by the (multiplicative) recursive equation $a_n = 2a_{n-1}; a_1 = 5$.

A variable with a subscript is used to identify a particular value in a sequence. Suppose you represent terms of the sequence with the letter "a." Then, for the sequence 5,000, 4,900, 4,800, 4,700, 4,600, 4,500, $a_1 = 5,000$ because 5,000 is the first term in the sequence. $a_2 = 4,900$ means the second term has a value of 4,900. The number 2 is the **term number**, or **index**—the integer indicating the location of the indicated term in the sequence. Indices generally begin at 0 or 1, and increase by 1 per term. $a_6 = 4,500$ means the sixth term has a value of 4,500.

Mathematicians are somewhat inconsistent in how they identify the first term of a sequence. Many times, when the focus is on "sequence," a_1 refers to the first term. Sometimes, however, sequences involve time and it makes more sense to identify the first term as a_0 to represent the initial value or the value when time equals 0. Then a_1 identifies the sequence value after 1 unit of time; a_2 after 2 units of time, and so on. Either method is acceptable, but you must always clearly communicate how you will identify the first term of the sequence. Regardless of your choice, the variable you use to identify the term is the explanatory variable in the recursive equation. Consecutive integers are used to identify term numbers.

c) Is the arithmetic sequence of balances 5,000, 4,900, 4,800, 4,700, 4,600, 4,500 discrete or continuous? Are all arithmetic sequences discrete or continuous? Explain your answer.

d) How long will it take for the balance to reach 0?

e) List the first eight terms of the sequence represented by the recursive equation

$p_n = p_{n-1} + 22; p_1 = 940$.

f) What is the index for the term 1,050 in part (e)?

5. a) A sequence of numbers does not have to follow a pattern. The term numbers must follow a pattern of consecutive non-negative integers.

    Let the numbers 16, 5, 32, 8, 50, 48, 71, 37, 7, 39 form a sequence of 10 numbers with $a_1 = 16$. What is a_8?

    b) What is the term number when $a_n = 50$ for the sequence in part (a)?

6. Given a recursive equation, write the first four terms (x_1, x_2, x_3, x_4) for each sequence. Identify whether the sequence is an arithmetic sequence or a geometric sequence.

   a) $x_n = x_{n-1} + 7.5$; $x_1 = 24$.

   b) $p_{next} = p_{current} + 40$; $p_{initial} = 1{,}654$.

   c) $a_n = 1.5a_{n-1}$; $a_0 = 800$.

   d) $y_n = 2y_{n-1} + 10$; $y_0 = 50$.

7. Each of the following sequences is the result of additive or multiplicative growth and can be represented by a simple recursive equation. Write a recursive equation that would produce each sequence. Remember to define your starting or initial value. Identify whether the sequence is arithmetic or geometric.

   a) 10, 18, 26, 34, 42, 50, 58.

   b) 5, 15, 45, 135, 405, 1,215.

   c) 2.64, 4.35, 6.06, 7.77, 9.48, 11.19, 12.90.

   d) 150, 165, 181.50, 199.65, 219.615.

8. a) In general, how do you know when a sequence of numbers represents additive growth?

    b) How do you know when a sequence of numbers represents multiplicative growth?

INDIVIDUAL WORK 2

9. This article from *USA Today* describes the 1995 Ebola epidemic that occurred in Zaire.

   You may have seen the film, *Outbreak*, which describes the outbreak of a similar virus. Coincidentally, the movie was produced just before the Ebola epidemic began.

   a) The outbreak started with one person and eventually died out. Sketch a graph to model the spread of a disease through a population. Identify your explanatory and response variables. Then describe one or two assumptions you are making and how your assumptions affect the graph.

   b) Which growth model, additive or multiplicative, most accurately describes the spread of a disease in a population? Give reasons for your answer.

Ebola Experts Count Epidemic's Days, Lessons

The clock is ticking in Kikwit, Zaire, counting down the days to Aug. 24, when the Ebola epidemic could be officially declared over. If no new cases arise, that date will mark 42 days from the last Ebola patient's discharge from the hospital, says Dr. Jim LeDuc of the World Health Organization in Geneva. WHO's criteria for declaring the outbreak over is a period of twice the 21-day incubation with no new cases.

Scientists have learned enough about the rare virus, which causes massive internal bleeding, that "we'll be in a better position to understand how the disease is transmitted and be better prepared to respond to the next outbreak that comes along," LeDuc says. Among discoveries: (1) The virus is identical to the microbes that caused the last major Ebola outbreak in 1976. "That tells us it stays in this part of the world over a long period of time in some silent cycle that only periodically erupts," he says. (2) Its primary route of transmission is from person to person. Doctors have tracked the passage of the virus among patients in 60%

of cases, says LeDuc. But "obviously one person got infected someplace to start the whole chain." That first source of infection remains a mystery. In hopes of finding the natural host, scientists have collected more than 32,000 bugs, including 18,094 mosquitoes and assorted bedbugs, ticks, sandflies, fleas and tsetse flies; and almost 1,660 animals, including rodents. Most will go to the Centers for Disease Control and Prevention in Atlanta for testing, LeDuc says, but don't expect a quick discovery. "It's slow work," he says, "and somewhat dangerous."

Manning, Anita. "Ebola experts count epidemic's days, lessons," *USA Today* (Tuesday, August 8, 1995) p. 1D.

ACTIVITY

STRAIGHTEN THE CURVE

3

In Activity 2, you examined the behavior of additive and multiplicative growth and generated a table of values for each type. Since you know the defining recursive equations, you already know which data are from an additive process and which are from a multiplicative process. But what are efficient ways to identify such data when you don't know in advance what type they are? In this activity, you will examine a number of representations to see which are useful (and which are not) in identifying these two types of growth. Now look back at your work in Activity 2.

1. Identify Plan A as additive or multiplicative, and justify your choice.

2. You made a table similar to the one in **Figure 6.6**.

Time n (years)	Total number of homes p_n	Total homes previous year p_{n-1}	Amount of growth $p_n - p_{n-1}$	Ratio of successive terms p_n/p_{n-1}	Relative rate of growth $(p_n - p_{n-1})/p_{n-1}$
0	21,000				
1	21,400	21,000	400	1.019	0.0187
...					

Figure 6.6.
Partial table from Activity 2.

Describe any patterns you see in your table for Plan A that would be easy to identify and that you think are valid for all additive processes.

ACTIVITY

3

STRAIGHTEN THE CURVE

3. You drew the time-series graph of Plan A in Activity 2. Use your table, a spreadsheet, the SEQ or LIST function on the graphing calculator, or pencil and graph paper to create very careful graphs of:

  a) p_n versus p_{n-1}

  b) $(p_n - p_{n-1})$ versus p_{n-1}

  c) $p_n - p_{n-1}/p_n$ versus p_{n-1}

  d) any other variables you think might help identify additive growth.

4. Describe the characteristics of each graph. Identify those that are perfectly straight (be sure). Which are curved? For those that are linear, find the slope and vertical intercept (algebraically); don't round your answers at intermediate calculations.

5. Based on your results in Item 4, what graph(s) should you construct to check for an additive process? Use the recursive equation for Plan A to explain why the graphs you selected do what they do.

6. Repeat Items 1–5 for Plan B.

INDIVIDUAL WORK 3

Growth Alignment

1. Is the population growth represented by the numbers in **Figure 6.7** more like additive growth or more like multiplicative growth? Explain your answer.

2. Refer to Item 2 of Individual Work 2.

   a) Prepare a six-column table (see **Figure 6.8**) to represent the growth of Mario's collection for 6 weeks.

   b) Describe any patterns you notice in your completed table. Is this consistent with what you observed in Activity 3?

Year	Population
1980	3.0 million
1985	5.2 million
1990	9.1 million
1995	16.0 million

Figure 6.7.
Population growth for Item 10.

Time n (weeks)	Total number of CD's a_n	Total number of CD's previous week a_{n-1}	Amount of growth $a_n - a_{n-1}$	Ratio of successive terms a_n/a_{n-1}	Relative rate of growth $(a_n - a_{n-1})/a_{n-1}$
0					
1					
...					
6					

   c) Graph the time-series graph a_n versus n. Describe the graph.

   d) Graph $(a_n - a_{n-1})$ versus a_{n-1}. Describe the graph.

Figure 6.8.
Table for Item 2(a).

3. Refer to Item 3 of Individual Work 2.

   a) Use a calculator or spreadsheet to complete a six-column table, like the one shown in **Figure 6.9**, to represent the growth of Gridville's home total for several years.

Time n (years)	Total number of homes a_n	Total homes previous year a_{n-1}	Amount of growth $a_n - a_{n-1}$	Ratio of successive terms a_n/a_{n-1}	Relative rate of growth $(a_n - a_{n-1})/a_{n-1}$
0					
...					
40					

Figure 6.9.
Table for Item 3(a).

INDIVIDUAL WORK 3

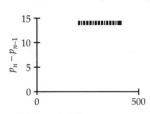

Figure 6.10.
Graph for Item 5(a).

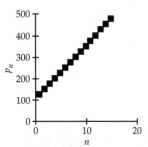

Figure 6.11.
Time series graph for
Item 5(b).

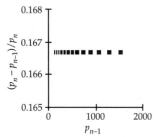

Figure 6.12.
Relative rate graph for
Item (5)c.

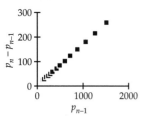

Figure 6.13.
Graph for Item 5(d).

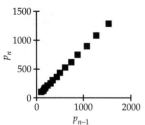

Figure 6.14.
Recursive graph for Item 5(e).

b) Describe any patterns you notice in your completed table. Is this consistent with what you observed in Activity 3?

c) Graph the time-series graph a_n versus n for $n = 0$ to $n = 25$. Describe the graph.

d) Graph $(a_n - a_{n-1})$ versus a_{n-1}. Describe the graph.

e) Graph the recursive graph a_n versus a_{n-1}. Describe the graph.

4. The increase in Mario's CD collection (see Item 2) is an example of additive growth. The regulated growth in the number of homes in Gridville (see Item 3) is an example of multiplicative growth. Compare your tables and graphs for Item 2 and Item 3 and describe the differences between additive growth and multiplicative growth for each of the following:

a) Amount of Growth (column 4 in the 6-column table)

b) Relative Rate of Growth (column 6 in the 6-column table)

c) Time-Series Graph (graph a_n versus n).

d) Graph of Growth (amount of growth $(a_n - a_{n-1})$ versus Population (a_{n-1})).

e) Graph of Relative Growth Rate versus Population (p_{n-1}).

5. Which of the graphs on the left might represent multiplicative growth? Which of these graphs might represent additive growth? Give careful reasons for each choice.

a) Refer to **Figure 6.10**.

b) Refer to **Figure 6.11**.

c) Refer to **Figure 6.12**.

d) Refer to **Figure 6.13.**

e) Refer to **Figure 6.14**.

6. a) For any sequence, amount of growth from term to term is also known as the **first differences**. In an additive recursive model the first differences are constant. Describe any patterns you notice for first differences in a multiplicative model.

b) In a multiplicative recursive model, you multiply the current term by the growth factor to get the next term. Describe the pattern that results when you divide each term by the previous term in a multiplicative model.

7. a) In general, how do you know when a sequence of numbers represents additive growth?

   b) How do you know when a sequence of numbers represents multiplicative growth?

8. The bird trainer at Sea World in San Diego, California said that if you put a male and a female mouse in a large field with plenty of food and no predators, there would be one million mice a year later. She explained that the hawk, a predator of the mouse, helped control the mouse population.

   a) Is the uncontrolled growth of the mouse population more likely an example of additive growth or multiplicative growth? Give reasons for your answer.

   b) Based on your answer to part (a) and numbers from the bird trainer, write an equation to represent the total number mice after n days.

   c) Use your equation to determine the number of mice in 5 years. Does you answer make sense? Explain.

9. The Sacramento "winter-run" is a distinct race of Chinook salmon, a species that is limited to the mainstream of the Sacramento River in California where it spawns every year. Around 1985 the average number of winter-run that came to spawn was 2,000. By 1989 the number had dropped to about 550, and by 1990 to about 450. The Chinook salmon winter-run was listed as an endangered species in California.

   a) Write an exponential equation to model the population between 1985 and 1989. Transform the explanatory variable by making 1985 = 0 and 1989 = 4. Write a closed-form exponential equation using the points (0, 2,000) and (4, 550).

   b) Use your exponential equation to obtain a predicted value for the year 1990 ($n = 5$). Compare the function value with the actual data value and write the residual value.

   c) Do you think your equation for part (a) is a good fit for all three points? Give reasons for your answer.

   d) If the decline continues to follow the same pattern, when will the level fall below 100? What equation do you solve?

10. On September 12, 1918, at Fort Devens, Massachusetts, one individual became ill with influenza. Six days later, in this contained environment, 6,674 people were stricken with the flu. By September 23 a total of 12,604 individuals at the camp had been diagnosed with influenza.

   a) Is this an example of additive growth or multiplicative growth? Give reasons for your answer.

   b) Write an equation to represent the total number of flu victims after n days.

   c) Use your equation to predict the total number of flu victims by January 1, 1919. Does you answer make sense? Explain.

You know that additive growth is described by

   (i) $p_n = p_{n-1} + k$,

where k is the additive constant. That means that

   (ii) $(p_n - p_{n-1}) = k$

for each value of n.

As you have seen, each of p_n, p_{n-1}, and $(p_n - p_{n-1})$ can be thought of as a variable quantity and listed in a table of values. So they can be graphed versus n or versus each other. You've done that already. When you graph p_n versus p_{n-1}, equation (i) says that it is exactly the same as graphing $y = x + k$, where k is still a constant.

The variables in equation (i) are p_n and p_{n-1} and act like y and x, respectively. In equation (ii) the only variable is $(p_n - p_{n-1})$; you could call it y if you wanted to. So it looks like $y = k$. Thus these graphs allow you to determine whether particular data come from an additive process, and if so, to identify its additive constant. For example, the graph of p_n versus p_{n-1} should be a line with slope 1 and vertical intercept k, for *every* additive growth sequence. So the vertical intercept of this graph tells you the additive constant for the process. Likewise, the graph of $(p_n - p_{n-1})$ versus n (or anything else) is a horizontal line with height k, the additive constant again. Check to see that your work in Activity 3 confirms these observations.

11. Multiplicative growth is described by

    (i) $p_n = kp_{n-1}$,

    where k is the growth factor. An alternate form is

    (ii) $(p_n - p_{n-1}) = rp_{n-1}$,

    where r is the relative growth rate.

    Following the example of the additive graphs discussed above, explain the graphical significance of each of the equations (i) and (ii) for multiplicative growth sequences. In particular, what graph does each equation help predict, and what important numbers can be obtained from such graphs?

12. First differences represent the amount of growth in a sequence. They are easy to compute and see in a table or spreadsheet. Another kind of graph that you may have encountered in an earlier course is a "signpost" graph. A signpost graph is one way to "see" first differences. **Figure 6.15** shows such a graph for an arithmetic sequence. **Figure 6.16** shows a signpost graph for a geometric sequence.

    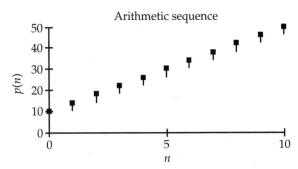

    Figure 6.15.
    Signpost graph for an arithmetic sequence.

    The signpost graph shows the amount of change from one term to the next in a time-series graph.

    a) What characteristics of a signpost graph do you identify with an additive growth model?

    b) What characteristics of a signpost graph do you identify with a multiplicative growth model?

    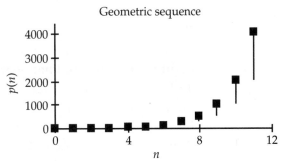

    Figure 6.16.
    Signpost graph for a geometric sequence.

c) Prepare a signpost graph for the recursive equation $p_n = 1.5 \times p_{n-1}$; $p_0 = 60$. Prepare the graph for values of n from 0 to 10.

d) Can you prepare a signpost graph for a closed-form equation such as $y = 3x + 1$?

e) Comment on how effective a signpost graph would be in identifying additive and multiplicative growth.

13. Instead of looking at the amount of change in a time-series graph, as in Item 12, you might also examine change in a recursive graph. Consider the arithmetic sequence defined be $a_0 = 5$, $an = a_{n-1} + 3$, and the geometric sequence defined by $g_0 = 4$, $g_n = 1.2g_{n-1}$.

a) Make separate recursive graphs for these sequences: graph term n versus term $(n - 1)$ for each sequence, a and g.

b) For each graph, describe the pattern of the graph. (Your description should agree with your observations in Activity 3 and in Item 11, above.)

c) In each graph, add vertical "signposts" below each point to indicate the amount of growth from one term to the next.

d) Now describe any pattern you see made by the "feet" of the signpost segments. That is, what can you say about the bottom points of the vertical segments? Be as specific as possible. (You might want to connect them to make any patterns more easily seen.)

e) Explain why the pattern you identified in part (d) exists and is the same for both graphs. Base your explanation on how the x- and y-coordinates of the "feet" are defined.

f) Comment on how effective this kind of graph would be in identifying additive and multiplicative growth.

With only a slight modification, the kind of graph you created in Item 13 is known as a **web diagram** or **web graph**. The usual modification is the addition of horizontal line segments joining each point at the top of each vertical segment to the foot of the corresponding next vertical segment. Thus, a web diagram may be drawn in three main steps. First, draw the line defined by the recursive equation, using p_{n-1} as x and p_n as y. Next draw the line $y = x$. Finally, starting with p_0 as the first x-value, draw a vertical line up to the recursive line, "bounce" horizontally over to the $y = x$ line, "bounce" vertically to the recursive line, and so on.

ACTIVITY

CONCENTRATION

4

In Activity 1 you probably graphed Eric's blood serum level versus time (a time-series graph). If so, you noticed the points followed an unfamiliar pattern. The graph curved (see **Figure 6.17**), but not in a way that you had seen before.

Because the graph is a curve, it may have seemed more like a geometric sequence than an arithmetic sequence. But is it really geometric? Activity 3 and Individual Work 3 have provided you with more tools to help confirm that decision. In addition, if the sequence turns out to be arithmetic or geometric, you now know how to determine the corresponding additive constant or growth factor.

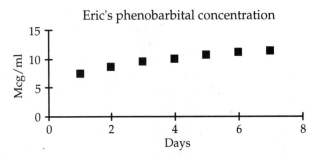

Figure 6.17.
Eric's phenobarbital concentration.

Constant first differences in a table of values, or a linear recursive graph (c_n versus c_{n-1}) with slope 1, identify arithmetic growth. Constant ratios of successive table values, or a linear graph containing (0, 0) of first differences versus c_{n-1}, or a linear recursive graph containing (0, 0), identifies geometric growth.

So, check it out. Return to the data in Figure 6.1 of Lesson 1, and apply one or more of your new mathematical tools to identify as many properties as possible for the sequence of values of Eric's concentration levels.

INDIVIDUAL WORK 4

Classified!

Your investigation of the medication levels in Eric's bloodstream should have indicated that those data are related to a recursive equation having both additive and multiplicative parts. The following items review your knowledge of arithmetic and geometric growth.

1. Write recursive and closed-form equations for each situation described below.

a) A fast-food restaurant kicked off a new promotion by offering six different figurines for sale with the children's meal. The promotion was an instant success, so the restaurant decided to release two new figurines each week. Write recursive and closed-form equations to represent the total number of figurines released by the restaurant in terms of the number of weeks since the beginning of the promotion.

b) Wild West Water Park opened five years ago and enjoyed total attendance of 93,000 that first summer. The management happily announces that total attendance has been increasing at a rate of 15% each year. Write recursive and closed-form equations to represent the total attendance in terms of the number of years since the park opened.

c) You borrowed $750 from a friend who does not charge you interest for the loan. Every week you make a payment of $20 to your friend until the debt is paid. Write recursive and closed-form equations to represent the amount you still owe in terms of the number of weeks since you borrowed the money.

d) Limit City has limits for driving, parking, eating, spending, and noise. The city currently has 5,000 homes and limits the building of new homes to 2% each year. Write recursive and closed-form equations to project the total number of houses in Limit City in terms of the number of years from now.

e) Which of parts (a), (b), (c), or (d), are examples of additive growth? Multiplicative growth?

2. Answer each question by writing and solving an exponential equation. Be sure to show your solution method clearly.

a) Twenty people at Clinton High School had the "bug" yesterday and were absent from school. Thirty people are absent today with the bug. The numbers are increasing by a factor of 50% each day. Assume the pattern continues until everyone has been infected. How long will it take until every one of the 1800 students has been infected by the "bug?" Assume that no student returns to school until everyone has recovered.

b) Inez still had 50 mg of a pain-reducing drug in her body when she took another dose of 450 mg. The drug is eliminated from her body at a rate of 28% each hour. How long until the level of the drug in her body drops below 100 mg?

c) Craig invested his money in a savings account that pays 9% interest compounded monthly (that is, 0.75% each month). He currently has $4,400 in his savings account. How many months will it take before he has $5,000 in his account?

d) The student population at East Valley High School is growing at a rate of 10% each year. The current population is 1,500 students. When will the population reach 2,500, the capacity of the school?

3. In this item you are given three different sequences. Determine whether each sequence is arithmetic, geometric, or neither. Give reasons for each answer.

a) 14, 17, 20, 23, 26, 29, 32.

b) 5, 12, 21, 32, 45, 60, 77.

c) 24, 12, 6, 3, 1.5, 0.75.

4. In this item you are given graphs for four sequences. Determine whether each sequence is arithmetic, geometric, or neither. Check the axes' labels carefully. If the sequence appears to be arithmetic or geometric, approximate the value of the corresponding additive constant or growth factor.

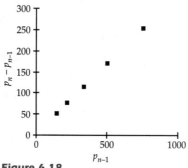

Figure 6.18.
Graph for Item 4(a).

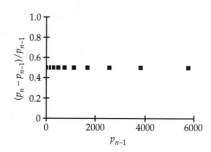

Figure 6.19.
Graph for Item 4(b).

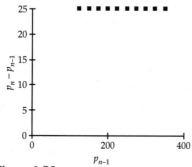

Figure 6.20.
Graph for Item 4(c).

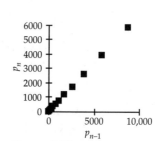

Figure 6.21.
Graph for Item 4(d).

a) Refer to **Figure 6.18**.

b) Refer to **Figure 6.19**.

c) Refer to **Figure 6.20**.

d) Refer to **Figure 6.21**.

5. a) **Figure 6.22** shows the graph of $(p_n - p_{n-1})$ versus p_{n-1} for a particular population. Based on the information provided by this graph, classify the type of growth and sketch the corresponding graph of $\dfrac{p_n - p_{n-1}}{p_{n-1}}$ versus p_{n-1}.

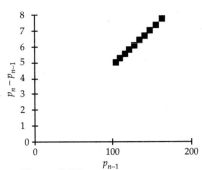

Figure 6.22.
Graph of growth versus population.

b) **Figure 6.23** shows the graph of $(p_n - p_{n-1})$ versus p_{n-1} for a particular population. Assume that $p_0 = 100$.

Based on the information provided by this graph, sketch the graph of p_n versus n.

Figure 6.23.
Graph of growth versus population.

6. The closed-form equation $p(n) = 40(1.25)^n$ represents the growth of a population.

a) Write the recursive form.

b) Write the first five terms for the sequence using the recursive form.

c) Determine the relative rate of growth.

d) Graph the amount of growth $(p_n - p_{n-1})$ versus the total population (p_{n-1}).

7. a) Dancing lessons at Marcy's School of Dance cost $5.00 per lesson. Marcy decides to increase her rates by 200%. What will she charge for a lesson after the increase?

b) You receive an allowance of $10.00 per week from your parents. They threaten to reduce your allowance by 200% if you bring home a bad report card. Discuss the effect of your parents' carrying out this threat.

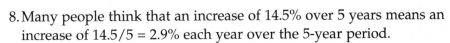

8. Many people think that an increase of 14.5% over 5 years means an increase of 14.5/5 = 2.9% each year over the 5-year period.

a) Show that this is false by determining the effect of 5 successive 2.9% annual increases on the hourly wage of an employee who is paid $10.00 per hour. What is the actual percentage increase that results from these 5 raises?

b) Find an annual percentage increase that does produce a 14.5% increase when applied over 5 successive years.

9. The tripling sequence is produced using the function
$y = 3^x$ or $p_n = 3p_{n-1}$; $p_0 = 1$.

   a) Produce a table (see **Figure 6.24**) with the first eight terms of the sequence. Include a third column and write the amount of growth (first differences).

n	p_n	$p_n - p_{n-1}$
0	1	
...		
7		

Figure 6.24.
Table for Item 9(a)

   b) Describe the relationship between the sequence of first differences and the original sequence.

Many applications of exponential decay are described using what is known as the **half-life**. In simple terms, the half-life of a substance is the amount of time needed for the decay to remove exactly half the material present when observations began. Of course, if half has been removed, then half still remains.

10. A certain drug is eliminated from the body in such a way that its half-life is 10 hours.

    a) Find the hourly "growth" factor.

    b) Write a closed-form equation for the percentage of the drug remaining in the body in terms of the number of hours since the drug entered the body.

11. Consider the case of a drug that decays exponentially so that 16% is removed from the body each hour. Suppose that initially (at time 0) there is 40 mg in the bloodstream.

    a) Write a closed-form equation for the amount remaining after h hours.

    b) What equation do you need to solve in order to determine the half-life for this medicine? Explain.

    c) Use a graph or table to find the approximate half-life for this drug.

Multiplicative growth models produce closed-form equations with exponents. Problems like those in Item 11 can be solved graphically, but you will need to work with exponents if you want to develop symbolic methods for solving such equations. Here is a review of the laws for exponents, which you may have studied in Course 1, Unit 6, *Wildlife*.

  I. Multiplication: $(b^n)(b^m) = b^{n+m}$

  II. Division: $b^n / b^m = b^{n-m}$

  III. Power: $(b^n)^m = b^{nm}$

To see why these rules hold, examine the rule for multiplication first. Note that the rule applies only when the base numbers for the product are identical. For this example, take $b = 1.5$. The closed-form growth equation $p = 1(1.5)^n$ represents growth for n time units, with a growth factor of 1.5. Now think of the result $1(1.5)^n$ as the starting value for a new growth having the same factor, this time for m units of time. The final amount is then $p = [1(1.5)^n](1.5)^m$. Of course, common sense tells you that the result for n time units followed by m time units is equivalent to $n + m$ time units using the same factor. Therefore, it must be true that $(1.5)^n(1.5)^m = (1.5)^{n+m}$.

Another way to help remember the multiplication rule is through a small, specific example not based on growth equations. For example, you know that $(2)(2)(2)$ is 2^3. Take the answer and double it four more times: $(2)(2)(2)(2)(2)(2)(2) = 2^3(2^4)$, which is equivalent to 2^7. Therefore, $2^3(2^4) = 2^7$, illustrating the multiplication rule.

The second rule, for division, is the reverse of the multiplication rule. If multiplying by b^m adds m to the exponent, then dividing by b^m should subtract m from the exponent. Again, of course, the base numbers must be identical.

To help remember this rule, think about how you reduce common fractions. For example, you can reduce the fraction $\frac{42}{72}$ by writing $(2)(3)(7)/(2)(2)(2)(3)(3)$ and dividing common factors of 2 and 3 from the numerator and denominator to get $7/(2)(2)(3)$ or $\frac{7}{12}$. Similarly, you can reduce the fraction $\frac{3^8}{3^5}$ by writing $(3)(3)(3)(3)(3)(3)(3)(3)/(3)(3)(3)(3)(3)$ and dividing five of the 3's from the numerator and denominator to get $(3)(3)(3)$ or 3^3, illustrating the use of the division rule above.

12. Practice rules (I) and (II) by writing equivalent exponential expressions for each of the following:

a) $x^4(x^8) =$

b) $y^3(y^{10}) =$

c) $n^2(n^5)(n^3) =$

d) $5x^4(2x^3) =$

e) $3x^5(x^5)(x^5) =$

f) $x^4(y^3) =$

g) $6x^3y^2(-3x^4y)$

h) $x^9/x^4 =$

i) $m^{10}/m^7 =$

j) $y^{12}/y^6 =$

k) $10x^8/20x^6 =$

l) $n^5/n^8 =$

m) $12x^2y^5/2xy^3$

To help remember the power rule, note that the expression $(n^2)^4$ means $(n^2)(n^2)(n^2)(n^2)$, which is equivalent to $(n)(n)(n)(n)(n)(n)(n)(n) = n^8$.

13. Write an equivalent expression for each of the following:

a) $(x^3)^2 =$

b) $(y^2)^5 =$

c) $(m^3)^4 =$

d) $(10^3)^5 =$

e) $(b^n)^m =$

Two additional rules for using exponents are:

IV. Zero Power: $b^0 = 1$ (0^0 is undefined; it is not a number)

V. Negative Exponents: $b^{-n} = 1/b^n$

The rule for division of exponential expressions involves subtraction of the exponents. For example, $x^5/x^2 = x^{5-2} = x^3$. Suppose you were to switch

the numerator and denominator to get x^2/x^5. Using the division rule, the result is $x^{2-5} = x^{-3}$. However, another way to simplify x^2/x^5 would be to write

$$\frac{(x)(x)}{(x)(x)(x)(x)(x)} \text{ , which becomes } \frac{1}{(x)(x)(x)},$$

or $1/x^3$. Thus the expression x^2/x^5 simplifies to both x^{-3} and $1/x^3$. These two expressions must be equivalent. This illustrates the fifth rule, of negative powers (exponents).

Similarly, x^5/x^5 can be used to illustrate the Zero Power law.

14. Simplify each of the following in two ways:

a) $m^3/m^5 =$

b) $x^1/x^6 =$

c) $10^4/10^9 =$

d) $y^6/y^6 =$

e) $8^3/8^3 =$

f) $b^0/b^n =$

Problems involving scientific notation frequently require applying the laws of exponents. For example, the number 57,000 may be written as 5.7×10^4 in scientific notation. Similarly, the number 300,000 may be written as 3.0×10^5 in scientific notation. Then you may use scientific notation to multiply 57,000 and 300,000: $(5.7 \times 10^4)(3.0 \times 10^5)$ becomes $(5.7)(3.0) \times (10^4)(10^5)$.

Apply the rules for multiplying exponential numbers with a common base (10) and you get

$17.1 \times 10^4 \times 10^5$

17.1×10^9

$1.71 \times 10^1 \times 10^9$

1.71×10^{10}.

15. Show how to use scientific notation to multiply each of the following.

  a) 150 × 8,000

  b) 9,000,000 × 7,000

  c) 0.0004 × 900

16. a) Compare the graphs of $y = 2^x$ and $y = (1/2)^x$. The graph of $y = 2^x$ represents a growth factor of 2. The graph of $y = (1/2)^x$ represents a growth factor that is 1/2, the reciprocal of $y = 2^x$. In what ways are the two graphs similar? In what ways are they different?

  b) Since $2^{-1} = 1/2$, $(1/2)^x = 2^{-x}$. Use this fact to explain your observations in part (a).

17. Remember from Item 7 of Individual Work 1 that parametric equations may be used to graph functions even when they are not originally expressed parametrically.

  a) Write parametric equations to graph the exponential function $y = 100(1.25)^x$.

  b) Explain how to rewrite these equations to shift the graph 6 units to the right.

  c) Explain how to rewrite the equations to shift the graph up 100 units.

  d) Given the closed-form equation $y = 100(1.25)^x$, write the closed-form equation to represent a shift 6 units to the right.

  e) Given the closed-form equation $y = 100(1.25)^x$, write the closed-form equation to represent a shift of 100 units upward.

LESSON THREE
Finding Time

KEY CONCEPTS

Inverse functions

Logarithms

Solving
exponential
equations

PREPARATION READING

Loggers

Throughout your study of algebra, you have written equations, graphed equations, and solved equations. In a very real way, graphing equations and solving them involve "opposite" directions of thinking. In building a graph, you use known values of the explanatory variable to compute corresponding values of the response variable in order to determine coordinates of points on the graph. To solve an equation, you need to start with a known value of the response variable and find the corresponding value of the explanatory variable.

In Lesson 2, you studied exponential functions and represented multiplicative growth with recursive and closed-form equations. Then you used tables and graphs to find a half-life or to solve equations such as $25{,}000 = 21{,}000(1.015)^x$ in which the variable is an exponent.

In this lesson, you examine the general process of solving equations analytically by revisiting the idea of an inverse operation. From that, you can develop a symbolic method for solving exponential equations (equations where the explanatory variable is an exponent). For any problem you face, though, you will still need to decide when it is best to use a symbolic method and when it is sufficient to use tables and graphs.

Developing a symbolic method for solving exponential equations introduces you to logarithms. The logarithm is more than just another mysterious button on the calculator; it is a powerful tool for investigating and solving a variety of problems.

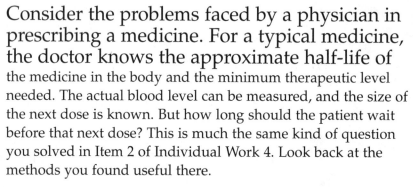

ACTIVITY

C'MON BACK

5

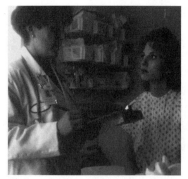

The Image Bank

Consider the problems faced by a physician in prescribing a medicine. For a typical medicine, the doctor knows the approximate half-life of the medicine in the body and the minimum therapeutic level needed. The actual blood level can be measured, and the size of the next dose is known. But how long should the patient wait before that next dose? This is much the same kind of question you solved in Item 2 of Individual Work 4. Look back at the methods you found useful there.

In Item 10 of Individual Work 4 you computed the half-life of a drug from its exponential equation—the same kind of problem. In each case, you knew the value of the response variable and needed to know the value of the explanatory variable. You have done this kind of algebra before, but for simpler kinds of equations. The key idea remains the same, though—to understand the inverse operation.

1. Suppose the equation $p = 4m + 80$ represents the regulated growth of a population of a particular animal over a period of m months.

   a) What kind of growth is represented by this equation?

   b) Make an arrow diagram to represent the process that begins with the explanatory variable m and produces the response variable p. Include the inverse process with your arrow diagram.

   c) Write an equation to represent the inverse process.

   d) Explain how the inverse process (arrow diagram and/or equation) could be helpful in solving the equation $2,104 = 4m + 80$.

   e) Graph the function and its inverse on the same graph, using a window in which horizontal and vertical units are identical. Note that the usual convention is to use the horizontal axis to represent the input for a particular process

ACTIVITY

C'MON BACK

5

and to show the output on the vertical axis. Thus, to graph the inverse process you need to swap the axes on which you plot values of m and p, plotting p along the horizontal axis. That is, the horizontal axis will represent different quantities for the two graphs (m for the original growth equation, and p for the inverse function). Describe the relationship between the completed graph of the function and the graph of the inverse. (Note: You will be able to judge relationships better if you first Zoom Square to obtain equal-sized units on the two axes.)

For the original function, the response variable, population, is the outcome of a process that begins with the explanatory variable, time, in months. The inverse of this function reverses the process and switches the roles of the variables. You used that idea in your graph in Item 1(e).

Another way to represent the inverse of a function is with a table. Just as the table provides sample pairs of values for a function, the same table may be used to provide sample pairs of values for the inverse of a function. You create a table for the inverse by exchanging the values for the response variables with the corresponding values for the explanatory variables.

For example, the table in **Figure 6.25** represents sample pairs of values for the population function above. Note that "x" and "y" are used only to indicate explanatory and response variables. They have no consistent "meaning" in the population context.

x Explanatory	0	1	3	4	7	10	25
y Response	80	84	92	96	108	120	180

Figure 6.25.
Sample values for the population function.

The table representing the inverse relation is shown in **Figure 6.26**.

Figure 6.26.
The inverse of a function is the result of exchanging the *x*- and *y*-values.

x **Explanatory**	80	84	92	96	108	120	180
y **Response**	0	1	3	4	7	10	25

Swapping x for y and y for x changes the roles of the values, but it does not change their meaning. Again, the letters "x" and "y" are used as generic abbreviations for explanatory and response variables within a particular setting. In the example above, x represents the variable m in Figure 6.25 and represents the variable p in Figure 6.26. In fact, it is the meaning of the variables that may make the exchange awkward. In general, you will probably find it easiest to use variable names based on the context so that their meanings will always be clear, rather than using the generic x and y. However, sometimes you examine mathematical equations without contexts. In those cases, x and y may be used.

ACTIVITY

C'MON BACK

5

2. a) Complete a table of sample pairs for the function
      $y = 2x + 4$ (see **Figure 6.27**) and its inverse (see **Figure 6.28**).

x Explanatory	−3	−1	0	1	2	4	5
y Response							

Figure 6.27.
Table for $y = 2x + 4$.

x Explanatory							
y Response	−3	−1	0	1	2	4	5

Figure 6.28.
Table for inverse of $y = 2x + 4$.

   b) Graph by hand the function $y = 2x + 4$ and its inverse on
      the same graph. Again, use equal scales on the two axes.
      Use your table from part (a) if necessary. Observe and
      describe the graphical relationship between the function
      and its inverse.

   c) Write an equation for the inverse of the function $y = 2x + 4$.
      Explain your method.

A table is a numeric way to represent the relationship between a
function and its inverse. The arrow diagram shows how an
inverse is the result of a reverse process. By hand, the process of
graphing an inverse function from a table is easy.

   3. Construct a table and graph for the exponential function
      $p = 1.5^t$ for t between −2 and 6. Use equal scales for the two
      axes. Then add the table and graph for its inverse and
      describe how the two graphs are related.

INDIVIDUAL WORK 5

Back and Forth

1. The formula $C = \frac{5}{9}(F - 32)$ is used to convert temperatures in degrees Fahrenheit to degrees Celsius. The inverse of this function may be used to convert from °C to °F.

   a) Prepare an arrow diagram to represent the process of converting from °F to °C.

   b) Add the inverse process to your arrow diagram from part (a).

   c) Write a closed-form equation to represent the inverse process.

   d) Prepare a spreadsheet you could use to convert temperatures from °F to °C in increments of 2°F from 70°F to 90°F. Explain how to use the resulting table for conversions from °C to °F.

2. a) The inverse of a function reverses that function and switches the roles of the explanatory and response variables. Create tables for the function $y = 2^x$ and its inverse. Use x-values between 0 and 5 for the first function.

   b) Graph the ordered pairs from your tables in part (a) on the same set of axes. Zoom Square to ensure equal scales. Connect the points with a smooth curve. This is the graph of the inverse of $y = 2^x$. It produces the exponents x from given y-values.

3. A particular medicine has a decay factor of about 0.955 when time is measured in hours. That is, the body eliminates about 4.5% of the medication present at any instant over the course of the next hour.

   a) Suppose the level for a certain patient is 9 mg. Write an equation to describe the amount of medicine remaining in the bloodstream if no more medication is taken.

   b) About when will the level drop below 6 mg? Explain what that question has to do with inverse functions.

You know how to graph the inverse of a function when you have a table of values from which to work. Simply reverse the roles of the two sets of values, making the explanatory variable into the response variable, and vice versa. That means that the domain of the original function becomes the range of the inverse, and the range of the original becomes the new domain. But how can you graph inverses of complicated functions?

There are three strategies. First, you could develop algebraic representations (equations) of the inverse functions. Second, you could develop other ways to use technology to graph inverse functions. Third, you could observe general properties of inverse graphs, properties that you could use for *any* function you encounter. Item 4 takes this third approach: investigate the graphical relationships between functions and their inverses. Items 5 and 6 use a new twist on a familiar calculator technique. Items 7 and 8 and Activity 6 develop an algebraic method.

4. Graph, by hand, each given function and its inverse on the same graph, using identical scales on both axes.

   a) $y = 0.5x + 8$.

   b) $y = x^2$ (just for positive x-values)

   c) $y = x^2 + 4$ (just for positive x-values).

   d) Examine the graphs you have constructed in this item and in earlier similar items. Describe the relationship between the graph of a function and the graph of its inverse.

5. This item explores one calculator technique for graphing inverse functions when the original function's equation is given. It is based on the idea of transformations. which you have studied throughout this course. (Look back at Item 7 of Individual Work 1 and at Item 16 of Individual Work 4.) The key to simple representations of transformations on the calculator is to be able to deal directly with x- and y-values separately; that means using parametric equations. So, change your calculator to parametric mode.

   a) The equation $y = 2x + 4$ may be expressed in parametric form (on the calculator) as $X_{1T} = T$, $Y_{1T} = 2T + 4$. Enter these equations into your calculator's function list, set a window of approximately $[-5, 15] \times [-5, 15] \times [-5, 15]$. Zoom Square to set your final window, but be sure that the T-window includes all of the X-window. Graph your equations to confirm that it really does produce the graph of $y = 2x + 4$.

   b) Now add a second function, defined by $X_{2T} = Y_{1T}$ and $Y_{2T} = X_{1T}$. Before graphing these new equations, describe what they tell the calculator to do. What should happen? After writing your description, graph the equations to confirm your predictions.

   c) Repeat this technique to graph each function-inverse pair you examined in Item 4.

6. Use the method of parametric equations to obtain a graph of the function $y = x^2$ (for x between -2 and 2, say) and its inverse. Check that you use equal-sized scales on both axes. Then use the graph of the inverse to explain the solution of the equation $4 = x^2$.

To graph the inverse of a function directly, using algebra, means to represent the inverse as an equation in the form $y =$, where y now represents the output for the inverse function (so, the input for the original function). Thus, your algebraic energy should be spent in solving for what was originally the explanatory variable in terms of what was originally the response variable; that is, isolate the original explanatory variable in the equation.

For example, to get the inverse of the linear equation $y = 3x + 40$, solve for the original explanatory variable x. That is, use algebra to rewrite the equation in the form $x = \ldots$:

Subtract 40 from both sides of the equation.

$$y - 40 = 3x + 40 - 40$$

$$y - 40 = 3x$$

Divide both sides by 3.

$$\frac{y - 40}{3} = \frac{3x}{3}$$

$$\frac{y - 40}{3} = x$$

For a function using the generic variables x and y, as this one does, exchange the x and y labels. Swapping x for y and y for x changes the roles of the variables so that they now represent the proper roles for the inverse function. (Note: For contextual functions, where the variable names represent particular quantities such as population and time, do *not* make this exchange of letters. Let the form of the equation indicate which quantity is the input and which is the output. Convention is to interpret the isolated variable as the response, or output, variable.)

The equation for the inverse has several equivalent forms:

$$y = \frac{x - 40}{3} \quad \text{or} \quad y = \frac{x}{3} - \frac{40}{3} \quad \text{or} \quad y = \frac{1}{3x} - \frac{40}{3}$$

7. Using function mode (not parametric), graph the equations $y = 3x + 40$ and $y = \frac{x - 40}{3}$ together to confirm that their being inverses is reasonable. Be sure to Zoom Square so that slopes "look right."

8. Use algebra to find an equation for the inverse of each of the following functions. Compare the results to those you obtain from arrow diagrams. (You may also wish to graph each equation with its inverse, using regular function mode instead of parametric, to confirm that your equations are reasonable.)

a) $y = 2x + 4$

b) $p = 4m + 80$

c) $y = 0.5x - 20$

d) $y = x^2$

9. Use your calculator to graph $y = 2^x$ and its inverse using "square" axes. Does your answer agree with what you obtained in Item 2? What equation represents the inverse of $y = 2^x$?

Items 6 and 8(d) illustrate an interesting point. Recall that this lesson began by reviewing the use of arrow diagrams to represent functions and their inverses. The squaring function, $y = x^2$, is one of many functions for which it is easy to "go forward" in the arrow diagram, but it is difficult to "go backwards." The reason is that there are *two* ways to get any particular positive number by squaring! For example, $(2)^2 = 4$ and $(-2)^2 = 4$. Thus, undoing the squaring produces two results, not just one. For this reason, mathematicians say that the inverse relation, the square root, is *not* a function.

ACTIVITY

LOG ON

6

Remember the town back in Activity 2? It has 21,000 homes and is considering two plans for managing growth. Plan A limits growth to 400 new homes each year. Plan B limits growth of new homes to 1.5% each year. Corresponding closed-form equations for the total number h of homes after y years are

Plan A:
$h = 21,000 + 400y$

Plan B:
$h = 21,000(1.015)^y$

1. Determine, symbolically if possible, when each plan will permit the total number of houses to exceed 42,000 (that is, when the town will have doubled). If you can not get an analytic solution, use and explain another method to obtain your answer.

Notice that the prediction requested in Item 1 involves the use of inverse operations. For example, using arrow diagram representations, the situation for Plan A looks like **Figure 6.29**.

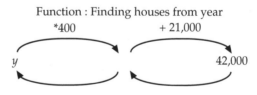

Function : Finding houses from year
*400   + 21,000

y 42,000

Inverse : Finding year from houses

Figure 6.29.
An arrow diagram relating houses and years for Plan A.

This function is fairly easy to reverse. You probably had no difficulty in doing the algebra to write its inverse.

LOG ON

ACTIVITY

6

Figure 6.30 shows the same information for Plan B.

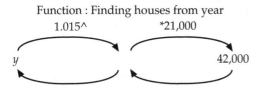

Function : Finding houses from year

1.015^   *21,000

y 42,000

Inverse : Finding year from houses

Figure 6.30.
An arrow diagram relating houses and years for Plan B.

The first step of finding the inverse here is easy, too. But reversing the "raise 1.015 to the power . . ." is tough! In fact, solving exponential equations requires the invention of an entirely new kind of function, called a **logarithm**. A logarithm is the exponent of the designated base, usually 10, that is equivalent to the given number. The term **base** is used in mathematics to describe a number being raised to a power.

The following guided exploration introduces logarithms (usually just called "logs") and several ways to evaluate them. Following the usual advice of "start simple," begin by learning how to solve simple exponential equations. Exponential equations may be simplified either by using more familiar base numbers or by looking first at special cases. You will do both. First, consider two familiar exponential functions, $y = 2^x$ and $y = 10^x$, before attempting to solve $42,000 = 21,000(1.015)^y$.

The function $y = 2^x$ is the "doubling function." In biology, for example, it represents the number of cells after x rounds of cell division. One example of a doubling sequence is:
1, 2, 4, 8, 16, 32, 64, ...

The function $y = 10^x$ is familiar from your previous work with scientific notation. You have written numbers such as 56,300 as 5.63×10^4.

ACTIVITY

LOG ON

6

2. Here are equations involving 10^x and 2^x. Solve each equation symbolically, without relying on graphs or other approximations. Show or describe the procedure you use to solve each equation. Hint: $10^2 = 100$.

a) $1{,}000 = 10^t$.

b) $100{,}000{,}000 = 10^t$.

c) $0.001 = 10^t$.

d) $2^x = 16$.

e) $64 = 2^x$.

Equations such as $1{,}000 = 10^t$ are easy to solve because $1{,}000$ can be written as 10^3. The value that makes $10^3 = 10^t$ becomes obvious ($t = 3$) when both expressions have the same base (see **Figure 6.31**). In an equation such as $2^x = 8$ you express 8 as 2^3 and the equation becomes $2^x = 2^3$. The answer, $x = 3$, is apparent when both expressions are written with the same base number.

One way to solve an exponential equation is to write the expressions in an exponential form so they have the same base number.

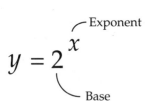

Figure 6.31.
Identifying the base and exponent.

3. What do you do when one or both expressions cannot be written easily in exponential form? For example, how do you solve the equation $2{,}500 = 2^x$ to determine the number of times a cell must split before there are 2,500 cells?

a) The number $2{,}048$ is 2^{11}. The number $4{,}096$ is 2^{12}. The number $2{,}500$ is between 2^{11} and 2^{12}. Your calculator gives you the number $2{,}893$ as the approximate answer for $2^{11.5}$. Use your calculator and a guess-and-check method to solve the equation $2^x = 2{,}500$. Find the answer to the nearest hundredth.

b) Estimate the solution to the equation $750 = 10^t$. Remember that $10^2 = 100$ and $10^3 = 1{,}000$. Then use a guess-and-check method to solve the equation to the nearest hundredth.

ACTIVITY

LOG ON

6

Keep a record of your guesses by making a table like the sample in **Figure 6.32**.

Guess	Evaluate	Decide
2.5	$10^{2.5} = 316$	Way too low!
2.75	$10^{2.75} = 562.34$	Still too low

Figure 6.32.
Sample record of guesses.

c) Confirm your answer by graphing $y = 10^x$ and tracing to find x when $y = 750$.

d) Solve the equation $0.037 = 10^x$ by graphing the equation $y = 10^x$. Confirm your answer using a guess-and-check method.

Looking at Item 3, you might guess that you can use the calculator to express any positive number as 2^x or 10^x with a guess-and-check method, or by graph and trace. However, writing numbers as 10^x turns out to be more practical because your calculator provides a way to do this quickly. Logarithms allow you to express any positive number as 10^x.

Find the logarithm key for your calculator; it is probably labeled LOG. Since the logarithm is the name of a function, it is written using function notation. That is, to indicate that you want the "logarithm *of* 1,000," you write log(1,000); the parentheses do *not* indicate multiplication.

John Napier (1550–1617) was one of the greatest mathematicians of the 17th century. A Scotsman, Napier is credited as the inventor of logarithms on the basis of his 1614 publication on the subject. Its title, translated, is "A Description of the Wonderful Law of Logarithms."

ACTIVITY

LOG ON

6

4. This item examines the use of the LOG function on your calculator.

   a) Use the LOG button on your calculator to find the values of log(1,000), log(100,000,000), and log(100). Compare the values to your work in Item 2. Describe any patterns you notice between each number and its logarithm.

   b) Predict the value of log(10,000). Give reasons for your prediction. Then check your answer using the calculator.

   c) Estimate log(500). Use the calculator to check the accuracy of your estimate.

The logarithm of 10, log(10), is 1 because $10^1 = 10$. The logarithm of 100,000, log(100,000), is 5 because $10^5 = 100,000$. The logarithm of 0.001, log(0.001), is –3 because $10^{-3} = 0.001$. The logarithm of 200, log(200), is 2.3010 (rounded to four decimal places) because $10^{2.3010} = 200$.

So what does the LOG button do? It evaluates the inverse of the function $y = 10^x$. It is the "backwards" arrow in the diagram in **Figure 6.33**. It gives the exponent (for the base 10) related to any number. Loosely, you might say that it undoes the "10-to-the . . ." function.

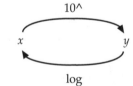

Figure 6.33.
Arrow diagram defining log.

Thus, the two sentences (equations) $y = 10^x$ and $x = \log(y)$ mean *exactly* the same thing. But there is nothing special about the number 10 as a base, at least in principle. For example, since you can think of an exponential function with the base of 2, as described by the arrow diagram in **Figure 6.34**, it should have an inverse, too.

ACTIVITY

LOG ON

6

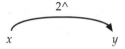

Figure 6.34.
The base-2 exponential function.

Its inverse is also a logarithm. To make clear that it uses base 2 instead of base 10, it is usually written as $\log_2$, and is read as "the logarithm to the base 2." But, based on the arrow diagram in Figure 6.34, the two sentences (equations) $y = 2^x$ and $x = \log_2(y)$ mean the same thing. And, there is nothing special about 2 or 10. Any number that can be the base of an exponential function is also the base of a corresponding logarithm. In general, for any positive base b (1), the following two sentences are equivalent:

$y = b^x$

$x = \log_b(y)$.

In fact, this symbolic representation of the relationship shown in the arrow diagrams is the definition of logarithm. You could write log as $\log_{10}$. However, since many logarithm calculations are done using base 10, it is customary to be "lazy" and leave the 10 off. Base-10 logs are called **common logarithms**.

Logarithms may be used to convert any positive number to an exponential expression with a base of 10. For example, in Item 3(c) you did the work to show that $\log(750)$ is about 2.88. (Check it with your calculator!) Likewise, because $\log(193)$ is about 2.2856, $193 = 10^{2.2856}$. The number 193 may be replaced with $10^{2.2856}$. The usefulness of this kind of substitution becomes apparent as you return to Item 1.

ACTIVITY

6

LOG ON

EXAMPLE:

Recall from Item 1 that the first step in solving the equation $42,000 = 21,000(1.015)^x$ is to divide by 21,000. The result is $2 = (1.015)^x$, which may be solved by changing 2 and 1.015 to exponential expressions. That is, use logarithms and change the form of the equation and solve for the exponent. (Note: The results of the following calculations are shown rounded to four decimal places. However, computations have been done using full calculator precision, with rounding taking place only at the end. Computations with rounded intermediate values will give slightly different and less accurate answers.)

$$2 = (1.015)^x$$

The logarithm of 2 is 0.3010 so replace 2 with $10^{0.3010}$.

The logarithm of 1.015 is 0.0065 so replace 1.015 with $10^{0.0065}$.

$$10^{0.3010} = (10^{0.0065})^x$$

Apply the power law for exponents, $(b^m)^n = b^{mn}$:

$$10^{0.3010} = 10^{0.0065x}$$

$$0.3010 = 0.0065x$$

$$x = 46.56.$$

Check the answer by replacing x with 46.56 in the equation $42,000 = 21,000(1.015)^x$.

The Earthquake-Logarithm Connection

In 1935, Charles F. Richter used mathematics and seismograms (visual records produced by instruments called seismographs) to develop the Richter Scale method for measuring the magnitude of earthquakes. Computing logarithms for different seismic wave amplitudes shown on seismograms was a key step in his calculations.

INDIVIDUAL WORK 6

Solutions

1. Write a logarithm equation using the same base as the indicated exponential base, and equivalent to the given exponential equation.

   a) $10^4 = 10,000$.

   b) $2^3 = 8$.

   c) $(0.2)^2 = 0.04$.

   d) $3.2^{1.3} = 4.54$.

   e) $8 = (1.75)^x$.

2. Write an exponential equation using the same base as the indicated logarithm's base, and equivalent to the given logarithm equation.

   a) $\log(0.01) = -2$.

   b) $\log_2(64) = 6$.

   c) $\log_5(25) = 2$.

   d) $\log_8(0.03125) = -\frac{5}{3}$.

Remember how logarithms may be used to solve the equation $50 = 2^x$.

Change each number in the equation to an equivalent form using base-10 logarithms. (Note: The calculations are shown here rounded to four decimal places, but full calculator precision is actually used in the final calculation.)

   The $\log_{10}(50) = 1.6990$, therefore, $50 = 10^{1.6990}$.

   The $\log_{10}(2) = 0.3010$, therefore, $2 = 10^{0.3010}$.

Substituting, the equation $50 = 2^x$ becomes $10^{1.6990} = (10^{0.3010})^x$ or $10^{1.6990} = 10^{0.3010x}$.

The expressions on each side of the equation are equivalent if the bases are the same and the exponents are the same. Since both bases are 10 you need only find what makes the exponents equal. Thus, solve the equation $1.6990 = 0.3010x$. The result is approximately $5.6439 = x$. The solution to the equation $50 = 2^x$ is approximately $x = 5.6439$.

Use the calculator to check your answer: $2^{5.6439}$ gives you 50.0015, which is reasonably close to 50. To do better, notice where the numbers that you divided came from; try entering log(50)/log(2) into your calculator.

3. a) Use logarithms to solve: $962 = 3^x$. Use the calculator to compute the power directly to verify your answer.

   b) Use logarithms to solve: $175 = 5^x$. Use the calculator to compute the power directly to verify your answer.

4. a) Solve $20 = 5^x$.

   b) Solve $14 = (1.005)^x$.

   c) Solve $2{,}000 = 400(1.015)^x$.

5. a) Graph $y = (1.75)^x$. Use the graph to find x when $y = 8$.

   b) Use base-10 logarithms to solve the equation $8 = (1.75)^x$. Compare your symbolic answer to the graphical answer from part (a).

   c) Which method is better for solving exponential equations, the graphical method using the exponential equation as in part (a), or the symbolic method using logarithms as in part (b)? Give reasons to support your answer.

6. Use logarithms to rewrite and solve each of the following equations:

   a) $400 = 10^x$.

   b) $1{,}250 = 25^x$.

   c) $838 = 6.52^x$.

7. Suppose you invest $8,000 in a savings account that earns 7.5% interest compounded monthly (that is, 0.625% each month). How many months will it take for your investment of $8,000 to grow to $10,000? Use logs to determine the answer.

8. There are 240 moose in a wildlife preserve. Suppose their population is growing naturally at a rate of 2% each year. Assume the growth pattern continues and determine the number of years it will take for the population in the preserve to reach 1,000 moose. Show your method.

9. Medication is frequently described by its half-life. (Recall that the half-life is the amount of time needed for half of the medicine to leave the bloodstream.) The phenobarbital that Eric took decays according to the formula $c = 15(0.99)^h$, where c is the concentration in the blood (in mcg/ml) and h is time since the dose (in hours). What is the half-life under these conditions?

10. Suppose a piece of paper is 0.004 inches thick. If you cut the paper in half and place the two halves together, the combined thickness is 0.008 inches. If you cut in half the stacked halves you get four pieces, which stack to a total thickness of 0.016 inches.

   a) How thick is the stack if you repeat this "cut-and-stack" process 20 times? 50 times?

   b) How many times would you need to cut and stack to produce a stack of papers that would reach from the floor to the ceiling of a typical house (8 feet)?

11. Chlorpheniramine maleate is the medical name for a drug found in some allergy medicines. A typical dose might be 25 mg each day of which only about 9 mg will make it into the bloodstream. Suppose the drug is eliminated from the body at a rate of 4.50% every hour.

   a) Write a closed-form equation to represent the quantity (in mg) of chlorpheniramine maleate in the body after t hours.

   b) What is the half-life of this drug?

   c) Use logs to determine when the drug level will drop to 6 mg.

   d) Complete a table like that in **Figure 6.35** to model the hourly level of the drug in the body for 24 hours. Explain the method you used to compute amounts.

   e) Graph the level over the 24-hour period.

Hour	Amount	Hour	Amount
0	9.00 mg	13	
1	8.58 mg	14	
...		...	
12		24	

Figure 6.35.
Level of chlorpheniramine maleate over a 24-hour period.

12. Model the growth of a guppy population in a pond. Assume that female guppies are able to produce one litter each month with 20 guppies in each litter surviving at birth. Assume that half of the surviving guppies are female. Begin the process with 1 male and 1 female.

a) Make a table or use a spreadsheet to model the growth of the guppy population in the pond assuming there are no deaths.

b) After 6 months, how many guppies are in the pond?

c) When will the guppy population reach a total of 10,000,000? (Assume no predators and sufficient food.)

13. a) Prepare a web diagram for $p_n = p_{n-1} + 8$; $p_0 = 20$.

b) Prepare a web diagram for $p_n = 1.5p_{n-1}$; $p_0 = 20$.

c) Describe how to use a web diagram to determine if a growth pattern is multiplicative or additive.

14. Decide whether the data in **Figure 6.36** approximate linear growth or exponential growth. Explain your choice. Produce an appropriate closed-form equation to fit the data and examine residuals.

15. Diseases often spread to nearest neighbors. Suppose a disease begins with a single home in the center of Gridville. The next day the disease is spread to the nearest neighbors of the one home. Five homes will have the disease. The next day the nearest neighbors of those five homes will become infected with the disease.

a) Draw a diagram and make a table or use a spreadsheet to model the spread of the disease through Gridville.

b) By the tenth day, how many houses will have been infected with the disease?

c) How long will it take until 2,000 homes are infected with the disease?

Year	California prison population
1980	24,569
1981	29,202
1982	34.640
1983	39,373
1984	43,328
1985	50,111
1986	59,484
1987	66,975
1988	76,171
1989	87,297
1990	97,309

Figure 6.36.
Prison Population in California from 1980 through 1990.
Source: California Department of Corrections.

BASE-TEN MONOPOLY

As you know, every exponential function is identified by its base, and each exponential function has a corresponding inverse function, a logarithm with the same base. But your calculator does not seem to have a log button for every base you can think of! This investigation looks at why that is so.

Recall that the base for the exponential expression 2^x is the number 2. The base of the logarithm function $\log_2(x)$ is 2 because this logarithm is the inverse for 2^x. The base for the logarithm function $\log_5(x)$ is 5 and $\log_5(x)$ is the inverse for 5^x. The base for the logarithm function $\log_{10}(x)$ is 10 and $\log_{10}(x)$ is the inverse of 10^x. The logarithm function in the calculator corresponds to base 10. In general, the base for the logarithm function $\log_b(x)$ is b and $\log_b(x)$ is the inverse of b^x.

Remember, $\log_b(c) = a$ means $b^a = c$.

You can use base-2 logarithms to solve the equation $32^x = 16$. Using powers of 2 and the above equation, $\log_2(32) = 5$, and $\log_2(16) = 4$, so $32^x = 16$ is equivalent to $(2^5)^x = 2^4$, so $2^{5x} = 2^4$ and $5x = 4$. Thus, $x = 0.8$.

You could also use base-10 logarithms to solve the equation $32^x = 16$. Replace 32 with $10^{1.5051}$ and replace 16 with $10^{1.2041}$. Solve the equation $1.5051x = 1.2041$ to get $x = 0.8$.

You can use base-2 logarithms or base-10 logarithms to solve the same equation. Can you use other bases? Is it easier to use base 10 or some other base? Will the answer be the same regardless of the base?

In this investigation you use tables and symbolic methods to compare base-10 logarithms to other bases. Begin by examining base-10 logarithms.

BASE-TEN MONOPOLY

1. a) Your calculator provides base-10 logarithms for all positive numbers. Use your calculator and complete a copy of **Figure 6.37**.

Number	Logarithm	Meaning
1	$\log_{10}(1) = 0$	$1 = 10^0$
2	$\log_{10}(2) = 0.3010$	$2 = 10^{0.3010}$
3		
...		
10		

Figure 6.37.
Table of sample values for base-10 logarithms.

b) Your calculator does not contain logarithms for base 2. Calculate their values yourself! Use graph and trace or guess and check to complete a table of sample base-2 logarithms as in **Figure 6.38**.

Number	Logarithm	Meaning
1	$\log_2(1) = 0$	$1 = 2^0$
2	$\log_2(2) = 1$	$2 = 2^1$
3	$\log_2(3) = 1.5850$	$3 = 2^{1.5850}$
4	$\log_2(4) = 2$	$4 = 2^2$
5		
6		
7		
8	$\log_{10}(8) = 3$	
9		
10		

Figure 6.38.
Table of sample base-2 logarithms.

BASE-TEN MONOPOLY

Why does the calculator not have logarithms for other numbers? Is there a faster way to find a logarithm to base 2 or base 3 or base 5 than by tracing a graph or using guess and check? Study the procedure in the following example.

Example: Find $\log_3(25)$.

Name the logarithm expression x: $x = \log_3(25)$.

Using the definition of logarithms, $x = \log_3(25)$ means $25 = 3^x$.

Use base-10 logarithms: $\log_{10}(25) = 1.3979$ and $\log_{10}(3) = 0.4771$.

Thus, $10^{1.3979} = 10^{0.4771x}$.

Solve the equation $1.3979 = 0.4771x$. (Remember, 1.3979 is the base-10 logarithm of 25 and 0.4771 is the base-10 logarithm of the base number 3.)

The solution to $1.3979 = 0.4771x$ is really just $\log(25)/\log(3)$, or about 2.93.

2. a) Use the example to find a general formula for finding the logarithm base b of any positive number. That is, find a way to calculate x if $\log_b(m) = x$.

   b) Use your formula from part (a) to complete a copy of **Figure 6.39**, a table of sample base-5 logarithms.

Number	Logarithm	Meaning
1	$\log_5(1) = 0$	$1 = 5^0$
2	$\log_5(2) = 0.4307$	$2 = 5^{0.4307}$
3		
...		
10		

Figure 6.39.
Table for sample base-5 logarithms.

   c) Explain why you need only one logarithm button on the calculator.

Perfect Ten

1. New life begins with a single cell. The cell divides to become two cells. The two cells divide to become four cells. Four cells become eight; eight become 16, and so on. Each time the cells divide the total number of cells doubles. The cells continue to divide and increase in number until a recognizable form appears. Cell division and the development of life forms is another context for the function $y = 2^x$.

   a) Complete a copy of the table in **Figure 6.40** for the function $y = 2^x$. Complete the table without using a calculator.

x	0	1	2	3	4	5
y						

Figure 6.40.
Table for $y = 2^x$.

   b) Use the values in your completed table to graph $y = 2^x$ by hand on graph paper in the window $[0, 6] \times [0, 40]$.

   c) Connect the points on your graph with a smooth curve to produce a continuous graph. Graph $y = 2^x$ on the graphing calculator and verify your table and your hand–drawn graph.

   d) What is the meaning of $2^{1.5}$ in the context of cell division? Does it make sense?

   e) Assume x represents time in hours and find the approximate time it takes for 1 cell to grow to 25 cells. Use your paper-and-pencil graph to solve the equation $25 = 2^x$. Check using logarithms.

For any base b, $\log_b(x) = \log_{10}(x)/\log_{10}(b)$. But some logarithms are easier to evaluate without a calculator than with one. For example, $\log_2(32) = 5$ since $2^5 = 32$.

2. a) Evaluate $\log_2(0.5)$ without using a calculator. Then use the formula above to check your answer.

   b) Find $\log_2(11)$.

3. The function $y = \log_{10}(x)$ is the inverse of the function $y = 10^x$. Therefore, the $\log_{10}(100) = 2$ because $100 = 10^2$. This relationship is generalized in **Figure 6.41**.

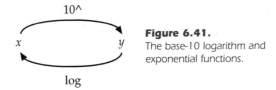

Figure 6.41.
The base-10 logarithm and exponential functions.

a) Use the definition of logarithm, without the calculator, to find: $\log_{10}(1,000)$; $\log_{10}(10,000)$; $\log_{10}(0.01)$.

b) If $\log_{10}(100) = 2$ and $\log_{10}(10) = 1$, what do you estimate is $\log_{10}(50)$?

c) Use the calculator to find: $\log_{10}(5)$; $\log_{10}(50)$; $\log_{10}(500)$; $\log_{10}(37)$; $\log_{10}(193)$.

d) Find the logarithms for several numbers between 10 and 100. What is similar about logarithms of numbers between 10 and 100?

e) What is similar about logarithms of numbers between 100 and 1,000?

4. Scientific notation uses exponents and base 10. Investigate the relationship between base-10 logarithms and scientific notation.

a) The number 67,400 is equivalent to 6.74×10^4 in scientific notation. Write 770, 38,300, and 582,496 using scientific notation.

b) Next, find the logarithms for 67,400, 770, 38,300, and 582,496. What similarities do you notice between scientific notation and the base-10 logarithms?

c) What is similar about the logarithms for 3.7, 37, 370, 3,700 and 37,000?

d) Find the logarithms for the following pairs of numbers. Use the calculator to find the logarithm for the first number in each pair. See if you can guess the logarithm for the second number. Then use the calculator to find the logarithm for the second number and verify your guess.

75 and 7,500;
8 and 80,000;
450 and 4,500;
9.25 and 925,000.

The laws of exponents apply to scientific notation and to logarithms, and are the reason that the logarithm for 6.74 is so similar to the logarithm for 67,400.

Recall the laws for exponents:

I. Multiplication: $(b^n)(b^m) = b^{n+m}$

II. Division: $b^n/b^m = b^{n-m}$

III. Power: $(b^n)^m = b^{nm}$

5. First, review these laws.

a) Simplify $(x^2)(x^5)$.

b) Simplify $(a^8)(a^3)$.

c) Write, in exponential form, $(10^5)(10^4)$.

d) Write, in exponential form, $(10^{0.4471})(10^3)$.

$\log_{10}(5.5) = 0.7404$ means the number 5.5 is equivalent to $10^{0.7404}$. The number 550, in scientific notation, is 5.50×10^2. Replace 5.50 with $10^{0.7404}$ and 5.50×10^2 becomes $10^{0.7404} \times 10^2$ That is, 550 is equivalent to $10^{2.7404}$ Therefore, the $\log_{10}(550)$ should be 2.7404.

Here are additional examples showing the relationship between scientific notation and base-10 logarithms.

$776 = 7.76 \times 10^2 = 10^{0.8899} \times 10^2 = 10^{0.8899 + 2} = 10^{2.8899}$

$54,000 = 5.4 \times 10^4 = 10^{0.7324} \times 10^4 = 10^{0.7324 + 4} = 10^{4.7324}$

$0.092 = 9.2 \times 10^{-2} = 10^{0.9638} \times 10^{-2} = 10^{0.9638 - 2} = 10^{-1.0362}$

6. a) Create three more examples of your own that demonstrate the relationship between base-10 logarithms and scientific notation.

b) Use the calculator to verify the logarithms for 776, 54,000, and 0.092 and for the examples you created in part (a).

7. Use a calculator to discover and verify certain rules for logarithms. Compare the rules for logarithms to the rules for exponents so you can answer the question, "What properties did logarithms inherit from exponentials?"

a) You know that $2 \times 3 = 6$. Find log(2), log(3), and log(6) and discuss the relationship among these three logarithms.

b) You know $5 \times 8 = 40$. Find $\log(5)$, $\log(8)$, and $\log(40)$ and discuss the relationship among these three logarithms.

c) You know $2^3 = 8$. Find $\log(2)$ and $\log(8)$ and discuss the relationship between these two logarithms.

d) You know $5^2 = 25$. Find $\log(5)$ and $\log(25)$. Discuss the relationship between these two logarithms.

e) You know $15/3 = 5$. Find $\log(15)$, $\log(3)$, and $\log(5)$ and discuss the relationship among these three logarithms.

f) Write an equivalent expression for $\log(ab)$.

g) Write an equivalent expression for $\log(a/b)$.

h) Write an equivalent expression for $\log(b^n)$.

i) What properties did logarithms inherit from exponentials?

j) Let $x = \log_{10}(a)$ which means $a = 10^x$.

Let $y = \log_{10}(b)$ which means $b = 10^y$.

Prove that $\log_{10}(a) + \log_{10}(b) = \log_{10}(ab)$.

8. Using logarithms sometimes allows you to reveal patterns not evident in a table of values.

a) Use a spreadsheet to make a four-column table for $y = x^2$. Use the first column for x-values and the second column for y-values. Put $\log(x)$ in the third column and $\log(x^2)$, which is the same as $\log(y)$, in the fourth. Then create the following three graphs: the time-series graph (y versus x), the graph of $\log(x^2)$ versus x (which is the same as $\log(y)$ versus x, and the graph of $\log(x^2)$ versus $\log(x)$ (which is the same as $\log(y)$ versus $\log(x)$).

b) What patterns do you notice when you examine the table and the graphs?

c) Explain part (b) using the laws you observed in Item 7.

LESSON FOUR
Sum Kind of Growth

KEY CONCEPTS

Arithmetic series

Geometric series

Quadratic sequences

The Image Bank

PREPARATION READING

Contract Settlements

Suppose two excellent companies offered you the same job doing the same type of work. One company offers you a beginning salary that is much higher than the other company. Do you take the job that pays more? Do you need more information?

Employment contracts can be complex. Incentives and raises complicate the decision. In some cases it may be financially better for you to accept the job offer that has a lower starting pay. Mathematics can help you decide which deal is the best for you.

In this unit you have been looking for a mathematical model to describe the level of a particular drug in the body. How is an employment contract similar to the level of a drug in the body? The blood level is a combination of the most recent dose and what remains from previous doses. The level is the result of accumulation. Each new dose adds to the amount already in the body.

Employment contracts may be compared based on accumulation. You may want to accept the job that pays the highest total over a certain period of time. The contract that pays the highest cumulative total may not be the contract with the highest beginning salary.

In this lesson you study sequences created by accumulating the values of arithmetic and geometric sequences. You look at patterns for functions that are known so you can recognize similar patterns in unknown situations. Could the serum level of phenobarbital in Eric's blood (Activity 1) be modeled by a function that is the accumulation of an arithmetic or geometric sequence?

CONSIDER:

1. Would it be better to have a salary raise based on a fixed percentage or a fixed amount? What conditions make one type of raise better than another?

2. A landfill is an example of accumulation. What is the remaining capacity in the landfill nearest you? How fast is it filling? How long until capacity is reached?

CUMULATIVE EARNINGS

Three excellent companies offer you the same job doing the same type of work.

Company A offers you a starting salary of $45,000 per year. The contract includes raises of $500 each year if you earn a satisfactory performance rating.

Company B offers you a starting salary of $40,000 per year with raises of $1,000 per year provided you earn a satisfactory performance rating.

Company C offers you a starting salary of $38,000 per year with raises of 4% guaranteed each year as long as your evaluations are satisfactory.

All other benefits are the same for Company A, Company B, and Company C.

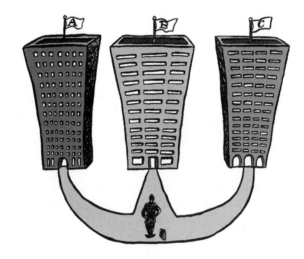

ACTIVITY

CUMULATIVE EARNINGS

8

1. Rank the offers in order from best to worst based on total earnings.

   Prepare to defend your ranking. Use mathematics to provide evidence supporting your decision. Remember to state your assumptions.

2. Suppose you worked for the same company for 40 years. Calculate the total amount of money you will have earned with Company A, Company B, and Company C after 40 years.

   Let the sequence $a_1, a_2, a_3, a_4, a_5, a_6, a_7, a_8, a_9, a_{10}$ represent the salaries for the first 10 years for Company A. The **sequence of partial sums**, $s_1, s_2, s_3, s_4, s_5, s_6, s_7, s_8, s_9, s_{10}$, is found in the following way:

   $s_1 = a_1$

   $s_2 = a_1 + a_2$

   $s_3 = a_1 + a_2 + a_3$

   . . .

   $s_{10} = a_1 + a_2 + a_3 + a_4 + a_5 + a_6 + a_7 + a_8 + a_9 + a_{10}$

   Thus, recursively, the next term, s_n, in a sequence of partial sums is found as:

   $s_n = s_{n-1} + a_n$,

   where a_n is the next term in the sequence that is being accumulated (added). Note that, though this relation is recursive, it directly involves three varying quantities, namely s_n, s_{n-1}, and a_n, not just s_n and s_{n-1}

3. How would the sequence of partial sums help you compare the offers from Company A, Company B and Company C?

4. Write recursive equations to represent the yearly salaries and the sequence of partial sums for each of the three contracts.

ACTIVITY

8

CUMULATIVE EARNINGS

5. Write the first ten terms of the sequence of partial sums for the contract with:

   a) Company A.

   b) Company B.

   c) Company C.

6. For each company, use your equations to determine the number of years it will take to reach the $1,000,000 club. (You reach the million dollar club when your total cumulative earnings surpass one million dollars.)

7. Examine each sequence of partial sums to determine whether an additive or multiplicative growth model describes it well. Defend your choice based on first differences, relative rate of growth, and patterns that appear in recursive graphs.

   (If needed, propose a new model and develop a checklist of characteristics for your new model.)

A sequence of partial sums is frequently called a **series**. When the sequence whose terms are being summed is arithmetic or geometric, then the sequence of partial sums is called an **arithmetic series** or a **geometric series**, respectively. Note that, since a series is formed by addition, an arithmetic series has first differences that form an arithmetic sequence. Likewise, a geometric series has first differences that form a geometric sequence.

INDIVIDUAL WORK 8

Pits

1. In 1992 the San Timoteo Landfill, which borders the cities of Redlands and Loma Linda in San Bernardino County, California, had room for 14.5 million cubic yards of compacted garbage. That year the surrounding communities contributed 583,000 cubic yards of garbage to the landfill. The following year the total added to the landfill was nearly 600,000 cubic yards. Each year the total amount added to the landfill increases by about 17,000 cubic yards per year. (Note: 17,000 is not the amount added to the landfill; it is the yearly increase in the amount added to the landfill.)

a) Write the sequence of partial sums for the first 10 years.

b) Develop a spreadsheet to complete **Figure 6.42**, representing the projected accumulation of garbage in the San Timoteo Canyon landfill.

Year ($n = 0$ for 1992)	New garbage	Current cumulative garbage (partial sums)	Previous cumulative garbage	First differences (amount of growth) of cumulative garbage	Relative rate of growth of cumulative garbage
0	583,000	583,000			
1	600,000	1,183,000	583,000	600,000	1.02916
2					
10					

Figure 6.42.
Table for landfill.

c) Describe any patterns you notice in the completed table.

d) Use your spreadsheet, or paper and pencil, to prepare four graphs:

- Cumulative Sums versus Years (column 3 versus column 1). This is the time-series graph of the accumulated garbage.

- Amount of Growth (first differences of accumulation) versus Previous Accumulation (column 5 versus column 4).

- Relative Rate of Growth of Accumulation versus Previous Accumulation (column 6 versus column 4).

• Cumulative Total versus Previous Cumulative Total (column 3 versus column 4). This is the recursive graph of the Cumulative sequence.

e) When (after how many years) will the landfill reach capacity?

f) Suppose that, instead of a 17,000 cubic yard increase in each year's garage, the San Timoteo Canyon Landfill projects the amount of garbage added each year to be 1.5% more than the previous year. Under this new assumption, when would the landfill reach capacity if 583,000 cubic yards are added the first year, and that amount increases by 1.5% each year?

g) Suppose 583,000 cubic yards are added to the landfill the first year and, due to recycling, that amount is reduced by 1.5% each year. How many years will this recycling program add to the "life" of the landfill as compared to your result in part (f)?

Landfill Heats School

Pattonville High School in Maryland Heights, Missouri, is the first public school in the United States to be heated by methane gas from decomposing landfill refuse. The gas, provided at no expense by the corporation that owns the landfill, arrives at the school for use in its hot water boilers by way of a 3,600-foot pipeline. In addition to saving the school about $40,000 per year in heating costs, the burning of methane gas improves air quality by eliminating potent greenhouse gas emissions.

Source: Pattonville School District

2. You have been working for the same company for five years and you are now earning $50,000 per year. You like your job, but you are unhappy with the 4% raises you have been getting each year. Another company offers you the same job with a salary of $55,000 and agrees to raise your salary by $3,000 each year. If money is the only factor, should you remain at your present job or accept the offer from the other company? Give reasons for your answer and supply mathematical evidence for your reasons.

3. Chess is a very old game, originating in India sometime around the seventh century. The game appears in many stories and legends. The following is a legend about its inventor.

When the king of India learned that his favorite game had been invented by one of his subjects, a man named Sessa, he summoned the inventor to appear before him.

When Sessa was brought before the king, the king greeted him and said, "I wish to reward you for your marvelous invention. Name your reward and you shall have it."

Sessa replied, "O, great king, I am but a poor teacher. To know that you find favor with my game is reward enough for me."

But the king was not pleased with this response. He wished to demonstrate his vast riches by granting whatever reward Sessa requested. He commanded Sessa to think of a proper reward.

After a moment's thought, Sessa made this request: "Sire, if it pleases you, I would like to have one grain of rice for the first square on the chessboard."

The king was astonished and not overly pleased. Was Sessa mocking him?

"And," Sessa went on, "two grains of rice for the second square, four for the third, eight for the fourth, sixteen for the fifth, thirty-two for the sixth . . ."

"Enough!" cried the king. "You shall have your grains of rice for all sixty-four squares of the chessboard, each doubling the amount of the square before.

But know that in naming such a small reward you have shown disrespect for me and my most generous offer. Go now! My servants shall bring you your sack of rice!"

a) How much rice did Sessa receive for only the 64th chessboard square?

b) What was the total amount of Sessa's reward?

Use the next questions to approximate a modern-day value for Sessa's reward:

There are approximately 730 gains of rice in one tablespoon. The price of rice varies on the commodities market. One example of a selling price for rice is between $9.77 and $10.37 per 100 pounds on the commodities market.

c) There are 3 tablespoons in 1/4 cup. If you use the old rule of thumb,"A pint's a pound the world around," you can estimate that 2 cups of rice weigh about 2 pounds. Approximately how many pounds of rice should Sessa receive?

d) If Sessa could sell his rice at today's prices (assume $10.00 per 100 pounds), how much would his reward be worth in dollars? Did he strike a good deal with the king by today's standards?

e) Two pounds of rice requires about 0.035 cubic feet of storage. If the inventor wanted to construct a storage facility to store his reward, how large would that facility have to be?

4. Foam padding and carpeting come in rolls. Suppose you want to estimate the length of a long piece of foam padding. It is an enormous amount of work to unroll the padding and measure the length with a tape measure. It would be more efficient to determine a mathematical relationship between the length of the padding and the number of layers in the roll. Since each layer is longer than the previous layer, the total length of the roll is an accumulation of increasing lengths.

Suppose a particular grade of padding compacts to a thickness of 0.25 inches when rolled and that the diameter of the cardboard "spool" in the middle of the roll of padding is 6 inches (see **Figure 6.43**).

a) Complete a copy of **Figure 6.44** and use it to launch your investigation.

Remember, the circumference, C, of a circle is $C = \pi d$, where d is the diameter. For example, when the spool has only one layer of padding wrapped around it, that layer is only 6π inches long, or about 19 inches long.

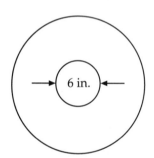

Figure 6.43.
Center "spool" is 6 inches in diameter.

Figure 6.44.
Table for total length of padding.

Total number of layers on spool	Diameter of outer layer	Length of outer layer (only)	Total length of roll
1	6.0	6π	6π
2	6.5		
...			
16			

b) Graph column 3 (length of outer layer) versus column 1 (number of layers), and graph column 4 (total length of roll) versus column 1.

c) What kind of sequence is column 3? How do you know?

d) Does the sequence of values in the fourth column represent arithmetic growth, geometric growth, or neither? Justify your answer. If it is either arithmetic or geometric, find the exact value of the corresponding additive constant or growth factor.

In Item 4, the sequence of Total Length is an arithmetic series since it is the accumulation of an arithmetic sequence. Item 5 explores one method for writing a closed-form expression for its values.

5. a) Suppose the roll has a total of 12 layers. Add the length of the first (inner) layer to the length of the twelfth (outer) layer. Add the length of the second layer to the eleventh layer. Add the length of the third layer to the length of the tenth layer. Continue this process and find the total length of padding.

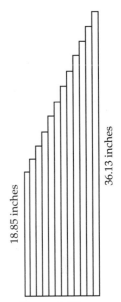

Figure 6.45.
Twelve strips showing the lengths of layers in the roll of padding.

b) The first layer is about 18.85 inches long and the second layer is about 20.42 inches long. Suppose you cut a strip of the padding so each strip is the length of the layer from which it is cut. For simplicity, think of the strips as being 1 unit wide. Place each of the 12 strips side by side as in **Figure 6.45**.

Stack a copy of Figure 6.45 (in reverse order) end to end with the original figure (see **Figure 6.46**). Describe how to use the resulting figure to find the total length of the roll of carpet padding.

c) Generalize the method of part (b) to write an equation to represent the total length of the roll as the accumulation of the layers. Use *n* to denote the number of layers (not just 12).

d) Use your equation to determine the length of the padding on the entire roll if the diameter of the outer roll is 32 inches.

6. Using the specific details describing the particular rolls, you showed in Item 5 that the arithmetic series in that problem had a quadratic description. The steps below outline how you can use the method of generalization to see that every sequence of partial sums based on an additive sequence is quadratic. That is, every arithmetic series has a quadratic closed-form description. You may wish to use Figure 6.46 as a guide to your thinking.

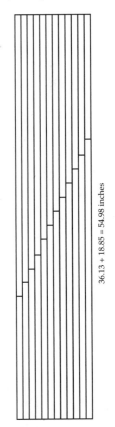

Figure 6.46.
Two copies of 12 strips stacked to form a rectangular shape.

Use the following variables in your work:

a_1 represents the first term of the arithmetic sequence.

a_n represents the nth term (the general term) of the arithmetic sequence.

d represents the additive amount.

s_n represents the sum of n terms of the arithmetic sequence.

a) The recursive equation $a_n = a_{n-1} + d$ represents the nth term of an arithmetic sequence. Write a closed-form equation to represent the nth term of an arithmetic sequence.

b) In generalizing Figure 6.46, how many strips "wide" is the figure? How "tall" is the figure? How many copies of s_n does the figure represent? Explain.

c) Use the generalized Figure 6.46 to write a formula for s_n. Note that your answer will involve n, d, and a_1.

d) Use your equation from part (c) to write a quadratic equation representing the sum of n terms for the arithmetic sequence in Item 4: $6\pi, 6.5\pi, 7\pi, 7.5\pi, 8\pi, \ldots$. How does it compare with your results from Item 5?

7. Marie lives in a city near a large vacant lot. Her neighbors formed a neighborhood group and approached the city council asking the city to clean and prepare the land on the lot so that it could be used to plant gardens.

The city council approved the project and provided the funds to prepare the land. Knowing the project would take a while to get established, the garden commission decided to start with a few individual garden plots and gradually increase the number. In this way, they would learn as the project developed and be able to try solutions before any problems became too large. The garden commission started with one square-shaped plot in the center of the empty lot. In this plot they placed both a water source with several long hoses and storage bins for the community gardeners to use. This center square-shaped garden plot is labeled "W" in **Figure 6.47**.

Figure 6.47.
Garden plot with water source and storage containers.

W $\|$ 5 yd.

Large-garden-square #0

The garden commission planned to form a ring of individual square-shaped garden plots around the center water-source plot (see **Figure 6.48**). Each individual square-shaped garden plot (including half of the footpath around it) would measure 5 yards on each side. These eight individual garden plots would be rented first.

Individual garden plot

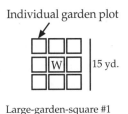

Large-garden-square #1

Figure 6.48.
Eight garden plots.

The second ring (Phase 2) would have 16 square-shaped individual garden plots and would be rented in the second phase (see **Figure 6.49**).

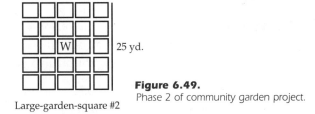

Large-garden-square #2

Figure 6.49.
Phase 2 of community garden project.

There are many different patterns of growth in the community garden situation. For example, you can investigate the sequence generated by the length of the side of the entire project as new phases (rings) are added, the number of individual garden plots in each new ring, the total number of individual garden plots in the entire project, or another pattern you observe.

a) Create a mathematical model (equation) of the length of a side of the outer ring in terms of the number of rings (phases).

b) Create a mathematical model (equation) for the number of new plots in a phase in terms of the phase number.

c) Create a mathematical model (equation) for the total number of plots in terms of the number of phases.

The sequence of partial sums for an arithmetic sequence is quadratic. Can the accumulation of a drug in the body be represented by a quadratic function? In Activity 1 you were challenged to determine the proper dose of phenobarbital for Eric. In Lesson 2 you determined that the sequence of blood level readings was neither arithmetic nor geometric. Here you will identify another characteristic of quadratic sequences so you can recognize the pattern in unknown situations. Perhaps then you can decide if Eric's blood serum level should be represented by a quadratic function.

8. The sequence of partial sums for the roll of carpet padding that you investigated in Item 4 is a quadratic sequence.

a) Return to the table you made in that investigation and add a column for **second differences**, the difference between the consecutive terms of a sequence of first differences. In that column write the first differences of the column of first differences. For example, the first entry will be 0.5π, representing the difference between 6.5π and 6.0π.

b) Describe the patterns in the First Differences and Second Differences columns in the table. Explain why these make sense.

9. In a previous course, you may have solved the quadratic equation $2 = 1 + 3p - p^2$. Show that the function $y = 1 + 3x - x^2$ fits the patterns for quadratic growth that you found in Item 8.

a) Complete a copy of **Figure 6.50** using integers for x. (Note the function is continuous. In order to form a quadratic sequence of values, the values for x must be uniformly spaced.)

x	Current y	Previous y-value	First differences	Second differences
1	3			
2	3	3	0	
3	1	3	−2	−2
4	−3	1	−4	−2
...				
10				

Figure 6.50.
Table for $y = 1 + 3x - x^2$.

b) Describe characteristics of a quadratic sequence that you notice in the table.

c) Prepare a time-series graph for $y = 1 + 3x - x^2$. Describe the shape of the graph.

10. In Items 8 and 9 you identified characteristics of particular quadratic sequences. How will you recognize growth patterns that are linear (produced by an additive process), exponential (produced by a multiplicative process), and quadratic (produced by summing the terms of an additive sequence)? In this item you prepare a checklist by summarizing your work from this and previous lessons. Analyze the following functions, using $x = 0, 1, 2, \ldots$.

$y = 2x + 5$ Linear Growth

$y = 100(1.50)^x$ Exponential Growth

$y = x^2 + 3x + 2$ Quadratic Growth

a) Write recursive equations to represent each function if you can. If you are unable to write a recursive equation for a particular function, explain any difficulties you encounter.

b) Write the first ten terms of each sequence and complete a table like that shown in **Figure 6.51**. Prepare a separate table for each function. You may want to use a spreadsheet to prepare the tables. You also may wish to add columns for other quantities that you wish to investigate further.

Term x	Current amount y	Previous amount	First differences	Second differences	Relative rate of growth
1					
2					
...					
10					

Figure 6.51.
Sample table for Item 10.

c) Prepare four graphs for each function. Graph (1) Current amount versus Term Number, (2) Amount of Growth (first differences) versus Previous Amount, (3) Relative Rate of Growth versus Previous Amount, and (4) Current Amount versus Previous Amount.

d) Examine first differences for each function. What patterns, if any, distinguish linear growth from exponential growth from quadratic growth? Explain your answer.

e) Examine second differences for each function. What patterns, if any, distinguish linear growth from exponential growth from quadratic growth? Explain your answer.

f) Examine relative rate of growth for each function. What patterns, if any, distinguish linear growth from exponential growth from quadratic growth? Explain your answer.

g) Examine the recursive graphs for each function. What patterns, if any, distinguish linear growth from exponential growth from quadratic growth? Explain your answer. For any graphs that appear linear, check that their slopes from point to point really are constant (do the arithmetic, without rounding any computations).

h) Decide whether the following "mystery" sequence of numbers is an example of linear growth, exponential growth, quadratic growth, or a type of growth you have not studied. Supply reasons for your answer.

Mystery sequence: 2, 5, 9, 14, 20, 27, 35.

11. Suppose you open a savings account with $100. The next month you add $150 to your account. The following month you add $200, then $250, then $300, and you continue the pattern of growth for m months. Check that the equation $s(m) = 25m^2 + 75m$ represents the total amount of money in your account after m months (assume that no interest is being added to your account). How long will it take to accumulate $5,000? Write and solve an appropriate quadratic equation.

12. A projectile is launched vertically with an initial velocity of 64 feet/second. The height of the projectile after t seconds can be approximated using the quadratic equation $h(t) = -16t^2 + 64t$. When does the object reach a height of 50 feet? Write and solve an appropriate quadratic equation.

13. The **additive identity** is 0. Add 0 to a number and the number is unchanged. The **multiplicative identity** is 1. Multiply any number by 1 and the number is unchanged.

    a) Describe an arithmetic sequence in which the number added is the additive identity.

    b) Describe a geometric sequence in which the growth factor is 1.

14. You have examined accumulations, both for arithmetic and geometric series. The sequence of partial sums for arithmetic series turned out to be quadratic. Geometric series don't seem to be quite so easily characterized. Compare the recursive graphs you obtained in Activity 4 of Lesson 2 to those you generated in this lesson's examinations of geometric series. What conclusions do you reach?

LESSON FIVE
Mixed Growth

KEY CONCEPTS

Mixed sequences

Web diagrams

Geometric series

Hulton-Deutsch Collection/Corbis

PREPARATION READING

Enough is Enough

Y ou're hired! You just landed your first job. What will you do with the money you'll be making? Perhaps you want to buy your own car or put away money for college. Suppose you decide to save money from each paycheck and put the money in a savings account that pays you interest. How much should you deposit each month so you will have enough money two or three years from now?

Investing money is one example of **mixed growth**, the combination of additive and multiplicative growth. Your savings account balance changes in two distinct ways each month. First, you make deposits in a constant amount each month; that's additive growth. In addition, you earn interest each month as a fixed percentage of the money in your account; that's multiplicative growth.

In Lesson 4 you investigated properties of series, both arithmetic and geometric. You may have noticed that the sequence of partial sums for any geometric series always has a linear recursive graph that misses (0, 0). That means its recursive equation can be written in the form $s_n = b \times s_{n-1} + a$, for some constants a and b. That's mixed growth.

Your analysis in Activity 4 (Lesson 2) of recursive representations of the blood level of phenobarbital in Eric's data indicates that such levels are also described by mixed-growth sequences. Other drug-dosing problems have suggested that appropriate processes are a percentage decay (multiplicative) and repeated doses (additive). Lesson 4's examination of accumulations shows that an arithmetic series (quadratic growth) is *not* equivalent to a mixed-growth sequence since its recursive graph is not actually linear.

In this lesson you examine mixed-growth sequences more carefully, determining identifying properties and the mathematical behaviors such sequences can produce. Study known situations, such as savings accounts, so you can recognize patterns that appear in unknown situations.

ACTIVITY

9

SAVINGS SITUATION

1. Suppose your uncle gives you $300 toward the purchase of a new car with the condition you put the money into a savings account on the first of the month. In addition, you agree to deposit $200 each month from your paychecks, with the first deposit to be made one month after the account is opened. The bank pays 6% annual interest (0.5% each month), compounded monthly on the last day of the month, on your balance for that month.

   a) Write a recursive equation to represent your balance at the beginning of each month (just after you make your $200 deposit).

   b) How long will it take for your account balance to reach $10,000? Does your answer make sense?

The recursive form of an arithmetic sequence is: $a_n = a_{n-1} + d$. The arithmetic sequence is produced by adding a constant amount d each month or interval.

The recursive form of a geometric sequence is: $a_n = b \times a_{n-1}$. The geometric sequence is produced by multiplying by a constant factor b each month or interval.

In the savings account problem you are multiplying by a constant factor each month (the growth factor is 1.005, representing the addition of the interest to the principal) and you are adding a constant amount each month (your $200 deposit). This is an example of a mixed sequence.

The recursive form of a mixed sequence is: $a_n = b \times a_{n-1} + d$.

2. Acetaminophen, the main ingredient in Tylenol®, is used by young and old to relieve aches and reduce fevers. Too much acetaminophen can be toxic and too little will be ineffective.

   The typical dose for an adult is 325–650 mg every 4 hours. Total amount per day should not exceed 4 grams. The half-life of the drug is 2.75 to 3.25 hours, which means about 21% of the drug is eliminated from the body each hour.

SAVINGS SITUATION

For a 70 kg person, if the amount in the body remains above 13,300 mg for 4 hours following an acute overdose, the patient may suffer liver damage.

a) Write a recursive equation for a 70 kg adult taking 500 mg doses every 4 hours. Note that the 21% rate is hourly and must be converted to a rate for 4-hour periods. Base your equation on the amount in the body immediately after each dose, and assume that all of each dose immediately enters the body. Thus, $a_0 = 500$ and a_1 represents the amount after 4 hours of decay and the taking of the next dose.

b) Use your equation to follow the peak concentration of acetaminophen over a period of three days. Describe any patterns you see.

3. Repeat Item 2, this time modeling the amount in the body just *before* taking each dose. Thus, $a_0 = 0$ and a_1 represents the amount after taking the first dose and then waiting for the 4 hours of decay.

4. The recursive equation $p_n = 1.5 \times p_{n-1} + 10$ is another example of a mixed-growth sequence. The growth factor is 1.5 and the additive amount is 10.

What are the characteristics of a mixed-growth model? How will you recognize that an unknown sequence of values represents a mixed-growth model? Prepare a table and several graphs for this mixed-growth sequence. Examine first differences, second differences, and relative rates of growth. Check ratios of successive terms. Check ratios of successive first differences. Look for other "check-able" patterns. Create and examine time-series and recursive graphs and web diagrams. Identify characteristics you suspect might identify mixed-growth sequences. Check your conjectures on the sequences from Items 1–3, above.

Prepare a presentation summarizing your conclusions.

INDIVIDUAL WORK 9

Steady States

1. The sequence below is a mixed sequence. Analyze first differences, second differences, growth factors, and the time-series and recursive graphs for the sequence. How can you determine whether a sequence of numbers represents a mixed-growth pattern? Verify the characteristics you conjectured in Activity 9.

   100, 160, 250, 385, 587.50, 891.25.

2. Hank likes to take his son to fish at the trout farm every Tuesday during the summer. Hank picks Tuesdays because he knows they re-stock the lake every Monday, so Tuesday is the easiest day to catch fish.

   Annisa operates the trout farm. She stocks the lake every Monday with 1,200 mature trout in order to keep the number of fish essentially constant from week to week. She estimates the fish are caught at a rate of 2% per day.

   a) How many fish are in the lake at the trout farm right after Annisa re-stocks the lake? Check your answer. Make sure that removing 2% of each day's stock results in removing a total of 1,200 fish for the week.

   b) Why isn't 2% each day the same as 14% per week?

3. Suppose that Annisa (see Item 2) did not know that 1,200 fish would keep her trout lake stocked at about the same level each week. In fact, suppose that she opens the lake in her first week of business with only 5,000 trout in the lake. (*You* know there are 5,000; she doesn't.) Assume that she still uses the "add 1,200 each week" plan under these new conditions, and that fishers still are able to catch about 2% of each day's available fish.

   a) Without doing any calculations, describe how you think this new set-up will behave.

   b) Write a recursive equation describing this system immediately after each re-stocking.

   c) Verify the conclusions of Activity 9 for this new mixed-growth situation.

d) Draw a web diagram representation of your recursive equation. Does it help you evaluate the accuracy of your prediction in part (a)? Does it matter (in the long run) that Annisa knew (in Item 2) how many fish were in the lake at the start of the week? Explain.

4. Phoenicia owes $5,000 on her credit card. She is charged 1.75% interest on the remaining balance every month, which means 1.75% of the amount she owes is added to the amount she owes at the end of each month. The minimum monthly payment is $200. Suppose that she makes no new purchases with her credit card and pays $200 on the first day of every month.

a) Write a recursive equation describing her balance just after she makes each payment.

b) How long will it take her to pay off her credit card debt? Show your reasoning.

c) How much will she pay, total, during that time?

d) Prepare a time-series graph to represent the remaining amount still owed over time. Describe the general shape of the graph.

e) Create a web diagram based on your recursive equation. (Note: you may wish to alter your scales in order to "see what's going on.") How does it verify your description in part (d)?

f) Verify the conclusions of Activity 9 for this new mixed-growth situation.

5. Charles deposits $400 on the first day of every month into a savings account that pays him 0.5% interest on the last day of each month. His balance today (just after his deposit) is $2,400.

a) Write a recursive equation for the balance in his account just after each deposit.

b) Write a recursive equation for the balance in his account just before each deposit.

c) How much will he have one year from today (after his deposit)?

d) How long will it take for his total to reach $10,000?

INDIVIDUAL WORK 9

e) Prepare a time-series graph for the two processes, one based on part (a) and the other based on part (b). Describe the two graphs.

f) Create a web diagram for your equation in part (a). (Alter scales as needed.) Describe it.

g) Verify the conclusions of Activity 9 for this new mixed-growth situation.

6. Consider the following "thought experiment." Suppose that a particular bucket has holes in it, so water always leaks out. Suppose also that you refill the bucket at a constant rate of 2 pints per minute. Thus, water is leaking out and water is being poured in, both at the same time.

a) Explain how it could be possible that this process could result in the bucket's filling up and overflowing.

b) Describe conditions under which the bucket would become empty through this same process.

c) Describe conditions under which the level of water in the bucket would not change, even while the leaks and refilling are going on.

The condition you described in Item 6(c) is known as a dynamic equilibrium. The word "dynamic" indicates that activity is taking place, and the word "**equilibrium**" means that the net effect is "no change," a kind of "balance" between opposing actions. An equilibrium is sometimes called a fixed point in the context of sequences. You have seen such behavior in other situations. For example, Annisa's trout farm maintains a kind of dynamic equilibrium between the fishing and re-stocking activities, resulting in essentially constant weekly inventories of around 9,100 fish.

7. Perhaps you know someone who has a heart condition. In critical cases a person with a heart problem may be admitted to a hospital and given intravenous (IV) injections of a cardiac drug. A doctor determines the correct dosage. If the dose is too low the person may experience a crisis. If the dose is too high, it could cause other problems.

For a particular heart medication, amrinone lactate, the minimum effective level for most patients is the equivalent of about 150 mg. You want the level to stabilize above 150 mg. The maximum safe level is about 400 mg. According to the *Handbook of Clinical Drug Data*, the drug is eliminated from the body at a rate of 16% per hour for the typical person.

a) Write a recursive equation to represent the level of the heart medication in the person immediately after each dose enters the system. Use any dose size you like for this part. Assume doses are taken hourly.

b) Create a web diagram for the equation you wrote in part (a). Use it to describe the long-term behavior of the amount of this drug in the body.

c) Write a recursive equation to represent the level of heart medication in the person immediately before each dose enters the system. Use the same dose size you used in part (a). Assume doses are taken hourly.

d) Adjust the dose size in parts (a) and (b) so that the level in the body reaches and remains above 150 mg (without exceeding 400 mg) after 8 hours.

e) Verify the characteristics of mixed growth that you described in Activity 9.

If the same dose is taken at regular intervals, the peak level of drug in the body reaches an equilibrium. Equilibrium is achieved when the new dose exactly replaces the drug eliminated by the body between doses. Mathematically, the equilibrium level is p_n and occurs when $p_n = p_{n-1}$. You can find this value algebraically, without graphing, just by substituting p_n for p_{n-1} in your recursive equation and solving for p_n.

f) What is the equilibrium level for the "after dose" amounts described in part (a)?

g) What is the equilibrium level for the "before dose" amounts described in part (c)? How is your answer related to that for part (d)? Is that reasonable?

Equilibrium is achieved when the dose of drug exactly replenishes the amount the body eliminates after the previous dose. A savings account is in equilibrium when, once a month, you withdraw an amount equal to the amount of interest you earned. Populations reach equilibria when the number of births and immigrants is equal to the number of deaths and emigrants.

INDIVIDUAL WORK 9

Mathematically, equilibrium is described by $a_n = a_{n-1}$. Geometrically, this is visible in a web diagram as the intersection of the recursive line and the line $y = x$. For example, if $a_n = 0.8681a_{n-1} + 1{,}200$, as in Item 2 above, then the set-up for the web diagram is as shown in **Figure 6.52**. The window is $[4{,}500, 10{,}000] \times [4{,}500, 10{,}000]$.

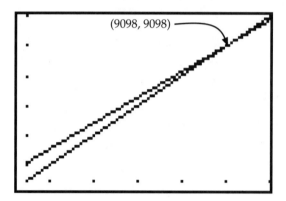

(9098, 9098)

Figure 6.52.
Lines defining web diagram for $a_n = 0.8681a_{n-1} + 1{,}200$.

8. a) Find the equilibrium level for a drug modeled by the equation:
$a_n = 0.35a_{n-1} + 60$.

　　b) Suppose your savings account pays interest of 0.75% each month. Your living expenses require you to withdraw $1,500 each month to pay your bills. What amount of principal would lead to equilibrium?

9. a) The maintenance dose for a prescription drug is a quantity that produces a steady state or equilibrium at a desired level. That is, the maintenance dose is a quantity d that keeps $a_n = a_{n-1}$ at the desired level, where $a_n = b \times a_{n-1} + d$. Find a symbolic expression, in terms of d and b, to represent the steady-state quantity a_n when the sequence reaches equilibrium for the given values of d and b.

　　b) Nothing about your work in part (a) depends on the fact that the equation represented drug dosing. Thus, its conclusions apply to any situation that can be modeled by a mixed-growth sequence. Interpret your result from part (a) in terms of web diagrams.

For drug dosing, there are always two recursive representations. One describes blood levels just after a dose is taken; the other describes levels just before doses. The equation given in Item 9(a) describes the "just after dose" levels, and the multiplicative constant is less than 1, representing elimination of the medicine from the body since the previous dose. The two control numbers available to medical personnel are b and d. The value of d is just the size of each dose. The value of b is related to the half-life of the drug (which can not be altered) and the time between doses (which the doctor can change).

10. Return to the dosing situation described in Item 7. Find a dosing scheme that results in equilibrium levels of 150 mg for the "before dose" level and 400 mg for the "after dose" level. Use the elimination information given in Item 7. Explain your method.

In Activity 9 and the items above you have examined and verified characteristics of mixed growth. Mixed-growth sequences have linear recursive graphs (p_n versus p_{n-1}) that do not contain (0, 0), so their web diagrams are built upon lines that intersect at some point other than (0, 0). In addition, the ratio of successive first differences is constant, and that constant is the multiplicative constant for the recursive equation. That same constant is the slope of the recursive graph.

The slope of the recursive graph (p_n versus p_{n-1}) is equal to the multiplicative constant in the recursive equation. This is most evident from the construction of a web diagram; the recursive equation is used to draw the starting figure, a line with slope defined by the multiplicative constant. The "web" lines that you add to indicate the generation of successive values of the sequence make the "change in y" and the "change in x" amounts clearly visible from point to point along the recursive line.

11. a) The slope of any line is defined as the "change in y" divided by the "change in x" from one point to another on the line. Suppose one point on the recursive graph of a mixed sequence is (p_{n-1}, p_n). What are the coordinates of the "next" point?

   b) Write expressions for the "change in y" and the "change in x" for these two points, then use those expressions to write a formula for the slope of the recursive graph.

   c) Now write the sequence of first differences for terms beginning with p_{n-1}. Calculate the ratio of successive first differences and compare to the results of part (b).

d) Parts (b) and (c) show why the ratio of successive first differences for a mixed-growth sequence is constant and why that ratio is the multiplicative constant. Illustrate these ideas on a web diagram.

12. Suppose there are 115 of a particular animal species in a wildlife preserve and the population is growing naturally at a rate of 12% each year. In order to manage the growth, wildlife biologists in the preserve decide to remove 15 of the animals at the end of each year and relocate them to an area better-suited for large numbers of them.

a) Predict the number of these animals in the preserve in 10 years; in 20 years. Determine the long-term effects of this policy.

b) Use the results of Item 9 to find the equilibrium value for this management scheme. Show the first few terms of the population sequence if the initial population is equal to the equilibrium value and this policy is maintained.

13. Both Items 2 and 12 of this Individual Work deal with managing "wildlife" of some sort. For the trout farm, there is a "natural" decline due to fishing, and the manager adds fish to replenish the loss. For the animal preserve there is a natural increase, and the manager removes animals to balance that gain.

a) Find the equilibrium level for Annisa's trout farm management policy.

b) Create web diagrams for Annisa's trout farm policy and for the animal preserve policy in Item 12(b). Use your diagrams to discuss the sensitivity of these two systems to fluctuations in the "uncontrolled" part of the process (the 2% fishing rate and the 12% growth rate). That is, if the management's estimates for these rates are slightly off, what is the effect if the policy is still used?

14. Suppose a professional ball player signs a new contract for $5 million per year to be paid at the end of each year for the next 5 years. Assume that the team's payroll account earns 0.5% per month, that this is the first of the year, and that the first payment to the player is not due until the end of the year.

a) How much money would the team need to set aside now in order to end up with exactly as much in the account at the end of the contract as when it starts?

b) How much money would the team need to set aside in the bank now in order to meet that contract, with the account balance being $0 after the final (fifth) payment?

15. Phenobarbital is one drug that is prescribed to patients who experience seizures. It raises the seizure threshold for the person. A typical patient may take regular doses of phenobarbital twice a day. The person will have his blood tested on a regular basis to make sure the level of phenobarbital is above the minimum effective level and safely below the maximum, or toxic, level.

Recall from Activity 1 that for phenobarbital, the minimum effective level is about 15 mcg/ml, which is approximately equivalent to 600 mg of phenobarbital in the body. The level must stabilize above this value. The maximum safe level is about 40 mcg/ml, which is roughly equivalent to 1,500 mg of phenobarbital in the body. (Note: the ratio of amount in the body (mg) to the concentration level in the blood (mcg/ml) depends both on the particular drug being administered and on the weight of the patient.) The bioavailability of the drug is 83%, which means if 100 mg are taken, then only 83% of the dose, or 83 mg, will be absorbed into the body.

   a) The drug is eliminated from the body with a half-life of 72 hours. Doses are typically given twice daily. Find the appropriate multiplicative constant for representing this decay process on a dose-by-dose basis.

   b) Determine an appropriate twice-daily dose so that a 70 kg patient reaches and stays safely above the 600 mg (15 mcg/ml) within 12 days. Remember that for each 100 mg taken, only 83 mg enter the system.

   c) Find the approximate equilibrium level achieved just after each new dose of 100 mg.

   d) What is the equilibrium level for the "before dose" amount?

16. (Just for fun.) The sequence with recursive equation $a_n = 5 \times a_{n-1} + 2$, $a_0 = 3$ begins like this: 17, 87, 437, 2,187, ... Each term ends in a 7. Make up a mixed sequence of your own that has all sequence values ending with the same last digit.

ACTIVITY

THE REST OF THE STORY

10

Mixed-growth sequences describe many different kinds of situations. Levels of medication in the body, amounts of money in investment accounts, balances owed on loans, and populations all are subject to these descriptions. The more you know about this kind of growth, the better you will be able to model situations and solve real problems you encounter.

You have learned some of the characteristics of this kind of growth. You know the form of the defining recursive equation. You know that mixed-growth sequences have recursive graphs that are lines that miss (0, 0). You know that the ratio of first differences is constant. In short, you are well equipped to identify data that fit the mixed-growth model, even when you do not know their equation. And the recursive equation allows spreadsheets or calculators to produce good graphs and predictions, at least for reasonable numbers of terms.

Can closed-form equations be developed to assist in predictions? Can the relationship between mixed-growth sequences and more familiar sequences be made more clear? Is there a simple description of the time-series graphs of mixed-growth processes? These questions are at the heart of this investigation.

1. A geometric sequence is a special mixed sequence, one with an additive growth constant of 0. That is, the mixed-growth sequence $a_n = b \times a_{n-1} + d$ is geometric when $d = 0$. What conditions are necessary for a mixed sequence to follow the pattern of an arithmetic sequence?

2. Sequence #1: $a_n = 0.4 \times a_{n-1} + 6$; $a_0 = 6$ is a mixed sequence.
   Sequence #2: $a_n = 0.4 \times a_{n-1} + 10$; $a_0 = 6$ is a mixed sequence.
   Sequence #3: $a_n = 0.4 \times a_{n-1} + 12$; $a_0 = 6$ is a mixed sequence.

   a) These three sequences are members of the same family of mixed sequences. Graph the time-series graphs for Sequences #1, #2, and #3, and describe the differences in their time-series graphs. Then describe their similarities.

ACTIVITY

THE REST OF THE STORY

10

b) Find a fourth member of the family for which $a_0 = 6$ is its equilibrium value.

c) Use web diagrams to explain why the sequences in this family are all so similar.

3. Use the recursive equation from Sequence #1 of Item 2. Create the time-series graph and web diagram for this equation if the starting term is changed to $a_0 = 14$. Compare the result to your answers for $a_0 = 6$ in Item 2. Describe the similarities of the two sequences.

4. Consider the recursive equation $a_n = 1.2a_{n-1} - 3$.

a) Create the time-series graph and web diagram for this equation if the starting term is $a_0 = 14$.

b) Repeat part (a) using $a_0 = 16$.

c) Describe the similarities of the two sequences in parts (a) and (b).

d) Is there an equilibrium value for this recursive equation? If so, find it and verify your answer. If not, explain why none exists.

The web diagrams of all mixed-growth sequences are formed by "bouncing" between two intersecting lines, one of which is $y = x$. The web diagrams of all geometric sequences are formed by bouncing between two intersecting lines, one of which is $y = x$. Those two descriptions (mixed growth and geometric growth) read remarkably alike! What's different? Only the location of the intersection; remember, that intersection is the location of the equilibrium value of the recursive relation. Thus, the equilibrium value for any geometric-growth process is 0.

5. The family of sequences in Item 2 is characterized by a growth factor of 0.4. Changing the additive constant changes the equilibrium value. Sequence #1 has an equilibrium value of 10. Both 6 and 14 are 4 units from that equilibrium value. Create time-series and web diagrams for the pure geometric

ACTIVITY

THE REST OF THE STORY

10

sequence defined by $a_n = 0.4a_{n-1}$, with $a_0 = 4$. Compare your results to those from Item 3. What is the equilibrium value for this recursive relation?

6. Create the time-series and web diagrams for $a_n = 1.2a_{n-1}$, with $a_0 = 1$. Compare your result to those from Item 4.

Based on the results of the preceding items, it seems reasonable to conjecture that mixed-growth sequences are nothing more than geometric sequences "in disguise." Specifically, mixed-growth sequences seem to be geometric sequences whose time-series graphs have been translated vertically, replacing the original equilibrium value $y = 0$ with the new equilibrium value you computed in Item 9 of Individual Work 9.

If that conjecture is true, then it should be possible to translate any mixed-growth sequence so that its translated equilibrium value is 0 and end up with a pure geometric sequence. For example, Sequence #1 in Item 2 above has an equilibrium value of 10. Subtracting 10 from each term would make the resulting (new) sequence have an equilibrium value of 0, and you already know how to identify geometric sequences.

7. a) If the conjecture about the nature of mixed-growth sequences is correct, what must be the form of the closed-form descriptions of mixed growth?

   b) Check the conjecture for Sequence #1 of Item 2 as suggested above. Then write an appropriate closed-form equation for Sequence #1.

8. Generalize your observations in Item 7 to describe a method for writing a closed-form description of any mixed-growth sequence.

9. Apply what you have learned here to determine a closed-form equation for the minimum amount of phenobarbital in Eric's bloodstream after n days of taking the medicine.

Mix and Mat(c)h

1. Explore the results of changing the constants in a mixed sequence. The constants in the sequence $p_n = 0.5 \times p_{n-1} + 20$; $p_0 = 30$ are the multiplicative growth factor 0.5, the additive growth amount 20, and the initial value p_0.

   a) Prepare a time-series graph for the sequence $p_n = 0.5 \times p_{n-1} + 20$; $p_0 = 30$.

   b) Predict how the time-series graph changes when you change p_0 from 30 to 15.

   c) Graph the sequence using a variety of different values for p_0. Generalize the effect of changing p_0.

   d) Predict how the time-series graph changes when you change the multiplicative growth factor from 0.25 to 0.5 to 0.75 to 1.0 to 1.5.

   e) Graph the sequence for a variety of different (positive) growth factors. Generalize the effect of changing the growth factor.

   f) Predict how the time-series graph changes when you change the additive growth amount from 20 to 30 to 15. What growth model would the sequence represent if the additive amount were 0?

   g) Graph the sequence using a variety of different additive growth amounts to test your prediction. Generalize the effect of changing the additive growth rate.

2. Item 1 explores changes in the time-series graphs of mixed-growth sequences. Now, use web diagrams to examine mixed-growth models and generalize the effect of changing the multiplicative growth rate. In particular, examine the behavior determined by the recursive relation: $a_n = b \times a_{n-1} + d$; $a_0 = 1$.

   a) What patterns do you notice when d is positive and $b > 1$?

   b) What patterns do you notice when d is positive and $0 < b < 1$?

   c) What patterns do you notice when d is negative and $0 < b < 1$?

   d) What patterns do you notice when d is negative and $b > 1$?

3. a) Assume the sequence 3, 11, 27, 59, 123 is geometric. Use exponential regression to determine the "best" exponential equation for these data. Check the fit of the equation by examining residuals.

Then check the "geometric sequence" assumption by computing ratios of successive terms.

b) Assume the sequence 3, 11, 27, 59, 123 is arithmetic. Use linear regression to determine the "best" equation under this assumption. Check the fit of the equation by examining residuals. Then check the "arithmetic sequence" assumption by computing first differences.

c) Assume that the sequence 3, 11, 27, 59, 123 is mixed. Test this assumption using ratios of first differences and a recursive graph. Use these tests to determine the appropriate growth factor and additive constant; that is, find the exact recursive equation for this sequence.

d) Part (c) should have checked. Now use the condition that $a_n = a_{n-1}$ at equilibrium to determine analytically the equilibrium value for this sequence.

e) Recall that the closed-form equation for mixed growth is $a_n = f + k(b)^n$, where b is the growth factor and f is the equilibrium value. Use the specific terms of the given sequence above to determine the appropriate value of k, and write a closed-form equation for the sequence. Verify your closed-form equation using the given sequence values.

4. a) Consider the mystery sequence, 10, 25, 47.5, 81.25, 131.875, ... Analyze these values to find a closed-form equation for this sequence.

   b) Recall that a geometric series is the sequence of partial sums of a geometric sequence. Verify by direct calculation that the geometric sequence $g^n = 1.5g_{n-1}$, $g_0 = 10$ generates a geometric series whose partial sums are exactly the mystery sequence of part (a). Thus, knowing that a geometric series is mixed can help find its closed-form representation.

5. Suppose you deposit $200 each month into an account that earns 0.5% interest per month. Then the value of your first deposit (without considering any other deposits) after n months is $200(1.005)^n$ dollars. Your second deposit will have been in the account for one fewer months, so its value at the same instant is $200(1.005)^{n-1}$. Similar statements may be made for later deposits. If you report the value of the account immediately after each deposit, then for $n = 1$, the total value is $200 + 200(1.005)$; that is, the new deposit plus the first deposit with its interest. For $n = 2$, the total value is $200 + 200(1.005) + 200(1.005)^2$.

Similarly, the total value after n months is $200 + 200(1.005) + 200(1.005)^2 + 200(1.005)^3 + 200(1.005)^4 + \ldots + 200(1.005)^n$. That makes the values of the account the terms of a geometric series, one with growth factor 1.005 and first term 200. That is, $s_0 = 200$, $s_1 = 200 + 200(1.005)$, etc. Notice that $s_1 = 200 + 1.005 \times s_0$.

a) Explain why $s_n = 200 + 1.005 \times s_{n-1}$.

b) Use the mixed-growth recursive equation in part (a) to determine the equilibrium value for this series and write its closed-form equation.

c) Show that the form of the recursive equation in part (a) applies to the geometric series in Item 4.

6. a) Show that every geometric series may be written recursively as $s_n = s_0 + b \times s_{n-1}$, where b is the growth factor for the generating geometric sequence.

b) Find a formula for the equilibrium value for this recursive relation, then write a closed-form expression for the geometric series.

7. In the episode "The Trouble with Tribbles" of the original *Star Trek* television series, small furry animals ("tribbles") are reproducing at a ferocious rate, threatening to over-run an entire space station. In discussing the growth of the tribble population, the main characters, Captain Kirk and Mr. Spock, have the following exchange.

KIRK: "There must be thousands of these things here—hundreds of thousands."

SPOCK: "Actually, sir, 1,771,561, assuming that each tribble has 10 offspring every 12 hours for the last three days."

Determine whether the writers of this script knew any math or not! Is Spock right?

ACTIVITY

BACK TO ERIC

11

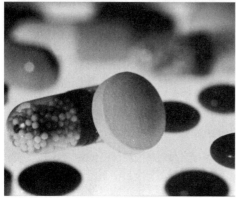

The Image Bank

Recall the problem of adjusting Eric's medicine doses, first introduced in Activity 1. In Activity 4 and Individual Work 9 you have learned a bit more about his situation. Now you are ready to solve the problem of Eric's dosing.

Assume that the increase in the concentration in his bloodstream after each dose is proportional to the size of the dose. For example, you know that the data in Activity 1 (repeated in **Figure 6.53**) are based on doses of 60 mg every 12 hours. If you were to compute the concentration of medicine in Eric's bloodstream just before a particular dose and again just after that dose, that difference would be relatively small—certainly not 60. The reason is that the 60 mg are distributed throughout the body; concentration measures the "density" of the medicine at any particular location. While the increase in amount (60 mg) and the increase in concentration (to be determined) are not equal, they are proportional. Doubling one doubles the other. Thus, the ratio of increase in concentration per dose to amount of dose is constant from dose to dose. (Note this is not true for the loading dose. It is administered directly into the bloodstream, so none of it is "lost" to metabolism before being absorbed. Therefore, the ratio of concentration to dose for the loading dose will be larger than for regular doses.)

ACTIVITY

BACK TO ERIC

11

Days	Blood level mcg/ml
1	7.5
2	8.5
3	9.4
4	10.0
5	10.5
6	11.0
7	11.3

Figure 6.53.
Daily minimum concentrations
of Eric's medicine.

Use the given data to determine equations that describe Eric's
current situation, both just before each dose (the given data are of
this type) and just after each dose. Remember, doses are taken
every 12 hours, but data are given only daily.

The minimum blood concentration for phenobarbital should be
above 15 mcg/ml to be effective, but it must also be below 40
mcg/ml to be considered safe. Phenobarbital levels above 40
mcg/ml are potentially toxic. Levels above 70 mcg/ml may lead
to a coma. Using your descriptive equations, adjust Eric's dose to
meet these safety and effectiveness conditions.

Wrapping Up Unit Six

1. Acetaminophen, the main ingredient in Tylenol®, is used by young and old to relieve aches and reduce fevers. Too much acetaminophen can be toxic, and too little will be ineffective.

   The typical dose for an adult is 325–650 mg every 4 hours. Total amount per day should not exceed 4 grams.

   The half-life of the drug is 2.75–3.25 hours, which means about 21% of the drug is eliminated from the body each hour (79% remains). For a 70 kg person, if the amount in the blood exceeds 13,300 mg after 4 hours following an acute overdose, the patient may suffer liver damage.

   a) Assume that 500 mg doses are taken every 4 hours, with all of the medicine reaching the bloodstream immediately. Write a recursive description or equation for the amount in the blood just after each dose enters the bloodstream. Classify the growth as arithmetic, geometric, quadratic, mixed, or other, and justify. (Hint: Convert all information to 4-hour time units first.)

   b) Use a table or spreadsheet to model the amount of acetaminophen in the individual's bloodstream over a period of 3 days.

   c) What is the maximum amount in the bloodstream after 3 days for a typical adult who takes 1,000 mg 4 times a day?

   d) What dosage four times a day would result in a maximum blood level of 13,300 mg in one day?

   e) How would the minimum-level sequence (for "just before doses") compare to the sequence you defined in part (a)?

2. Start with a $1,000 deposit in a new savings account that adds 0.6% interest each month.

   a) Suppose you deposit $200 at the end of each month, just after interest is posted. Write a recursive description or equation for the balance in the account just after each deposit, and classify the growth as arithmetic, geometric, quadratic, mixed, or other.

   b) How much money will you have accumulated in two years if you continue to deposit $200 each month?

   c) When will your account balance reach $10,000 if you continue to deposit $200 each month?

d) How long would it take to reach $10,000 if you deposited $300 each month?

e) How long would it take to reach $10,000 if you deposit $200 each month and the interest rate is lowered to 0.5% each month?

3. In 1995, the total population of the United States was about 263,034,000 people. Suppose the population continues to grow naturally (births exceed deaths) at a rate of 1% each year.

a) Write a corresponding recursive description or equation for the total U. S. population and classify the growth as arithmetic, geometric, quadratic, mixed, or other.

b) Write a closed-form equation for the U. S. population and use it to project the total population for the year 2020. Show your procedure.

c) When will the population double in number for a total of 526,000,000 people? What equation do you solve to determine the answer?

d) Project the population for the year 2020 if the United States allows 820,000 people to immigrate each year.

e) Compare the prediction for 2020 using the following highest projection for new immigrants each year with that based on the lowest projection.

Highest: 1,370,000 per year.

Lowest: 300,000 per year.

4. You are offered two identical jobs with different contract options. Which job should you take? Use mathematics to support your answer.

Job 1: Salary $35,000 and 5% raises.

Job 2: Salary $40,000 and $2,500 raises.

5. A landfill has space remaining for 25 million cubic yards of garbage, trash, and cover. The landfill expects to add 700,000 cubic yards this year and increase that amount by 1.5% each year. Note that the accumulated garbage is described by a geometric series.

   a) In how many years will the landfill reach capacity?

   b) Suppose recycling efforts reduce the amount added each year by 1.5%. That is, instead of having 1.5% more each year, 1.5% less is added. How many years would be added to the "life" of the landfill?

6. Newton's Law of Cooling describes the cooling (or heating) of an object left alone in a room or tank held at a constant temperature. It is given by the recursive equation:

   $t_{n+1} = t_n + r(t_n - t_R)$, where

   t_n is the temperature of the object in question at time n,

   t_{n+1} is the temperature of the object at time $n + 1$,

   r is a proportionality constant for the particular situation (liquid, time period, . . .). and

   t_R is the temperature of the surroundings (assumed here to remain constant).

   Suppose investigators need to calculate the time of death in a homicide investigation and that the body was discovered at 9:30 a.m. The coroner found the temperature of the body at 10:15 a.m. to be 71.0°F. One hour later the body temperature had fallen to 69.5°F. The temperature in the room remained steady at 65.0°F.

   a) Simplify the expression for Newton's Law of Cooling so that you can easily determine whether it is arithmetic, geometric, mixed, or neither.

   b) Treat the first observation, 71.0°F, as t_0 and 69.5°F as t_1. Note that $t_R = 65.0$°F. Evaluate the proportionality constant r, then compute the equilibrium value of this sequence. Does this value make sense?

   c) Use your results from part (b) to write a closed-form equation for the temperature of the body in terms of the time since it was found.

   d) Use your equation from part (c) to determine the time of death, namely, when the body temperature was 98.6°F. (Note that the time should be negative since it was before the body was found!)

7. Identify the type of growth—arithmetic, geometric, or mixed—for each of the following. Practice changing from recursive to closed form, and from closed form to recursive form.

Change to closed form:

a) $p_n = 1.75p_{n-1}$; $p_0 = 40$.

b) $a_n = a_{n-1} + 6.5$; $a_0 = 17.2$.

c) $a_n = 0.8a_{n-1} - 3.5$; $a_0 = 10$.

Change to recursive form:

d) $p(n) = 5.4n - 127$.

e) $a(n) = 3,800(1.04)^n$.

f) $a(n) = 500 + 1,000(1.04)^n$.

8. Use scientific notation and the laws for exponents to simplify each of the following expressions. Write your answer in scientific notation.

a) $(40,000)(700) =$

b) $(600,000)(200)(5,000) =$

c) $(0.005)(90) =$

9. Use the laws for exponents to simplify each of the following expressions.

a) $x^5(x^3) =$

b) $y^{10}/y^2 =$

c) $3^2(3^4)(3^5) =$

d) $x^8/x^5 =$

e) $(n^3)^4 =$

f) $(2x^3)^5 =$

10. Practice skills related to logarithms (assume the base is 10 unless otherwise stated):

a) If $\log_{10}(92) = 1.9638$, what is the $\log_{10}(9200)$? (Do not use a calculator.)

b) $\log(8) + \log(5) = \log(x)$. Find x.

c) If $\log_{10}(140) = 2.1461$, what is the log of 0.0014? (Do not use a calculator.)

d) What is similar about logarithms for numbers between 100 and 1,000?

e) What is similar about log(5.8), log(580), log(58,000)?

11. The logarithm function is an important example of an inverse. Practice representing inverses in this item.

a) Suppose $c = 2.50x + 200$ represents the cost c to produce x items. Cost, the response variable, is a function of the number of items. Write an equation to represent the inverse where cost is the explanatory variable and the number of items is the response variable.

b) Use your equation for the inverse in part (a) to determine the number of items when the cost is $2,000.

c) Complete the arrow diagram in **Figure 6.54** to represent the inverse process.

Add 50   Multiply by 3

x y

Figure 6.54.
Arrow diagram for Item 11(c).

d) Use your answer to part (c) to find x when y is 600.

e) Graph the inverse of $y = 2^x$.

12. a) The Consumer Price Index uses the average cost of goods for 1982, 1983, 1984 as its base value of 1.00. Ten years later, in 1993, the CPI was 1.45, which means goods that cost $1.00 in 1983 cost $1.45 in 1993. The linear equation $CPI = 0.0442x + 0.858$ approximates the CPI since 1980. Thus $x = 1$ represents 1981. What equation do you solve to predict when the CPI rises to $2.00? Use your equation to find the year.

b) The population of Earth was about 6 billion in the year 1997 and is growing at a rate of about 1.7% each year. Write a closed-form equation to represent the population after n years.

c) When will the population reach 10 billion, the expected capacity of Earth? What equation do you solve?

d) A company estimates its profits $p(x)$ based on the number of items sold x as $p(x) = -0.002x^2 + 20x - 30,000$. How many items must the company produce to make a profit of $5,000? What equation do you solve?

13. Below you are given some terms of sequences. Find the missing terms and write a recursive rule to describe each sequence. Assume that each sequence is either arithmetic, geometric, quadratic, or mixed.

a) $a_1 = 1{,}200$, $a_2 = 900$, $a_3 = 675$. Find a_4, a_5, a_6.

b) $a_4 = 33.3$, $a_5 = 38.6$, $a_6 = 43.9$. Find a_1, a_2, a_3.

c) $s_1 = 3$, $s_2 = 11$, $s_3 = 24$, $s_4 = 42$ represent terms for a sequence of partial sums. Find s_5 and s_6, and a_1 through a_6.

14. **Figure 6.55** shows the number of houses in a community. Create a mathematical model to represent the growth of the community. Identify past pattern(s), then project the number of houses in the year 2020 if the present pattern continues.

15. A four-year-old asthma patient was given a 3.2 mg/kg intravenous dose of theophylline. **Figure 6.56** records the level of theophylline in the blood plasma (in micrograms per milliliter).

Year	Houses
1990	1700
1991	1785
1992	1874
1993	1968
1994	2066
1995	2170
1996	2285
1997	2400
1998	2515
1999	2630

Figure 6.55.
Number of houses in a community.

Time (hr.)	Concentration	Time (hr.)	Concentration
0.25	10.62	4.00	3.98
0.50	9.15	6.00	2.78
1.00	7.42	8.00	1.94
2.00	5.78	10.00	1.35
3.00	4.77		

Figure 6.56.
Theophylline concentration. Source: *Biopharmaceutics and Clinical Pharmacokinetics*, Robert E. Notari, page 54.

Note that these time intervals are not uniformly spaced. You can ignore some of the data and use the remaining data to develop a model. Then the data that were initially ignored can be used as "predictions" to check your model.

a) Let times $t = 2$, 4, 6, 8, and 10 be represented as terms $n = 0, 1, 2, 3$, and 4. Use this sequence to determine whether the "growth" is additive, multiplicative, or neither. Give reasons for your answer.

b) How is this drug situation similar to the problem involving Eric in Activity 1? How is it different?

c) Find a closed-form equation for the "amount versus n" data in part (a). Then substitute $t = 2n+2$ to convert to times and check your equation using the data that were not included in your original analysis.

Mathematical Summary

T he main reason you analyze the mathematical characteristics of known growth models is so you can identify the characteristics in an unknown sequence of numbers and propose a growth model based on those patterns. A second reason is that knowledge of the processes actually controlling a situation may provide information about its mathematical description. Mathematical models permit prediction, both of future or missing values and of behavior. All can be useful.

Four types of mathematical models of growth were studied in this unit: linear (arithmetic), exponential (geometric), quadratic, and mixed. Each model has graphical, tabular, and symbolic characteristics that distinguish it from other models. Each can be represented by equations in either recursive or closed form.

An additive-growth model produces a linear time-series graph and is known as an arithmetic sequence of values. It is represented by recursive equations of the form $a_n = a_{n-1} + d$ and closed-form equations of the form $a(n) = dn + k$. The control number d denotes the difference between consecutive terms (known as first differences) of the sequence. It is constant for the additive-growth model and adjusts the slope of the line in the time-series graph.

A multiplicative-growth model produces an exponential time-series graph and is known as a geometric sequence. It is represented by recursive equations of the form $a_n = b \times a_{n-1}$ and closed-form equations of the form $a(n) = a_0 b^n$. The control number b denotes the ratios of consecutive terms (known as growth factor) of the sequence. It is constant for the multiplicative growth model, and adjusts both the steepness and direction (increasing/decreasing) of the time-series graph. That graph is decreasing and bounded for $0 < b < 1$ and increasing and unbounded for $b > 1$. The graph is a horizontal line for $b = 1$, and is neither exponential nor geometric.

A sequence of partial sums may be formed using any sequence. The sequence of partial sums for an arithmetic sequence produces a quadratic model with a parabolic time-series graph. The differences between consecutive terms for the sequence of first differences (known as second differences), is constant for the quadratic sequence. The sequence of partial sums for a geometric sequence is a mixed sequence. For the geometric sequence $a_n = b \times a_{n-1}$ and first term a_0, the sequence of partial sums is represented recursively as $s_n = b \times s_{n-1} + a_0$.

Mixed sequences are represented by recursive equations of the form $a_n = b \times a_{n-1} + d$. The ratio of consecutive first differences is constant for mixed sequences; that ratio is b. The control number b determines whether the curve is bounded as n increases. The values b and d determine the equilibrium level, $d/(1 - b)$.

Recursive graphs are used frequently in this unit to reveal patterns that may not be obvious in unknown sequences. In particular, the graph of a_n versus a_{n-1} is useful for identifying arithmetic, geometric, and mixed sequences. For arithmetic sequences, the graph of a_n versus a_{n-1} is linear with a slope of 1. For geometric sequences and mixed sequences, the graph of a_n versus a_{n-1} is linear with a slope equal to the growth factor. For geometric sequences, that line contains $(0, 0)$. Web diagrams, used to see the successive changes from term to term more clearly, are enhanced versions of the recursive graph. They include the line $y = x$ for reference and show the changes from term to term as horizontal and vertical segments joining successive points in the recursive graph.

Inverse processes are used to solve equations in which the value of the response variable is known and the value of the explanatory variable is sought. Inverse processes for most functions, including linear and exponential functions, can be represented with arrow diagrams, tables, graphs, or equations. Inverses are represented with reverse processes in an arrow diagram, the interchanging of explanatory and response variable roles in a table, and the reflection across the line $y = x$ in a graph. The range of the function becomes the domain for its inverse; the domain of the function becomes the range for its inverse.

The symbolic expression for the inverse of an exponential function is called the logarithm. The inverse of the exponential function $y = 10^x$ is $y = \log_{10}(x)$. Thus, $\log_{10}(x) = 0.6990$ means $10^{0.6990} = x$. Logarithms are identified according to their bases. The most common base is 10. The LOG button on a calculator produces the base-10 logarithm of a number.

Glossary

ADDITIVE GROWTH:
A growth process described by recursive equations of the form $p_n = p_{n-1} + k$ where k is a constant.

ADDITIVE IDENTITY:
The additive identity is 0. Add 0 to a number and the number is unchanged.

AMOUNT OF GROWTH:
The difference between successive values in a growth sequence. Use the expression $p_n - p_{n-1}$ to calculate the amount of growth between successive terms.

ARITHMETIC SEQUENCE:
A sequence of numbers that can be modeled by a recursive additive process.

ARITHMETIC SERIES:
A sequence of partial sums of an arithmetic sequence.

BASE:
The number that is raised to a power in an exponential expression. For example, in the expression b^x, b is the base. A base-10 logarithm is the exponent of the number 10 where 10 is the base.

COMMON LOGARITHM:
A base-10 logarithm (log).

EQUILIBRIUM:
A fixed point of a sequence. The quantity p_n such that $p_n = p_{n-1}$ for all values of n above a fixed number. In the context of monitoring drug levels in the body, the term "steady-state" refers to the equilibrium amount that occurs when the dosage exactly replaces the amount of drug eliminated by the body since the previous dose.

EXPONENTIAL GROWTH OR DECAY:
Behavior described by a closed-form exponential function of the form $y = ab^x$. Exponential growth describes situations where the growth factor b is greater than 1 and exponential decay describes situations where b is between 0 and 1.

FIRST DIFFERENCES:
Amount of growth. The difference between successive values in a sequence.

GEOMETRIC SEQUENCE:
A sequence of numbers that can be modeled by a recursive multiplicative process.

GEOMETRIC SERIES:
A sequence of partial sums of a geometric sequence.

GROWTH FACTOR:
The amount by which an exponential quantity is multiplied in one recursive period.

HALF-LIFE:
The amount of time it takes for half of a quantity of a drug to be eliminated from the body. For example, if the half-life of a drug is 2 hours, then a 100 mg quantity is reduced to 50 mg in 2 hours and the 50 mg quantity is reduced to 25 mg in 2 more hours.

HOME SCREEN ITERATION (HSI):
The feature of a calculator that allows you to repeat a process (several times) by simply pressing an ENTER or RETURN key.

LOGARITHM:
The exponent of the designated base, usually 10, that is equivalent to the given number. Logarithm is often abbreviated as "log." Logarithms using base 10 are called common logarithms. $\text{Log}_{10}(2)$ is 0.3010 because 2 is equivalent to $10^{0.3010}$.

MIXED GROWTH:
The combination of additive and multiplicative growth. $p_n = b \times p_{n-1} + d$ where b and d are constants.

MULTIPLICATIVE GROWTH:
A growth process described by recursive equations of the form $p_n = b \times p_{n-1}$ where b is a constant.

MULTIPLICATIVE IDENTITY:
The multiplicative identity is 1. Multiply any number by 1 and the number is unchanged.

QUADRATIC GROWTH:
A sequence of numbers that can be modeled using a closed-form equation of the form $y = ax^2 + bx + c$, where x represents the term number.

RECURSIVE GRAPH:
The graph of a_n versus a_{n-1}.

RELATIVE RATE OF GROWTH:
The change in growth divided by the total population before that growth.
The expression
$(p_n - p_{n-1})/p_{n-1}$ or $(p_{current} - p_{previous})/p_{previous}$
represents the relative rate of growth.

SCIENTIFIC NOTATION:
An equivalent expression for a number written as the product of a number between 1 and 10 and the number 10 raised to a power. The scientific notation for 53,800 is 5.38×10^4.

SECOND DIFFERENCES:
The difference between the consecutive terms of a sequence of first differences.

SEQUENCE:
An ordered list of values.

SEQUENCE OF PARTIAL SUMS:
Also known as a series. A sequence of numbers characterized by the recursive expression $s_n = s_{n-1} + a_n$.

SERIES:
A sequence of partial sums.

TERM NUMBER (INDEX):
The integer indicating the location of the indicated term in the sequence. Indices generally begin at 0 or 1, and increase by 1 per term.

WEB DIAGRAM OR WEB GRAPH:
A recursive graph to which has been added the line $y = x$ and horizontal and vertical segments joining each point on the recursive graph to its successor by "bouncing" off the graph of $y = x$. (See **Figure 6.57**.)

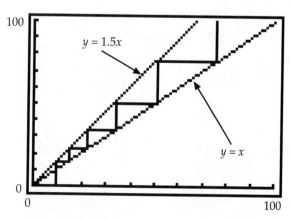

Figure 6.57.
Web diagram of $a_n = 1.5a_{n-1}$, $a_0 = 10$.

UNIT

7

Motion

The context of this unit involves the staging of high-risk stunts such as motorcycle and car jumps. In Course 1, Unit 5, *Animation* you learned how to describe the motion of an object as it moves along a line so that you could simulate its motion on your calculator. In this unit you will develop methods for analyzing real motion such as the motion of toy cars scooting across your classroom floor, the motion of a falling ball, or the motion of a stunt performer. You will use a motion detector to gather data and draw distance-versus-time graphs to represent the motion. From your data and graphs, you will be able to determine the velocity and acceleration of the moving object. To apply what you have learned, you will design and model several high-risk stunts and then use small-scale versions of these stunts to test the validity of your models.

THRILLS AND SPILLS IN MOTION

Have you ever watched stunt drivers in action? Touring stunt shows are not as popular today as they were in the early 1970s, but adults old enough to remember that period will know of Evel Knievel. He and auto drivers such as Joie Chitwood became celebrities by stunt jumping with motorcycles and cars. In fact, according to Evel Knievel's website, 52% of U.S. households viewed his 1975 televised King's Island, Ohio, jump over 14 Greyhound buses. That continues to be an *ABC Wide World of Sports* record.

Of course, such stunts require careful planning or the consequences can be tragic. In 1996, Butch Laswell was killed during an attempt to set a Guiness World Record for ramp-to-ramp motorcycle high jumps. He had planned to jump the Oasis Resort Hotel's 38-foot tall Skywalk Bridge but missed the landing ramp. Wind was thought to be a factor. Prior to this jump, Laswell had completed more than 5,000 jumps in 20 years of stunt jumping and, according to a hotel spokesperson, Laswell had a "100 percent safety record."

In this unit you won't be performing any motorcycle ramp-to-ramp jumps, but you will be asked to plan a scaled-down version of such a jump for a toy car.

LESSON ONE
Learning Your Lines

KEY CONCEPTS

Interpreting
graphs

Average
velocity

Slope

Linear
equations

Piecewise-defined
graphs

Used with permission of the Evel Knievel Fan Club

PREPARATION READING

Jack be Nimble, Jack be Quick

You've seen this, or something like it in the movies: A car careens down a street, narrowly missing a truck crossing an intersection. For stunt drivers, trial-and-error alone is not a viable method for planning such stunts. A mistake could cost them their lives. Careful planning, frequently supported by a knowledge of mathematics and the laws of physics, is essential to the creation and subsequent execution of successful stunts.

Early in his career, Evel Knievel often relied on "gut-level instincts" to help him create stunts. The results were sometimes more spectacular than intended. In one show, for example, Evel sandwiched a row of open crates containing rattlesnakes between two ramps. A hush fell over the crowd when he signaled for the start. He revved up his motorcycle's engine, sped up the first ramp, sailed over the rattlesnakes . . . and fell short of the landing ramp. He landed on a crate, taking down one of its sides and freeing the snakes. Unharmed, Knievel sped off into the sunset as the crowd quickly dispersed. (An application of mathematics might have saved the show.)

Undeterred by occasional mishaps, and tiring of a steady diet of motorcycle jumping, Evel decided that he would stand on the ground and leap over an oncoming motorcycle. At Evel's signal, a member of his thrill show raced his cycle straight toward him. At the appropriate moment Evel jumped. He cleared the front wheel of the oncoming cycle . . . but not the handle bars. Evel was flung 15 feet into the air. When he landed he broke most of his ribs. (Knowledge of the laws of physics might have saved his ribs.)

Jeff Lattimore's specialty in Chittwood's Thrill Show is "The Leap for Life." While this stunt might be considered a variant of Evel's failed leap over a speeding motorcycle, Jeff has performed it successfully for years. In his stunt, Jeff climbs a ladder and stands atop an eight-foot stool. Then the ladder is removed, leaving Jeff stranded just as a car comes speeding toward him. Jeff jumps a moment before impact, the stool is snapped out from under him, the car whizzes beneath . . . and Jeff lands safely on the ground.

Near collisions, ramp-to-ramp jumps, and leaps over oncoming vehicles are staples of motorcycle and car thrill shows. In this unit, you'll have the opportunity to plan and, in some cases, execute small-scale versions of these stunts. However, in order to avoid needless injury due to miscalculation (or other unexpected factors), you'll stage your stunts using toy vehicles or you'll simulate your stunts using a calculator.

CONSIDER:

> Which of the three stunts described above, the car-truck, near-collision stunt, Knievel's ramp-to-ramp motorcycle jump, or Lattimore's "The Leap for Life," do you think was the easiest to design? Which was the most difficult? Why?

The Preparation Reading introduced three types of stunts: a two-vehicle near-collision stunt, a ramp-to-ramp jump, and a leap-over-an-oncoming-vehicle stunt. Items 1–3 ask you to think about how *you* would design such stunts. (Refer to the Preparation Reading for examples of each of these stunt types.)

1. Suppose that it is your job to design a two-vehicle near-collision stunt.

    a) What information would you specify about the vehicles themselves and about your design of the physical layout of the stunt area?

    b) What kind of instructions would you give to the stunt drivers?

2. Next, suppose that you have been commissioned to design a ramp-to-ramp motorcycle jump.

    a) Draw a sketch of the set-up for the take-off and landing ramps.

    b) What information would be helpful in planning your stunt?

    c) How might you obtain this information?

3. Jeff Lattimore's "The Leap for Life" is a successful leap-over-an-oncoming-vehicle stunt. Evel Knievel's attempted leap over an oncoming motorcycle nearly cost him his life. Clearly, this type of stunt needs careful planning.

    a) What made Jeff's Lattimore's stunt design successful while Evel Knievel's was not?

    b) What important elements would you need to consider if you were designing a leap-over-an-oncoming-vehicle stunt?

ACTIVITY

COORDINATED EFFORTS—STUNT DESIGN

1

So far, you've considered three types of stunts and you'll have the opportunity to study each one as you work through the unit. This lesson will focus on near-collision stunts. Lessons 2–4 will handle leap-over-an-oncoming-vehicle stunts. Lesson 5 will be the grand finale. You'll use what you've learned about designing the first two types of stunts to design and then perform ramp-to-ramp jumps.

A major goal of this lesson is to design a two-vehicle near-collision stunt using battery-operated toy vehicles. Look back at your answer to Item 1(b). In your instructions, you probably told the drivers how fast to drive. However, your toy vehicles won't respond to such instructions. Instead, you'll have to design your stunt based on the velocities of your toy vehicles when they are turned on. Before planning a stunt involving the simultaneous motion of two toy vehicles, you'll find it helpful first to analyze the motion of each of your toy vehicles individually.

4. Imagine that a battery-operated toy car is moving along a straight line across the floor of your classroom. How could you obtain information about how fast the car is traveling? How could you tell whether or not the car is moving at a constant speed?

ACTIVITY

2

RECORDING THE MOTION

In Activity 1, you and the members of your group probably thought about several different methods for determining how fast a toy car moves and whether it travels at a constant speed or not. Did anyone think of making a time-series graph to describe the car's motion? Probably not! You first encountered such graphs in Course 1, Unit 5, *Animation* and used them there to describe the motion of an object along a line.

But to make such a graph and for it to be useful, you need accurate measurements of the car's location. How can you obtain precise moment-by-moment information about a toy car's distance from you as it moves away?

You may never have used an ultrasonic motion detector before. However, this device is exactly what you need to gather distance-versus-time data from a moving object. An ultrasonic motion detector sends out a beam of ultrasonic sound. If an object is in the beam, this sound reflects off the object back to the sensor, which detects the returning signal. The distance between the object and the motion detector can be determined from the time lapse between emitting the signal and detecting the return signal because the sound emitted by the detector travels at a known speed.

The motion detector must be connected to a calculator or a computer. You'll need to use an intermediate device to link a calculator to a motion detector unless you are using a "smart" motion detector (such as the TI-CBR). **Figure 7.1** illustrates a typical calculator-motion detector set up.

Motion detectors have a limited range for detecting motion. The beam of the motion detector pictured in Figure 7.1 makes an angle of 15° at the center of the detector. Unless your teacher tells you otherwise, assume that your motion detector accurately measures the distance from itself to an object that is between 1.5 and 24 feet away from its front face (in the beam).

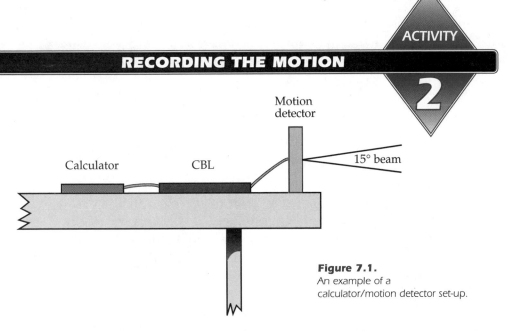

ACTIVITY

RECORDING THE MOTION

2

Figure 7.1.
An example of a
calculator/motion detector set-up.

During this activity you will run a program that collects and
records the distance from an object to the front of the motion
detector and the time at which this measurement is taken. In particular, you will want to know how long the program runs, how
many readings it takes, and whether the measurements are in
feet, meters, or some other unit of measurement. For example,
suppose that you can set your program so that it takes and
records 60 distance measurements (feet) and their corresponding
times (seconds) and displays the data in a scatter plot. Your program is fast—it's over in just six seconds! Any program similar
to the one just described will be called HIKER.83P (or .82P). Your
teacher will give you more specific details about the HIKER
program you will be using.

Once you have set up your motion detector apparatus, you'll
need to place a moving object in the motion detector's beam.
Suppose, for example, a battery-operated toy car is three feet in
front of a motion detector and the car is turned on the instant the
motion detector begins taking readings. Two seconds later, the
motion detector locates the car five feet away. Based on the readings from the motion detector, your calculator plots two data
points: (0, 3) and (2, 5). In addition, the times, 0 and 2, are stored
in one list and the distances, 3 and 5, in another. Based on the
information provided by the motion detector, you know that the
toy car has traveled from a location three feet from the detector to
a location five feet from the detector, a displacement of two feet
in two seconds.

ACTIVITY

RECORDING THE MOTION

2

Using a motion detector to gather information about the motion of a student as he walks in front of the detector is no different than using it to track the motion of a toy car. The advantage of gathering information about your own walk is that you actually experience the motion that produces the distance-versus-time graph. So, for the remainder of this activity you will use the motion detector to gather information about the walks of students in your class.

WALKING

Your group should set up the motion detector as described in Handout H7.2 or as described by your teacher. Your group should have approximately five copies of Handout H7.3 on which to record your observations.

Next, members of your group should take turns walking in such a way that the motion detector apparatus produces a graph. Remember to keep in mind the length of time your HIKER program runs. For example, if your program runs for six seconds, you may wish to have a member of your group count, "Go, 1, 2, 3, 4, 5, 6" during the data collection.

1. Select a member of your group to be the first walker. Watch carefully as the first walker completes his walk. On Handout H7.3, describe what you see the walker do. Then sketch the graph that is produced on the calculator screen. If possible, repeat this process so that each group member has a chance to be the walker.

TALKING

2. With your group, select two graphs from Handout H7.3 to present to the class. Once you have agreed on those two graphs, transfer your sketches and descriptions of the walks onto large sheets of paper. Label them with your group's name. Hang your presentations on a wall in your classroom.

3. After all the graphs have been posted, take a few minutes to look at them. Sort these graphs into groups. Record how you decided which graphs belong in the same group. Explain your reasoning.

INDIVIDUAL WORK 1

Equations of Motion

Recall from Course 1, Unit 5, *Animation*, that the **average velocity** of an object between time 1 and time 2 is determined by the following ratio:

average velocity from time 1 to time 2
= (change in location)/(change in time)
= (distance 2 – distance 1)/(time 2 – time 1).

Notice this is the same formula as the one used to calculate the slope of the line joining the points (time 1, distance 1) and (time 2, distance 2).

1. Imagine that a toy car is moving away from you along a straight line and that the time-series graph in **Figure 7.2** displays the distance between you and the car each half-second.

   The time-series graph in Figure 7.2 can provide much information about the car's velocity during its nine-second trip.

   a) What is the average velocity of the car from $t = 0$ to $t = 9$? (Be sure to include the units for the velocity.)

   b) What is the average velocity over the two-second time interval from $t = 0$ to $t = 2$? What about from $t = 2$ to $t = 4$? Is the car traveling faster during the first two seconds of its trip or the next two seconds? How could you tell just from the graph in Figure 7.2?

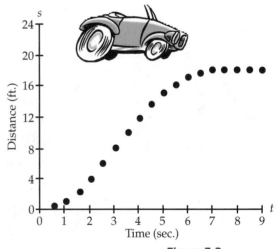

Figure 7.2.
Distance-versus-time graph for a toy car.

   c) Find a two-second interval over which the car appears to be traveling at a constant velocity. What is this velocity? Can you find more than one such two-second interval?

   d) Describe in words what is happening to the toy car's velocity during its nine-second trip.

2. Scott's motion was recorded by a motion detector. He started moving as soon as the motion detector began to record data. At $t = 0$ he was five feet from the detector. At $t = 6$ he was 23 feet from the detector. Assume he walked at a constant velocity.

a) What was Scott's velocity?

b) Complete the table in **Figure 7.3**.

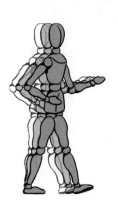

Elapsed time (sec.)	Scott's distance from the motion detector (ft.)
0	5
1	
2	
3	
4	
5	
6	23

Figure 7.3.
Table recording Scott's motion.

c) Draw a distance-versus-time graph that represents Scott's motion.

d) Write a closed-form equation to model Scott's distance from the detector in terms of the elapsed time as he walked away from the motion detector. Interpret the numbers in your equation in the context of Scott's motion.

e) Using your model from part (d), how far was Scott from the detector at $t = 2.5$ seconds?

3. Imagine that you have set up a motion detector to track the motion of your friend as she walks at a constant velocity *toward* the motion detector.

a) Draw a distance-versus-time graph for your friend's imaginary walk. Assume that you monitor her walk for six seconds.

b) What is your friend's average velocity during her six-second walk? Your answer should be negative. Why?

c) Write a closed-form equation describing the relationship that you graphed in part (a).

4. The graph in **Figure 7.4** is a time-lapse graph of the motion of a toy car over a six-second interval. This graph is similar to ones that you drew in Course 1, Unit 5, *Animation*. Unlike the lines in distance-versus-time graphs, the solid line here indicates the actual path of the car. Sample times have been added to the graph to show when the car passed particular locations.

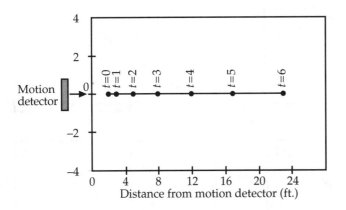

Figure 7.4.
Time-lapse graph of the motion of a toy car.

a) Is the car traveling at a constant velocity? How can you tell?

b) What is the average velocity of the toy car over the six-second trip? Over what one-second interval do you think the car's average velocity was closest to its average velocity for the entire trip?

c) Draw a distance-versus-time graph for the toy car.

d) What type of a function describes the relationship between distance and time in this situation? How do you know? (If you have trouble answering this question, recall from Unit 6, *Growth*, what first and second differences can tell you about a function.)

5. Suppose the first graph in **Figure 7.5** was recorded when Carol walked in front of the motion detector. The program used to collect the data recorded time in seconds and distance in feet. TRACE was used to find the points that are shown in the other graphs.

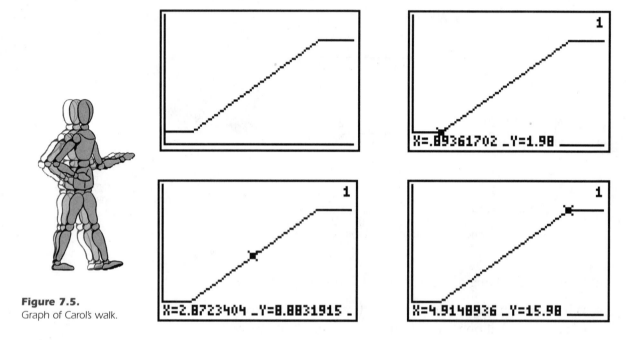

Figure 7.5.
Graph of Carol's walk.

a) How far was Carol from the motion detector when the program began running?

b) What was the length of time that Carol stood still before moving?

c) How much time did Carol spend walking?

d) How far did she walk?

e) How fast did she walk?

f) Write an equation that models the relationship between distance, d, and time, t, during the period that Carol was moving. State the domain of your model.

g) Does the d-intercept in the equation you found in part (f) have meaning in the context of this situation? What about the slope? Explain.

6. Select two of your walks your group made during Activity 2. For each walk, describe as carefully as possible the average velocity over some interval. If you can compute their numerical values, do so.

7. In previous units you fit a line to data and then examined the residuals to see if your linear equation was a good model for the data. What are residuals? How are they calculated? What does a residual plot from a "good-fitting" model look like?

8. Look back at your work in Item 5. **Figure 7.6** shows an attempt by one of Carol's classmates to use a line to describe the distance-versus-time data from her walk.

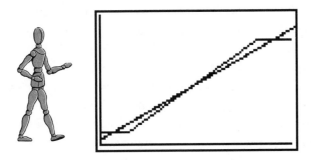

Figure 7.6
An attempt at fitting a line to Carol's data.

a) Is the line a good fit? (Does it describe the essence of her walk?)

b) Does the line's slope over- or underestimate Carol's velocity?

c) Make a rough sketch of what the residuals for this line would look like. Explain how the residual graph verifies your answer in part (a).

WALKING THE WALK

As part of Activity 1, you and your classmates produced a variety of distance-versus-time graphs by walking in front of a motion detector. Someone may have produced a horizontal line for a graph. How would you walk so that your distance-versus-time graph is a horizontal line?

In this activity you'll try to match walks to graphs and graphs to walks. In Part I, your group will specify the graphs and then the designated walker will attempt to "walk like one of the graphs." In other words, the walker will attempt to control his walk so that it produces a distance-versus-time graph that matches one of the specified graphs. In Part II, the process is reversed. Your group will write instructions for walks and then attempt to match them to distance-versus-time graphs that would result from such walks.

Set up the motion detector and get ready to conduct the following experiments!

Part I:
GRAPH TO WALK

1. Walk in such a way that your distance-versus-time graph is a straight line with a positive slope. When you are satisfied with your graph, copy the data to lists that are not used by your program (such as L5 and L6, or named lists). Then link calculators with the other members of your group so that everyone has a copy of these data.

   a) Make a sketch of your graph. On your sketch, label the approximate distance between the walker and the motion detector at the instant the motion detector began recording data. Also label the approximate distance between the walker and motion detector the instant the motion detector stopped recording data.

   b) Determine a linear model that describes the relationship

ACTIVITY

WALKING THE WALK

3

between distance and time for the walk in part (a). Then make a residual plot for your model. Does your model do a good job describing the data? Explain.

c) If your model in part (b) does not do a good job describing the data in part (a), either adjust your model (perhaps by eliminating outliers and refitting the model) or gather new data. Then repeat part (b). Was your walker able to walk a "good" line?

d) Interpret the slope and vertical intercept of the model you wrote in part (c). That is, what do these two numbers tell you about the walk itself?

2. Next, draw distance-versus-time graphs, provide instructions to walkers on how to walk, and then have the walkers follow the instructions. How good will your walkers be at matching the planned graphs? Perform this experiment to find out.

a) Plan four more graphs, with different shapes. Sketch the shape of each planned graph on a separate set of axes on the left side of Handout H7.4. Add appropriate scales and labels to the axes.

b) Next, write instructions to a walker describing how the walker should move in order to produce each graph in part (a). Include details such as where to start, which direction to move, how fast to walk, and any other instructions that you want the walker to follow.

c) Members of your group should take turns following the "walking instructions" corresponding to each of the graphs in part (a). Use a motion detector to record data from each walk. Sketch the actual graph on the axes to the right of your sketch of the planned graph. Again, add appropriate scales and labels.

d) Identify any differences between the planned and actual graphs, and discuss possible explanations for those differences.

HOW to WALK

ACTIVITY

3

WALKING THE WALK

Part II:

WALK TO GRAPH

In this experiment, you will provide written instructions for the walk. Next, you will predict how the distance-versus-time graphs should appear. Then, you will test your prediction against the actual graphs produced by walkers.

3. a) Plan four more "walks." Describe in written instructions how you want a walker to move. Include things like where to start, which direction to move, how fast to walk, and any other information you wish the walker to have. Record your instruction on a second copy of Handout H7.4.

   b) Predict what the graphs will look like. Sketch each predicted graph on a separate set of axes on the left side of Handout H7.4. Add appropriate scales and labels.

   c) Try the walks. Follow your "walking instructions" and record the actual graph on the same axes as your sketch of the predicted graph. Label which graph is the predicted one and which graph is the actual one.

   d) Identify any differences between the predicted and actual graphs, and discuss possible explanations for those differences.

After this activity, you should be able to look at a moving object's distance-versus-time graph and describe its motion; conversely, given a written description of an object's motion, you should be able to make a rough sketch of its distance-versus-time graph.

INDIVIDUAL WORK 2

Interpreting Motion Graphs

Figures 7.7(a) through (e) show graphs recording student walks.

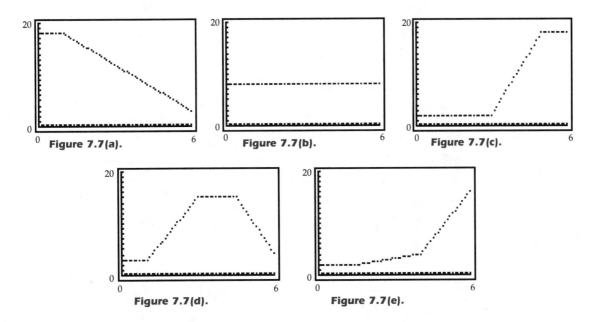

Figure 7.7(a).   Figure 7.7(b).   Figure 7.7(c).

Figure 7.7(d).   Figure 7.7(e).

1. The graphs in Figure 7.7 resulted from students walking in front of a
   motion detector. In all cases, the *y*-axis represents distance (feet) from
   the detector to the object and the *x*-axis represents time (seconds).
   For each graph, describe the students' motions.

2. The two graphs in **Figures 7.8 and 7.9** were recorded by a motion
   detector when a student walked in front of it. Assuming the scales
   are the same for both graphs, in which case was the student walking
   faster? Explain your answer.

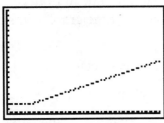

Figure 7.8.
Student #1's walk.

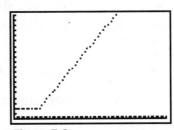

Figure 7.9.
Student #2's walk.

3. Suppose the graph in **Figure 7.10** is the distance-versus-time graph for the "walk-a-positive-slope" data from Item 1 of Activity 3.

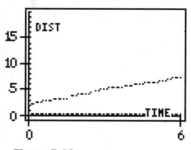

Figure 7.10.
"Walk-a-positive-slope" data.

a) What was the approximate distance between the student and the motion detector at the instant the motion detector began taking readings?

b) How far did the walker walk during the time that the motion detector was reporting readings?

c) What was the walker's average velocity over the entire time that the motion detector was taking readings?

d) Is the velocity you found in part (c) a good measure of the walker's velocity at each instant during his walk? Explain.

e) Find a linear model that describes the relationship between distance, d, and time, t.

f) What is the significance of the d-intercept in the equation you found in part (e)? What is the significance of the slope?

In preparation for the near-collision stunt in the next activity, it's time to apply what you have learned from your walks to the motion of toy cars.

4. Suppose that a toy car moves along a straight line in front of a motion detector and that its distance-versus-time graph turns out to be a line.

a) What does this tell you about the toy car's velocity?

b) How could you determine the toy car's velocity?

c) Suppose that the distance-versus-time graph has a negative slope. What does that tell you about the car's motion?

5. Suppose the distance-versus-time graph for a toy car traveling along a straight line in front of a motion detector is curved rather than straight. What does this tell you about the car's velocity? Explain.

6. Henry and George each felt that he had walked the better line. Data from their walks appear in **Figure 7.11**.

Elapsed time (sec.)	Henry's distance from the motion detector (ft.)	George's distance from the motion detector (ft.)
0.5	6.0	6.0
1.0	7.5	7.0
1.5	7.9	8.1
2.0	8.8	9.2
2.5	9.5	10.6
3.0	11.0	12.2
3.5	12.5	13.2
4.0	13.5	15.0
4.5	15.5	16.8
5.0	16.5	18.6
5.5	16.8	20.5
6.0	17.4	22.4

Figure 7.11.
Data from Henry and George's walks.

a) Use the data from Figure 7.11 to make a distance-versus-time scatter plot representing each of Henry's and George's walks.

b) Fit a least squares line to each walker's data. Sketch the least squares lines on the appropriate scatter plots drawn for part (a).

c) Assume for the moment that your equations in part (b) are reasonable models for both Henry's and George's walks. According to these models, approximately how fast was Henry walking? What about George? (Remember to include the units as part of your answer.)

d) Make a residual plot for each model. Based on the residual plots, who do you think walked a better line? Explain.

ACTIVITY

I MISS YOU NEARLY

4

Now it's time to apply what you have learned to the problem of stunt design. Recall the description of the two-vehicle, near-collision stunt in the Preparation Reading. You'll first plan, and then stage, this stunt in your classroom using battery-operated toy vehicles. The idea is to design just such a stunt.

1. **The Roadway.** Assume that both roads are straight and that they intersect. Decide on the configuration of the intersection: right angles or some other angles. Make a rough sketch of the intersecting roads.

2. **The Vehicles.** Next, you will need two battery-operated vehicles (cars, trucks, robots).

   a) Describe the toy vehicles that you will be using for your stunt.

   b) Use the motion detector to gather information about how your battery-operated vehicles move. Describe with an equation the distance-versus-time data collected by the motion detector. Justify your equation. Include sketches of the graphs produced by the motion detector readings as part of your justification.

   c) Based on your work in part (b), what is your best estimate for how fast each vehicle moves?

3. **The Stunt.** Describe how you will stage this stunt. (Where on the intersecting roadway will you place the vehicles? Will you start both vehicles at the same time? If not, how do you plan to stagger the starts?) Remember, the key to this stunt is to cause some anxiety for the spectators—the vehicles should pass through the intersection as closely as possible without colliding. Include in your description the mathematics supporting your design of this successful stunt.

4. **The Proof.** Execute the stunt. Did the stunt go according to your plans? If not, what do you think accounts for any deviations between what you planned and what actually happened?

INDIVIDUAL WORK 3

Moving Cars

1. The two-vehicle, near-collision stunt that you designed in the previous activity used battery-operated toy vehicles. How would you characterize the distance-versus-time graphs for battery-operated toy cars? What does this tell you about the velocities of these toy cars?

2. Suppose a battery-operated toy car is allowed to travel down a straight highway until its batteries run out.

   a) Draw a sketch of how you think its distance-versus-time graph would look. Assume that distance is measured from the car's starting position. Be sure to label the units on your axes.

   b) Describe in words what your graph indicates about the car's velocity during its drive.

3. Suppose that a battery-operated toy car moving in front of a motion detector produced the distance-versus-time graph in **Figure 7.12**.

   a) Approximately how far was the car from the motion detector when it began recording data? How far was the car from the motion detector when it stopped recording data?

   b) After the motion detector began recording data, how much time elapsed before the car began moving?

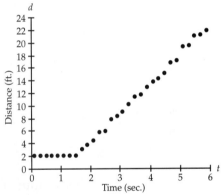

Figure 7.12.
Distance-versus-time graph for a toy car.

   c) Suppose that Jason wanted to determine a model for the car's motion and decided to fit a least squares line to the data in Figure 7.12. A graph of his model appears in **Figure 7.13**. According to Jason's model, what was the approximate velocity of the car? How did you get your answer?

   d) Do you think your approximation of the car's velocity in part (c) is too high, too low, or about right? Explain.

   e) Do you think Jason's model does a good job in describing the car's motion? If not, what would you have done differently?

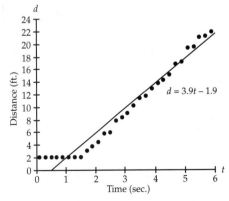

Figure 7.13.
Least squares line superimposed on scatter plot.

4. The screens in **Figure 7.14** show a distance-versus-time graph for a wind-up toy car powered by a spring. The times, x, in seconds and corresponding distances, y, in feet are indicated for three different times.

   a) Is the car speeding up, slowing down, or traveling at a constant velocity? How can you tell from the first graph in Figure 7.14?

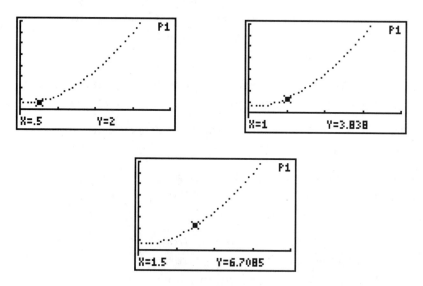

   Figure 7.14.
   Distance-versus-time graph for wind-up car.

   b) Compute the average velocity of the car the first half-second that the motion detector recorded data. What about the average velocities for the second and third half-seconds? Based on these average velocities, is the car speeding up, slowing down, or traveling at a constant rate?

   c) Explain how you could use the TRACE feature (if you had the actual data on your calculator) to compute an approximate velocity for a specific time. That is, how could you determine what a speedometer might read at a particular time?

LESSON TWO
Falling in Line

KEY CONCEPTS

Quadratic functions

Translations

Instantaneous velocity

Slope of a curve at a point

Symmetric difference quotient

Piecewise-defined graphs

The Image Bank

PREPARATION READING

The Death-Defying Leap

*A*fter completing Lesson 1, you should have some specific ideas about how mathematics can be used to plan successful two-vehicle, near-collision stunts. Before you staged your own stunt of this type, you analyzed distance-versus-time graphs produced by toy vehicles moving across a classroom floor. These graphs gave you information about each vehicle's velocity and allowed you to model the motion of individual toy vehicles. In particular, if a vehicle's distance-versus-time graph turned out to be approximately linear, you knew that your vehicle moved approximately at a constant velocity. Then you could determine its velocity from the slope of the graph.

In Lesson 1, all motion took place on a horizontal surface—the floor. Now you're ready to consider stunts that include vertical motion as well as horizontal motion. Recall, for example, Jeff Lattimore's specialty stunt "The Leap for Life" (Preparation Reading, Lesson 1). Jeff stands atop an eight-foot stool and jumps a moment before an oncoming car crashes into his stool. After the car passes beneath, Jeff lands safely on the ground. In this stunt, the car's motion is horizontal, and Jeff's is vertical. The thrill of this stunt is still "near collision," so Jeff must carefully synchronize his jump with the impact of the oncoming car.

In another variation of the falling-object-synchronized-with-moving-vehicle stunt, success is a planned "collision." For example, you've probably seen movies in which the hero, with no other route of escape, jumps off a building and miraculously lands in the back of a passing pickup truck loaded with mattresses (or an open garbage truck loaded with soft, squishy garbage).

In the next two lessons, you'll investigate mathematics useful in modeling falling-object-synchronized-with-moving-vehicle stunts. From Lesson 1, you already have developed some expertise in modeling the motion of the moving vehicles. All you need to add in order to plan similar stunts is experience in modeling falling objects.

CONSIDER:

1. Which stunt do you think would be more difficult to plan, a leap off a stool to avoid collision with an oncoming pickup truck or a fall from a building into the back of a pickup truck? Why?

2. How would you gather information about the motion of a person falling from a building or the motion of Jeff Lattimore leaping from his stool?

FALLING BODIES

ACTIVITY

5

CONSIDER:

After reading the Preparation Reading, how would you answer the following two questions?

1. Do you think that Jeff Lattimore would be able to execute "The Leap for Life" safely if a prankster sawed 12 inches off the legs of his stool? Why or why not?

2. Suppose that the stunt design calls for the hero to step off a rooftop the instant a pickup truck reaches a white line painted across the road. Do you think the hero would still land safely in the back of the pickup if the truck's velocity is 2 mph slower than called for in the stunt's design? What information would help you answer this question with more confidence?

Both stunts referred to in the Consider questions above involve falls. Jeff Lattimore's fall seems a bit more complex than the hero's fall because it is preceded by a jump upward. However, if you simplify Jeff's stunt by ignoring his initial jump (assume for the moment that he steps off his stool), then you can use the same form of model to describe both the hero's and Jeff's motion during their stunts. (All that's different is the initial height from which they fall.) Later, in Lesson 3, you can revise your model for Jeff's stunt to account for his initial jump.

Part I: SPECULATING ABOUT FALLS

1. Imagine that the hero steps backwards off a rooftop 25 feet above the ground and falls vertically to the ground.

   a) Sketch a possible distance-from-the-ground-versus-time graph for the falling hero. Add appropriate scales and labels to the axes. What leads you to believe that the hero will fall as you have described in your graph?

   b) How might you collect data that would help you decide whether your graph does a reasonable job describing the hero's fall?

ACTIVITY

5

FALLING BODIES

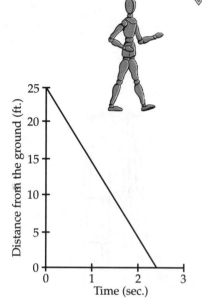

Figure 7.15.
Jackie's graph.

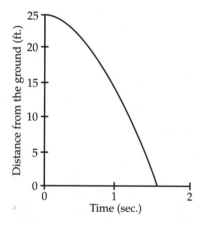

Figure 7.16.
Darryl's graph.

2. Jackie and Darryl drew the graphs in **Figures 7.15 and 7.16** as part of their answers to Item 1.

In Lesson 1, you experimented with having walkers walk toward or away from a motion detector (in a straight line) to produce similar distance-versus-time graphs.

a) Write instructions to a walker so that his distance-versus-time graph would look as much like Jackie's graph as possible.

b) Write instructions to a walker so that her distance-versus-time graph would look like Darryl's graph.

Whose graph, Jackie's or Darryl's, better describes what would happen if a person fell off a 25-foot rooftop? What kind of experiment could you perform in a classroom that could shed light on this question? Since dropping people is out of the question, you'll drop a book (or some other inanimate object) and use a motion detector to gather data as it falls.

You have two choices for the experimental set-up:

(1) You could put the motion detector on the floor and drop the book over the motion detector. Of course, you don't want the book actually to hit the motion detector. So, you will need to put the motion detector in a protective frame or have a member of your group catch the book before it hits the motion detector. (The frame is safer.)

(2) You could secure the motion detector above the book and then drop the book directly beneath the motion detector.

FALLING BODIES

3. Suppose that two students plan to use motion detectors to gather information about what happens when objects fall. The first student places the motion detector on the floor and drops the object onto the sensor; the second secures the motion detector above the object and drops the object beneath.

   a) Which student's graph will look most like the one you drew for Item 1?

   b) If both students drop the same book from the same height, what similarities and differences do you expect in their distance-versus-time graphs? You may want to illustrate your answer with sample graphs.

Part II: RECORDING THE FALLS

Enough speculation. What you really need are data. Of course, as was already noted, you can't drop a person. But different objects may fall differently. People are heavy, but not all equally so. It might help to experiment with a number of different books or other "heavy" objects. Therefore, if possible, not every group should use the same object.

Keep in mind your ultimate goal, too. In order to plan and carry out a successful "fall" stunt, you will need to be able to make accurate predictions for falling objects. Equations will probably be helpful, just as they were with the near-miss stunt. Likewise, knowing how the features of falls (such as starting location and velocity) relate to their equations can help you fine-tune your stunt for the most dramatic effects.

Recall that the HIKER program collected data over a period of six seconds. How long do you think it would take for a book to fall five or six feet? Certainly much less than six seconds. HIKER, which records one data point each 0.1 second, will be too slow to capture the important features of the motion of a falling book. Instead, you'll have to shorten the time that your program runs. Any program that can be used effectively to record the motion of objects falling from heights less than 12 feet will be called BALL-DROP. This type of program records data too quickly to graph

the data in real time. So, the program will record all the data first, then produce the graph after all data are collected.

4. a) Drop a book, and use a motion detector to collect distance-versus-time data on its fall. Your teacher will supply the directions. When you are satisfied with your data, transfer it to a location in your calculator where it won't get erased. Then link calculators and make sure that everyone in your group has the data.

   b) Make a rough sketch of the distance-versus-time graph produced by your data. Be sure to add appropriate scales and labels to the axes. Explain what each piece of the graph means. (If you followed directions exactly, you should have three distinct pieces. Don't worry if you have an extra small piece here or there.)

   c) Look at the portion of your graph that represents the motion of the object while it is falling. Was the object falling at a constant velocity? If not, did the object speed up or slow down as it fell? How can you tell from the graph?

Part III: ANALYZING THE DATA

5. You will need the results from this item for use in Activity 6. In particular, save your equation from part (d).

   a) Edit your data so that you have only the portion of data that represents the object's fall. Continue to edit (remove) any stray points so that the graph of the edited data is fairly smooth. Make a quick sketch of the graph of the remaining data. Be sure to specify the window settings for your graph.

   b) Look at the list containing the times for your edited data. Select two times from this list as follows: select time 1 from the times near the beginning of this list (but not the first entry) and select time 2 from the times near the end of this list (but not the last entry). What are your selected times?

ACTIVITY

FALLING BODIES

5

c) Now, imagine that you had a speedometer attached to the book during its fall. Use your data to estimate the speedometer reading at time 1. What if you had a "velometer"—a fictitious device that records **instantaneous velocity**—the rate of change of location with respect to time at the instant $t = t_0$. How would the velometer reading at time 1 be related to the speedometer reading? Explain. Use your data to estimate the book's velometer reading at time 2.

d) What type of equation do you think might describe the edited portion of the distance-versus-time data? Fit an equation to the edited data. Then make a residual plot. Does your equation appear to be a reasonable model for these data? (If not, select another type of function and try again.) Save your equation for use in later activities.

e) You learned in Unit 6, *Growth*, how to use first and second differences to identify certain types of graphs. Apply those methods to your edited data to confirm your results from part (d). Comment on the effectiveness of these methods here.

f) How might you use the equation from part (d) to get better approximations for your answers in part (c)?

6. If time remains, go back to Item 2. Select one of your group members to be the walker. Your "designated walker" should take two six-second walks and try to create distance-versus-time graphs that resemble the shapes of the graphs in Figures 7.15 and 7.16. (Don't worry about matching the scaling on the *x*-axes.) After the walker has completed both walks, identify any differences between the shapes in Figures 7.15 and 7.16 and the distance-versus-time graphs of the walks. Discuss possible explanations for those differences.

INDIVIDUAL WORK 4

Walk Like a Parabola

T he equation you found to describe your falling-body data from Activity 5 was probably quadratic. You are familiar with parabolas, the graphs of quadratic equations, but you may never have tried to match the features of such graphs (and their equations) with features of motion. For linear distance-versus-time graphs, for example, the slope provides useful numerical information about the velocity of the motion. How can you get similar information from situations such as those in Activity 5? This assignment explores velocity in new settings.

1. The calculation of average velocity uses the formula (distance 2 – distance 1)/(time 2 – time 1). In your work with quadratic functions in Unit 6, *Growth*, you found it useful to compute first differences. For distance data, the first differences are of the form (distance 2 – distance 1).

    a) Explain how to convert a table of first differences of distance data into a table of average velocities.

    b) Compute a table of first differences and average velocities for your edited data from Activity 5. Graph each quantity versus time and describe the graphs.

    c) Based on what you know from Unit 6, *Growth*, if the distance-versus-time data are perfectly quadratic, then the first differences (thus, the average velocities) will be linear versus time. Comment on how well this description applies to the graphs you made in part (b). How good is a quadratic model in describing the fall?

2. Imagine that you have the fictitious velometer that shows your velocity at each instant. (This will be your instantaneous velocity, which will be the same as a speedometer reading when your velocity is positive.)

    a) If you walk so that your distance-versus-time graph is a straight line, what can you say about your velometer readings during your six-second walk? Explain.

    b) What does that mean about the velocity in your Activity 5 experiment?

3. Recall the distance-versus-time graph of the toy car in Individual Work 1. A copy of that graph appears in **Figure 7.17**.

   a) Determine the toy car's average velocity between $t = 3$ and $t = 9$. Next, determine the car's average velocity between $t = 5.5$ and $t = 6.5$. Note that $t = 6.0$ is at the center of each of these intervals. Which of the two average velocities do you think comes closer to approximating the car's instantaneous velocity (its velometer reading) at $t = 6$ sec.? Why?

   b) Estimate the car's velocity at $t = 2.5$ sec.

   c) Would you be more confident estimating the toy car's instantaneous velocity at $t = 2.5$ or $t = 6$? Explain.

   d) At what time (or times) do you think the toy car's instantaneous velocity would be greatest? At what time (or times) would it be least?

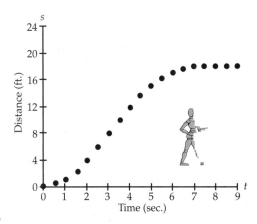

Figure 7.17.
Graph of toy car's motion.

Ms. Montgomery's math class had a contest to see who could walk the best parabola. Each group planned how their designated walker would have to walk in order to produce a distance-versus-time graph that looked like a downward-opening parabola. Items 4–6 refer to this (fictitious) experiment.

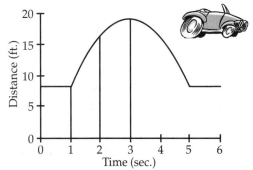

Figure 7.18.
Graph of Anita's walk.

4. Suppose that Anita was the designated walker for her group. The distance-versus-time graph for Anita's walk is displayed in **Figure 7.18**. The coordinates corresponding to points A, B, and C are (1, 8), (2, 16.25), and (3, 19), respectively.

   a) What was Anita's average velocity from $t = 1$ to $t = 3$? What was her average velocity from $t = 1$ to $t = 2$? From $t = 2$ to $t = 3$?

   b) What do the velocities in part (a) tell you about how Anita walked between $t = 1$ and $t = 3$?

   c) Does the average velocity you compute for the interval from $t = 1$ to $t = 2$ best represent the velometer reading for $t = 1$, $t = 2$, or some other instant? Explain.

d) Suppose you had data for every half-second instead of just for each full second. Explain how you could use first differences from those data to compute average velocities over shorter time periods.

5. Now, put yourself into Anita's shoes and imagine that you are walking her walk for Figure 7.18.

   a) Is your average velocity for the first half of your trip the same as your instantaneous velocity (the velometer reading) at each point along that part of your walk? Explain.

   b) Imagine that you look down at a speedometer (or a velometer) at $t = 2$. Which of your three answers to Item 4(a) do you think is closest to what the speedometer showed? Explain why you think you are correct.

   c) Now suppose that you look down at the velometer at $t = 3$. What is your instantaneous velocity? What are you doing at this point in your walk?

   d) What can you say about your velometer readings between $t = 3$ and $t = 5$?

6. Anita's group used quadratic regression to fit an equation to the data corresponding to times between $t = 1$ and $t = 5$ (the times that Anita was actually walking). After looking at the residual plot, they decided that the model $d = -2.75t^2 + 16.5t - 5.75$ did a good job in describing these data.

   a) Anita's average velocity between any two given times, say time 1 and time 2, can be calculated from the ratio (distance 2 – distance 1)/(time 2 – time 1) or (distance traveled)/(elapsed time). What would happen if you tried to use this same ratio to find her instantaneous velocity at $t = 2$ seconds, using both time 1 and time 2 as 2 seconds?

   You can't find average velocities over intervals of length 0. You *can* use very short intervals, though. The basic idea of computing an "instantaneous" velocity by finding the average velocity for a very short interval of time is sound. However, as you saw in Item 1, velocity calculations based on data from a motion detector have enough "noise" that the calculated velocities do not really represent the "smooth" nature of the fall very well. The equation you wrote in Item 5 of Activity 5, however, does capture the essence of the motion of your falling object; its graph *is* smooth, so calculations based on that model should actually be better than those from the real data. That's one advantage of finding a mathematical model!

The same applies to Anita's data. You can approximate Anita's instantaneous velocity at $t = 2$ by finding her average velocity over a very small time interval centered at $t = 2$. Although your brief time interval will be longer than an instant, you can still get pretty good results. And results based on the group's model (equation) should be better than those from the raw data.

b) Select an interval centered on $t = 2$ and use the quadratic model found by Anita's group to estimate her instantaneous velocity at $t = 2$. Is your estimate consistent with your answer to Item 5(b)?

c) Use the quadratic equation determined by Anita's group to estimate Anita's instantaneous velocity at $t = 3$. Repeat for $t = 4$.

d) One of the velocities that you calculated in part (c) should be negative. Interpret the meaning of a negative velocity in the context of Anita's walk.

e) If a distance-versus-time graph is linear, you can determine the instantaneous velocity at any time from the slope of the line. However, the graph of $d = -2.75t^2 + 16.5t - 5.75$ is curved. How might you define the "slope" of this graph at $t = 0.5$?

7. Return to your data from Activity 5. **Figures 7.19 and 7.20** display distance-versus-time graphs typical of data collected in that experiment. Figure 7.19 shows the graph of all the data recorded by the motion detector. When those data are edited to exclude all points not relevant to the book's motion while falling, the result is similar to that shown in Figure 7.20. Overlaying the graph of a quadratic model completes Figure 7.20. If you have not already done so, complete graphs like Figures 7.19 and 7.20 for your data.

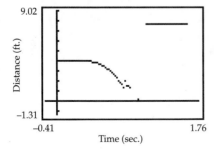

Figure 7.19.
Falling book's
distance-versus-time graph.

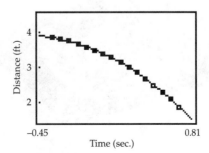

Figure 7.20.
Portion of graph representing the
fall, with quadratic model overlay.

INDIVIDUAL WORK 4

a) After your book was released, did it travel at the same velocity throughout its fall? How can you tell from your graphs?

b) Explain how you might use your quadratic model to estimate the velocity of the book at the instant $t = 0.5$ seconds (assuming that instant is in the "fall" part of your data). What is your estimate?

c) Use a calculator to graph your model in a window appropriate for that experiment. Sketch and describe that graph. Then trace to the point on your graph that corresponds to the time $t = 0.5$ (or, if that time is not during the "fall," to some other time when the book was still in the air) and zoom in on this point several times in order to magnify a small section of graph containing your point. Sketch and describe this magnified section of your graph.

d) Use your answer to part (c) to re-answer Item 6(e); that is, define the "slope" of a curve at a particular point on the curve. Explain your method and the idea behind it.

e) Explain what measuring the slope of your graph at time $t = 0.5$ tells you about the fall of the object in Activity 5.

BOOK DROP

Return to the problem of designing a "falling-
object-synchronized-with moving-vehicle" stunt
such as the simplified version of Jeff Lattimore's
"The Leap for Life" (in which Jeff steps off his stool into a fall).
Imagine performing the following scaled-down variation on
Jeff's stunt.

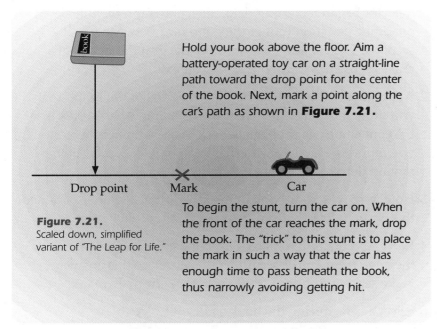

Hold your book above the floor. Aim a
battery-operated toy car on a straight-line
path toward the drop point for the center
of the book. Next, mark a point along the
car's path as shown in **Figure 7.21.**

Drop point   Mark   Car

To begin the stunt, turn the car on. When
the front of the car reaches the mark, drop
the book. The "trick" to this stunt is to place
the mark in such a way that the car has
enough time to pass beneath the book,
thus narrowly avoiding getting hit.

Figure 7.21.
Scaled down, simplified
variant of "The Leap for Life."

Think back to your planning for the two-vehicle near-miss stunt
in Lesson 1. There you needed to know the size of each vehicle,
where each started, and how fast each moved. Velocity allowed
you to determine how long each would take to get into (and out
of) the intersection.

In some ways you can think of the book-drop version of "The
Leap for Life" as a vertical version of a similar stunt. You still
need to know sizes, but now one of the "how long to reach the
intersection" times is the drop time for the book. It is unlikely
that the motion detector recorded the exact instant that you

ACTIVITY

BOOK DROP

6

released the book, and the range of your detector prevents it from "seeing" the very end of a drop. Therefore, neither the start nor the end of the drop are part of your original data!

1. In Item 5 of Activity 5, you determined a quadratic model that described the relationship between d, the book's distance from the motion detector, and t, the elapsed time since the detector began recording data. Sketch a graph of your quadratic equation (not of the data) for t between 0 and 2 seconds. Then shade the portion of the graph that actually makes sense for your book-drop context.

2. Approximate the coordinates of the vertex of the graph that you drew for Item 1. Interpret the meaning of the vertex in the context of this problem. Check to see whether the vertex you found is exactly one of the data points recorded by the motion detector during the Activity 5 experiment.

3. How long was your book in the air? Could you have obtained this answer directly from your data? Why or why not?

4. Use your equation for d to determine the average velocity of the book from $t = 0.1$ sec. to $t = 0.2$ sec. Does this answer make sense in the context of your book drop? Explain.

5. Select a time during the book's fall.

   a) Without calculating the book's velocity for the time you selected, is the velocity positive or negative? How do you know?

   b) Approximate the instantaneous velocity of the book at your selected time. Describe the method that you use.

BOOK DROP

For now, assume that you'll perform the vertical near-miss stunt holding the book at the same height as you did when you collected your group's book-drop data in Activity 5. In fact, you can use your quadratic equation from Activity 5 (the one that you graphed in Item 1) as the basis for a model for this scaled-down stunt. Recall that your equation relates the distance from the motion detector to the amount of time since the detector was turned on. Note that distance from the motion detector is not the same as distance from the floor; the motion detector has thickness!

6. a) Write a model that describes the relationship between h, the height of the book from the floor, and t, the elapsed time since the motion detector began recording data.

   b) How is the graph for h related to the graph that you drew for d in Item 1?

   c) Explain why the first and second coordinates of the vertex of the graph of h are important in planning the near-miss, book-drop stunt.

   In Item 2, you approximated the coordinates of the vertex of the d-versus-t graph. You also know how that graph is related to the h-versus-t graph. Here is a method for finding the vertex of *any* quadratic model quickly, easily, and exactly.

   Course 1, Unit 8, *Testing 1, 2, 3* introduced the quadratic formula for solving quadratic equations. Recall that an equation of the form $ax^2 + bx + c = 0$ has solutions given by

   the formula $\dfrac{-b - \sqrt{b^2 - 4ac}}{2a}$ and $\dfrac{-b + \sqrt{b^2 - 4ac}}{2a}$.

ACTIVITY

BOOK DROP

6

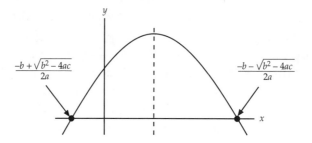

Figure 7.22.
Graph of a general quadratic function $y = ax^2 + bx + c$, $a < 0$.

You may also recall that the quadratic formula has a graphical interpretation. Suppose you have a quadratic model whose graph crosses the x-axis. For example, the graph might be similar to the one in **Figure 7.22**. Then the quadratic formula gives the values of the x-intercepts.

d) Use the quadratic formula and your model from part (a) to determine the "hit-the-floor" time for your book. You should get two answers. One of them is unrelated to the actual fall, but both make sense for the equation itself.

e) Based on Figure 7.22 and your two answers to part (d), what is the first coordinate of the vertex for your model?

Every parabola is symmetric about the vertical line drawn through its vertex. Thus the two x-intercepts of any parabola are equally distant from the vertex.

f) Use the information in Figure 7.22 and the symmetry of parabolas to find an expression in terms of a, b, and c for the first coordinate of the vertex of the graph of $y = ax^2 + bx + c$. Since your formula involves a, b, and c, it will apply to *any* quadratic model.

g) Use your answer to part (f) to determine the value of t at the vertex of the graph of your model for h. (Write your answers to four decimal places for use in later work.)

ACTIVITY

BOOK DROP

6

h) What does the number $\dfrac{\sqrt{b^2 - 4ac}}{2a}$ tell you about the book drop situation?

7. Use the information you developed in Item 6 to design a book-drop, toy-car, near-miss stunt. Do not carry out your design yet, though; just plan it.

8. a) What do you get when you substitute $t = 0$ into your equation from Item 6? What does this tell you about the motion of the book during the stunt? Based on your answer to this question, why might it be more appropriate to start timing the stunt once the book is released?

   For your model in Item 6(a), the explanatory variable, t, is the elapsed time since the motion detector began recording readings. Thus, it might be called t_{record}. This is not, however, the most appropriate explanatory variable to use. Discussions using this time measurement are not particularly informative. After all, you really need to know how long the book took to fall, from when it was released, not from when you turned on the detector. For example, if you compare the value of t for when your book hit the floor to the value of t for when another group's book hit the floor, does the comparison make any sense? A better variable might be t_{fall}, the elapsed time since the book was released.

   b) Select values for t separated by 0.05 seconds during the book's fall and record them in a table similar to the sample in **Figure 7.23**. Then use your equation to approximate the height of the book at these times.

t, elapsed time (sec.) since the detector began recording data	t_{fall}, elapsed time (sec.) since the book was released	h, height above the floor (ft.)
0.45		4.02
0.50		3.93
0.55		3.75
0.60		3.50
0.65		3.18

Figure 7.23.
Times and corresponding heights.

BOOK DROP

c) Now look at the data in your table from part (b). The time data in column 1 refer to the time that has elapsed since the motion detector began recording readings. You used these times to calculate the corresponding values for h. How would you adjust these times to complete column 2?

d) If you plot the data in column 3 versus the data in column 1, the points on your scatter plot will lie on the graph of h versus t and will be modeled by the equation you used to get those data in the first place. How would a scatter plot of column 3 versus column 2 relate to the first scatter plot? (You may wish to plot both scatter plots to be sure.)

e) As stated at the beginning of this problem, the old explanatory variable, t, which measured the elapsed time since the motion detector began recording data, is not the most natural variable to use in the context of this stunt. A better explanatory variable would be t_{fall}, the time elapsed since the book was released. Adjust the model (equation) for h so that the book's height can be determined from values for t_{fall}. (In determining this adjustment, use all four decimal places from your answer to Item 6(g). If you have trouble, recall how you translated absolute-value equations, as well as other types of equations, in Unit 1, *Gridville*.)

Save your model from Item 8(e) for use in Activity 7.

9. Use your model from Item 8 to verify your plan from Item 7. Which model is easier to use?

10. Execute your planned stunt. Comment on your success.

More Walkers

Recall from your work with the quadratic formula in Activity 6 that the first coordinate of the vertex of any parabola whose equation is of the form $y = ax^2 + bx + c$ is always $-b/(2a)$.

1. Sonia dropped the book for her group in Activity 5.
   To model the book's motion, her group used the equation
   $h = -15.32t^2 + 5.53t + 3.13$, where h is the height above the floor and t
   is the elapsed time since the motion detector began recording data.

   a) Sketch a graph of their model.

   b) Based on this model, how long after the motion detector began
      recording data did Sonia release the book? How did you deter-
      mine your answer?

   c) How high was the book the instant Sonia released it?

   d) If the motion detector had not been in the way, how long would it
      have taken from the time of release until the book hit the floor?
      Could you determine this value if you did not already know the
      coordinates of the vertex? How fast was the book moving when it
      hit the floor (hint: zoom)?

   e) Use your answers to parts (b) and (c) to re-express Sonia's model
      in vertex form (in other words, in "$y = a(x - h)^2 + k$" form, but you
      may use t instead of x if you prefer).

   f) The time variable, t, in Sonia's model refers to the elapsed time
      since the motion detector began recording data. Suppose Sonia's
      group wants to describe the relationship between h and t_{fall},
      where t_{fall} is the elapsed time since Sonia released the book. How
      can they do this?

As you have seen in this lesson, falling objects are reasonably well
described by quadratic models. In order to make the most of these
models, though, you will need to know how the control numbers for the
models relate to the physical features of the motion. Your work with
"fall" data is a start, but other motions may provide insight, too. For
example, think back to the walks you and your group took in Lesson 1.

INDIVIDUAL WORK 5

Time (sec.)	Distance from sensor (ft.)	Velocity (ft./sec.)
0	3	
1	6	
2	9	
3	12	
4	15	
5	18	
6	21	
7	24	
8	27	

Figure 7.24.
Distance-time data for Wanda's walk.

Time (sec.)	Distance from sensor (ft.)	Velocity (ft./sec.)
0	2	
1	3.15	
2	6.8	
3	12.95	
4	21.6	
5	32.75	
6	46.4	
7	62.55	
8	81.2	

Figure 7.25.
Distance-time data for Rhonda's walk.

2. Suppose that Wanda wanders along a line and that her motion is recorded by a motion detector. **Figure 7.24** shows her distance from the sensor every second.

   a) Plot the distance-versus-time graph from the table of Wanda's motion. Be sure to add labels and scales to your axes.

   b) Describe the shape of the graph of Wanda's motion.

   c) Based on your experience with such walks in Lesson 1, what should be true of Wanda's velocity if your description is correct?

   d) Copy Figure 7.24 and complete column 3. Check your conjectures from part (c). To estimate the instantaneous velocity at a particular time, use the average velocity for the smallest time interval centered at that time. For example, use the average velocity from $t = 2$ to $t = 4$ to approximate the instantaneous velocity at $t = 3$. This is called a **symmetric difference quotient** because the points used to calculate the slope are equidistant left and right from the point at which you want the instantaneous slope.

   e) Find an equation for Wanda's distance from the motion detector in terms of time.

   f) Plot the residuals for the equation you found in part (e) and interpret the plot.

3. Wanda's best friend, Rhonda, wanted to help. She also was recorded by the motion detector. Some of her data, for one-second intervals, are shown in **Figure 7.25**.

   a) Plot the distance-versus-time graph for Rhonda, including labels and scales.

   b) Describe the shape of Rhonda's graph.

c) Copy Figure 7.25 and complete the velocity column. Use symmetric difference quotients to do the calculations.

d) Did Rhonda walk at a constant velocity? Explain.

e) Based on your answer to part (d), it might make sense to examine a graph of velocity versus time for Rhonda's walk. Make such a graph and comment on what it tells you about the shape of her distance-versus-time graph.

f) Find an equation for Rhonda's distance from the motion detector in terms of time.

g) Make a residual plot for your equation in part (f) and interpret its meaning.

h) Use the equation you wrote in part (f) to zoom in on several of the points you included in your table in part (c). Compare the velocities based on zooming with the velocities you computed earlier using symmetric differences, and comment on the agreement between the two methods.

Save your tables and equations from Items 2 and 3 for use later in Individual Work 6.

4. What force or forces act on the book and cause its velocity to change during its fall? What would happen if these forces were not present when the book was released?

Talk About a Walk!

When Ffyona Campbell arrived in John o'Groat's, Scotland, on October 14, 1994, she deserved a rest. She had just finished walking 19,586 miles over a period of 11 years. Her trek took her through 20 countries on four continents and earned her a place in the Guinness Book of Records for the most miles walked by a woman.

Source: *The Guinness Book of World Records* (1957–1997 edition)

LESSON THREE
It Feels Like Fall

KEY CONCEPTS

Acceleration

Constant-acceleration motion

PREPARATION READING

See You in the Fall

Remember the hero perched atop the building with no route of escape? If he steps off now, will that approaching pickup truck loaded with mattresses be good news or bad?

Your work in Activities 5 and 6 has provided you with a good bit of experience with quadratic models for the falling objects you used in those experiments. But what about other objects? Do they fall according to the same equation? If each group in your class did a separate "drop" experiment, did you all get the same equations? Probably not.

Unfortunately, the data you've gathered from your one "drop" experiment only lets you solve a motion equation for the particular object you dropped. However, if you work with classmates from other groups, you now have data about a variety of falling objects. Those data can provide more insight into the general question of how things fall. Then you can build models that will allow you to design a safe "fall-into-the-back-of-a-moving-truck" stunt.

CONSIDER:

You will soon be asked to design a stunt similar to the hero's fall from a building into the back of a pickup truck.

1. Explain why it would be useful to be able to model the relationship between the height of a falling object and how long it has been falling.

2. Explain why it would be useful to be able to model the relationship between the object's velocity during the fall and how long it has been falling.

ACTIVITY

DO DROP IN

7

You've determined why this information is useful. Now it's time to put your models together and find the information you need.

First, take a close look at velocity. A constant velocity results in a linear distance-versus-time graph. A changing velocity results in a curved distance-versus-time graph. In fact, since quadratic models seem to do a reasonable job of describing falling objects, you can probably guess just how the velocity of a falling object changes. Your first task, then, is to see if the data support your guess.

1. Look back at the model you wrote in Item 8(e) of Activity 6. It represents the height of the falling object in terms of the time since it was released. If you have not already done so, rewrite your equation so that it contains no parentheses, and combine terms. Then round the constants to two decimal places. Write your simplified equation.

2. a) Use your model (equation) from Item 1 to complete a copy of the table in **Figure 7.26**. You may wish to share the calculations among the members of your group. (Round your final results to two decimal places.)

   b) Make a scatter plot for the velocity-versus-time data in your completed table. Then find an equation that describes the relationship between velocity and time.

Time t_{fall} (sec.)	Instantaneous velocity v (ft./sec.)
0	
0.1	
0.2	
0.3	
0.4	
0.5	

Figure 7.26.
Table of velocity-versus-time values (using translated times).

The rate of change of velocity with respect to time is known as **acceleration**. Acceleration occurs only when a force acts on an object. Since acceleration is a rate of change, you can calculate the **average acceleration** from time 1 to time 2 much as you calculate the average velocity. However, for average acceleration use the ratio

(velocity 2 – velocity 1)/(time 2 – time 1) = (change in velocity)/(change in time).

Since velocity has units of distance per unit of time, the units for acceleration are distance per unit of time per unit of time. For example, acceleration units for Item 2 would be feet per second per second, or feet per second^2.

ACTIVITY

7

DO DROP IN

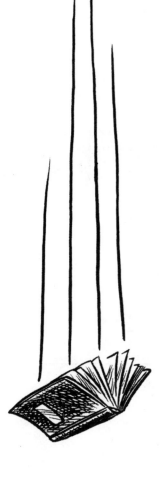

3. Use your table from Item 2 to find the average acceleration of the book over the following intervals.

a) What is the average acceleration from $t = 0$ to $t = 0.1$? Be sure to include the units as part of your answer.

b) What is the average acceleration from $t = 0.1$ to $t = 0.2$?

c) What is the average acceleration from $t = 0.2$ to $t = 0.3$?

d) Your velocity equation in Item 2 is linear. What is its slope? How does the slope of the velocity equation compare to the average accelerations that you computed in parts (a)–(c)? Explain why that makes sense.

e) What does it mean when the acceleration is negative?

Your results so far apply only to stunts involving dropping *your* object from the height *you* used. That's pretty restrictive! Generalize your work by pooling it with that from other groups. Your teacher will add your information to a transparency showing all the class data. Copy the class information onto Handout H7.6 to help organize the information you have.

4. What patterns, if any, do you see in the information on Handout H7.6?

5. Each group's height-versus-time model should be of the form $h = at^2 + c$. Interpret the values of a and c in the context of the falling book.

6. What is the connection between acceleration and your model (equation) for velocity?

7. Suppose that you drop a book from a height of 20 feet. Use ideas from Items 4–6 to predict a model for this situation.

a) Write a model describing the height of the book in terms of the elapsed time since the book was released. (Be sure to specify the domain of your model.) What assumption(s) did you make?

ACTIVITY

DO DROP IN

7

b) How long will it take for the book to hit the ground? First solve this problem by writing an equation and solving it using algebra (you should get two solutions to your equation, only one of which makes sense in the context of a falling book). Then explain how you could approximate the solution using a graph.

c) Write a model for the velocity of the book in terms of time. According to your model, how fast is the book traveling when it hits the ground? (For this item, assume that the book's fall is not broken by the motion detector and that it actually hits the ground.)

By this time, you probably have a good idea about how to model the motion of a falling book. When you released the book, its velocity was 0 ft./sec. The force of gravity caused the book to fall. Acceleration is always the result of force. The acceleration due to the force of gravity caused the book's velocity to change. Remember, if your motion detector was on the floor when you collected your data, then distance was measured "up" from the detector. So, negative velocity corresponds to falling. Since gravity pulls down, the velocities become more and more negative (the book falls faster), so acceleration is negative.

8. The usual value of acceleration due to gravity listed in physics books is –32 ft./sec^2 (or –9.8m/sec^2).

a) The constant that appears in your velocity equation is probably not –32 so you probably did not end up with a height-versus-time equation that looks like $h = -16t^2 + c$. Explain why that might have happened.

b) Suppose that you drop a balloon or a large, lightweight, beach ball. What force other than gravity might be acting on these objects as they fall? Write what you think would be a believable equation for the fall of such an object.

DO DROP IN

7

9. Set up the motion detector and try dropping a large beach ball or a balloon with several drops of water inside. Test your theory from Item 8(b).

a) What is your height-versus-time model for the falling beach ball (or balloon)? Remember to include the domain as part of the description of your model. Choose your time variable so that $t = 0$ corresponds to the release time.

b) What is your model for the beach ball's velocity in terms of time?

c) What is the acceleration of the beach ball?

Motion in Space: Free Rides and Aerobraking

Mathematicians and aerospace engineers have developed a way to use the gravity of the planets to direct the trajectories, or paths, of spacecraft. Essentially, a spacecraft can save fuel and assist its journey by flying close to a planet and using the planet's gravitational forces to speed up (get a free ride). To slow down, a spacecraft can dip lower into a planet's atmosphere and use the atmospheric drag in a technique called aerobraking. The first successful aerobraking occurred in 1993 when the Magellan spacecraft's orbit was lowered closer to Venus, positioning the spacecraft in a more circular orbit closer to the planet. From that position, Magellan could map more of Venus and gather more complete data on the planet's gravity.

INDIVIDUAL WORK 6

More Parabola Walkers

1. Look back at your data tables from Figures 7.24 and 7.25 in Individual Work 5. Use those data to complete the following analysis.

   a) Add a fourth column, labeled "acceleration," to each table. Use the average velocity for the smallest interval centered at a particular time to approximate the acceleration at that time; that is, use symmetric differences. For example, use the average acceleration from $t = 2$ to $t = 4$ to approximate the instantaneous acceleration at $t = 3$.

   b) What does the sign (+ or −) of Wanda's velocity tell you about how she walked?

   c) Now look at acceleration. After she started walking, what was Wanda's acceleration? What does her acceleration tell you about her velocity as she walked?

   d) What was Rhonda's acceleration? What does her acceleration tell you about how she walked? What does the sign of her acceleration tell you about her velocity?

2. Imagine that you are driving a car and traveling down the highway at 65 mph.

   a) Is your acceleration positive, negative, or zero? Explain.

   b) You decide to pass a truck, so you step on the gas pedal. Is your acceleration positive, negative, or zero? What effect does your acceleration have on your velocity?

   c) After passing the truck you let your velocity return to the 65 mph speed limit. What can you say about your acceleration as you return to 65 mph? What effect does your acceleration have on your velocity?

3. After collecting data from the book-drop experiment in Activity 4, Sonia's group found that $d = -15.46t^2 + 3.93$ did a good job in describing its motion. (Recall that d was the distance (feet) from the motion detector and t was the elapsed time (seconds) since the book was released.)

   a) Sketch a graph of this model in an interval that makes sense in the context of the falling book.

b) Your graph in part (a) records the relationship between the distance of the book from the motion detector and time. It does not record the path of the book. Sketch a time-lapse graph that describes the motion of the book along its path. Identify points on the path separated by 0.10-second intervals. Assume that Sonia is standing at the zero x-location and that she holds the book one foot away from her body as she drops it.

c) In Course 1, Unit 5, *Animation*, you used parametric equations to describe the motion of objects. In order to write a parametric equation for your graph in part (b), you need an equation for x that specifies the x-coordinate of the book's location at time t and an equation for y that specifies the y-coordinate of the book's location at time t. Write a set of parametric equations that describes your graph in part (b).

d) Change the mode settings on your calculator from function to parametric and change the format from connected to dot. (For example, if you are using a TI-83, you would press MODE and change the settings from Func to Par and from Connected to Dot.) Graph your parametric equations from part (c). Does your calculator-produced graph resemble the one that you drew for (b)?

e) How can you tell from the time-lapse graph that the book's velocity is changing?

4. Juanita is driving on the Interstate at 55 mph. She comes up behind a slow car and decides to pass. To pass, she steps on the gas pedal and goes around the slower car. Once around the car she slows back to 55 mph. **Figure 7.27** shows a graph of her velocity versus time for this event.

a) Describe her acceleration during this trip.

b) Sketch a graph of her acceleration versus time.

5. Michon walked in front of the motion detector producing a perfect parabola for her distance-versus-time graph. Her classmates found that the equation $d = -0.555t^2 + 7t + 2$ was a perfect fit for her data. Distances were measured in feet and time was measured in seconds.

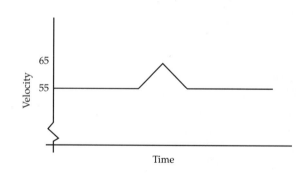

Figure 7.27.
Graph of velocity versus time.

a) Based only on this equation, predict whether Michon's acceleration was constant. If you think it was, predict its value (just from the equation). If you think it was not, explain why.

b) Complete a copy of the table in **Figure 7.28** using the equation that Michon's class found. Use symmetric difference quotients to approximate the velocity and acceleration entries.

Time (sec.)	Distance from sensor (ft.)	Velocity at time *t* (ft./sec.)	Acceleration (ft./sec.^{2})
0		xxxx	xxxx
1			xxxx
2			
3			
4			
5			xxxx
6		xxxx	xxxx

Figure 7.28.
Data from Michon's walk.

c) What does the sign of Michon's velocity tell you about how she walked?

d) What was Michon's acceleration?

e) What does her acceleration, particularly the sign of her acceleration, tell you about how she walked?

f) Extend the table to 10 seconds. Then, if necessary, revise your answer to part (e) based on this additional information.

6. John's movement is being recorded by a motion detector. He starts to move exactly when the detector starts to make readings. When he starts, he is 2 feet from the motion detector. He moves away rapidly at first, but slows to a stop and then moves back toward the sensor, speeding up as he goes.

a) Sketch a rough graph of his distance-versus-time graph.

b) Sketch a rough graph of his velocity-versus-time graph.

c) Is John's acceleration positive, negative, or zero? Why?

7. Suppose that John's average velocity for the first second (time 0 to time 1) was 7 feet per second. For the next second, it was 5 feet per second. After that, successive one-second average velocities were 3 feet per second, 1 foot per second, −1 foot per second, and −3 feet per second. Assume that he started 2 feet from the motion detector.

a) What does it mean to say that his average velocity for an interval was negative?

b) How far did he walk in the first second?

c) Make a careful graph of John's location versus time.

d) Calculate and describe John's acceleration. What does it tell you about his motion?

e) What kind of curve must describe John's distance-versus-time graph? Why?

f) Determine an equation that describes John's distance-versus-time graph. (You may use the regression capabilities of your calculator to determine this equation.)

g) Make a residual plot for your equation and use it to verify your prediction in part (e).

h) What do the coefficients in your equation tell you about John's motion?

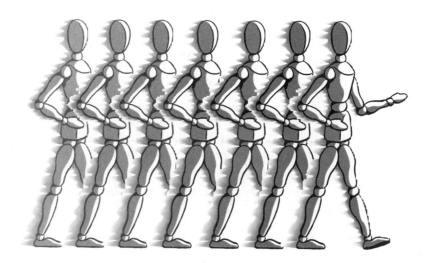

8. Renee attempted to walk away from the motion detector so that her distance-versus-time graph would be a parabola. Her data are repeated in **Figure 7.29**.

Time (sec.)	Distance from sensor (ft.)	Velocity at time t (ft./sec.)	Acceleration (ft./sec.^{2})
0.0	2.0		
0.5	2.7		
1.0	3.6		
1.5	4.8		
2.0	6.5		
2.5	8.7		
3.0	11.7		
3.5	15.6		
4.0	21.0		

Figure 7.29.
Data from Renee's walk.

a) Make a careful graph of Renee's data. Be sure to scale and label your axes.

b) Based only on your graph, do you think Renee was slowing down or speeding up during her walk? Describe the feature of your graph that helps you decide, and explain your reasoning.

c) Given your answer to part (b), was Renee's acceleration positive, negative, or zero? Explain.

d) Copy and complete the table in Figure 7.29. Use symmetric difference quotients to calculate the entries in the velocity and acceleration columns.

e) Using the table that you just completed, do you think Renee's distance-versus-time graph is a parabola? Explain your reasoning.

f) Use the quadratic regression to fit a quadratic equation to Renee's data. Then make a residual plot for your equation, and use it to check your answer in part (e).

ACTIVITY

8

DESIGNING THE HERO'S FALL

For this assignment, you will plan the details of a stunt in which the hero falls off a rooftop into the back of a pickup truck driving parallel to the building. In your plans you'll need to specify the height of the building, where to mark the road to signal the hero that it's time to fall, the speed of the pickup when it reaches this mark, and the relevant dimensions of the pickup.

1. Specify the following details. (Be as realistic as possible.)

   a) Building height.

   b) Length and height of the truck's bed as well as the length and height of the truck's cab.

   c) Truck's speed when it reaches the signal mark (converted to feet per second).

2. Develop a model that describes the motion of the hero during his fall from the building. State any assumptions that you make.

3. How long will it take the hero to fall to the height of the cab? How long will it take him to fall to the height of the back of the truck?

4. Next, you need to determine where to place a mark on the road. The hero should begin his fall the instant the front of the truck reaches this mark. Where should you place this mark so that the hero will land safely in the back of the truck without hitting the cab of the truck? (Specify the distance between the mark and the hero's drop point.) Make a sketch indicating where the hero will land in the back of the pickup. Explain how you determined your answers.

5. Another concern related to the survival of the hero is his vertical velocity the instant that he lands in the back of the truck. Approximate that velocity and explain how you determined your answer. Convert your answer to miles per hour. Do you think the hero is likely to "feel" the impact?

6. How sensitive is the success of this stunt to the hero's timing? Investigate how close to "on time" he must jump in order to hit the truck safely.

The Image Bank

PREPARATION READING

The Return of Jeff Lattimore

In Jeff Lattimore's stunt "The Leap for Life," Jeff leaps upward
into the air an instant before an oncoming car rams into his
stool. In Lesson 2, you simplified Jeff's stunt by ignoring his
initial leap. This allowed you to apply to Jeff's fall what you
had learned about other falling objects. Take a moment to
reflect on what you've learned from Lessons 2 and 3 about
how things fall.

Recall that when an object is released from a height, its fall produces a parabolic distance-versus-time graph. You can model the object's fall using a quadratic equation of the form $h = at^2 + c$, where h is the height of the object and t is elapsed time since the object's release. The constants a and c also have meaning in the context of the falling object: a is one-half the acceleration and c is the initial height of the object. Constant acceleration and linear velocity with respect to time further characterize this type of motion.

If no forces other than gravity act on a falling object, its acceleration is -32 ft./sec^2. Air resistance, however, acts in a direction opposite to the motion, causing a decrease in the magnitude of the acceleration. On a heavy, compact object, air resistance is probably negligible, especially if the object is dropped from a relatively low height. However, on a light object with a relatively large surface area (such as a balloon or a piece of tissue paper), air resistance is not negligible.

In this lesson you will add one additional element to your models of the motion of falling objects. You will model the motion of an object that is tossed upward and allowed to drop. Then you can allow Jeff Lattimore to leap upward from his stool, thus removing the restriction that he simply step off the stool, and you can adjust your model to account for this more complex motion. After adjusting your model, you can decide for yourself whether sawing 12 inches off Jeff's stool puts his life in danger.

CONSIDER:

1. Would you expect air resistance to be much of a factor in Jeff Lattimore's jump? Explain.

2. How high above the stool do you think Jeff can jump?

IT'S A TOSS UP

You know how to model motion when it starts with a stationary release. Now, in order to model Jeff's motion as he leaps from his stool and falls to the ground, you need to know something about how an initial vertical jump affects the motion.

You need to gather more data about leaps and falls. For this lesson, assume that Jeff leaps vertically upward and falls vertically to the ground. In reality, of course, there may be some forward momentum to Jeff's jump. You probably should ignore that possibility until after Lesson 5, where the analysis of a car's motion in a ramp-to-ramp jump will give you additional experience with motion that has both horizontal and vertical components.

As was the case with falling objects, you really can't experiment on people. So a good way to investigate the motion involved in Jeff's stunt is to study smaller objects that are tossed upward and then allowed to fall. That's exactly what you'll do in this activity. You'll use a motion detector to track the motion of a basketball (or some other similarly sized ball) that has been thrown upward. If you choose not to use basketballs, be sure to use other objects that are large enough for the motion detector to "see."

Recall that dropping a variety of different objects in Activity 5 produced data that allowed you to see how equations of motion varied for different falling objects. Since different groups will toss objects with different initial velocities, you may be able to see the effect of those different tosses through new data.

ACTIVITY

9

IT'S A TOSS UP

Part I: PREDICTIONS

1. Before you actually throw the basketball straight up into the air and use a motion detector to record its motion, think about the motion.

   a) What forces will act on the ball during its flight?

   b) What do you think its distance-versus-time graph will look like? Why?

Part II: THE EXPERIMENT

Now prepare the equiment for the toss. Attach the motion detector to your calculator. Place the motion detector on the floor so that its beam is pointing straight up.
Refer to **Figure 7.30**.

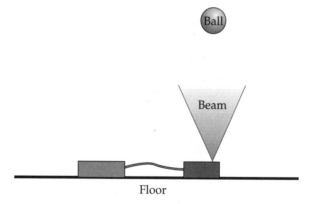

Figure 7.30.
Motion detector set-up.

Perform the following experiment.

One member of your group should kneel and hold a ball over the motion detector.

The program operator should execute a BALLDROP program. He should say "go" the moment that the motion detector begins recording data.

IT'S A TOSS UP

When the person holding the ball hears the word "go," she should toss the ball straight up.

Designate some member of your group to catch the ball just before it reaches the motion detector. Watch out for the tosser's (or catcher's) hand or head getting in the sensor's beam.

If this sequence is done properly, the motion detector will record the height of the ball not only during its flight, but also before it is thrown upward. You may need to repeat this experiment several times before you get a set of data that produces a good graph.

Once you have a good graph, link all the calculators in the group and share the data.

2. Make a sketch of your graph. Add appropriate scales and labels.

Part III: MODELING

Before beginning work with your data, save the original data for use in Individual Work 7.

3. If your group followed the instructions for the experiment carefully, your graph in Item 2 should show three distinct stages. First, the tosser held the ball stationary. Next, the ball was moved upward while it was still being held. Finally, the ball left the tosser's hand and traveled "freely." Identify the locations of the ball and the times at which each of these stages began. Explain how you determine your answers.

4. Use your results from Item 3 and the method of your choice to determine a model (equation) for the distance d above the motion detector to the ball, in terms of the time t_{fall} since the ball was released. Note that for this choice of time, you will need to use a translation of the data or a translation of a preliminary equation.

Save your data and results for use in Individual Work 7.

INDIVIDUAL WORK 7

Three-Point Shots

Your work in Item 4 of Activity 9 may have been somewhat more difficult than that for corresponding equations for Activity 5. Data from the ball-toss experiment may not be as "clean" as the data that you collected when you simply dropped a book. In particular, if you are not able to toss the ball perfectly vertically, the motion detector may fail to track the ball at each reading, so it may include some "stray" readings. This assignment examines several ways to avoid being sidetracked by such readings.

1. Edit your data. Keep only those data relevant to the ball toss; exclude any stray points. Describe any difficulties you have in carrying out this task.

You may have found in Item 4 of Activity 9 that stray points affected your ability to fit a model using regression. Depending on how many such points there are, their removal could be tedious and time consuming. Items 2 and 3 develop methods for fitting a quadratic equation to data that avoid the need to remove stray points (data that are not related to the actual motion). You will have a chance to use these methods to refit a model to your data.

2. Consider the data in **Figure 7.31**.

Figure 7.31.
Table of data.

x	0.1	0.2	0.3	0.4	0.5	0.6	0.7	0.8
y	9.84	9.36	8.56	7.44	6.00	4.25	2.16	−0.24

a) Without fitting a model to these data, how could you tell if these points lie on a parabola? Do they lie on a parabola?

Remember that the standard form of a quadratic equation is $y = ax^2 + bx + c$. Since you know that each point is from the same equation, each pair of numbers (x, y) in the table must make this equation true. If the first point is selected and placed in the general quadratic equation, the result is:

$9.84 = a(0.1)^2 + b(0.1) + c$, or $9.84 = 0.01a + 0.1b + 1c$.

Therefore, this equation says that (0.1, 9.84) is on the given parabola.

If two more points are selected and substituted into the general quadratic equation, you will have a total of three equations written in terms of the unknowns a, b, and c. If the resulting system is solved, you will determine values for a, b, and c. Substituting these values into the general quadratic equation will give a quadratic model for the data in Figure 7.31.

b) Have each member of your group pick three different points from Figure 7.31 and write the corresponding system of equations.

c) Use the method of your choice to solve your system of equations, and write the corresponding quadratic equation for the data using the values of a, b, and c that you found.

d) Compare the quadratic equations obtained by each group member. What do they tell you about the data?

e) Enter all eight points into your calculator's lists. Make a scatter plot of these data. Then enter your equation from part (c) and overlay its graph on your scatter plot. Describe the fit.

f) Compute the residuals for this equation. Do the residuals support your conclusion from part (d)?

3. Look back at the data in Figure 7.31. Do you know how many points you must have in a data set for your calculator to carry out a quadratic regression? You know it can be done with eight points because you just did it. Experiment to determine the fewest number of points needed. Then use your work in Item 2(c) to explain why your conclusion is correct.

OBSERVATION: In each of the last two items, you have found the equation of a parabola by using just three points. You have long known that any two points completely determine a line, and you know how to find such a line's equation from the two points. You now know a similar fact about parabolas: any three points (not in a straight line) will completely determine a parabola described by an equation of the form $y = ax^2 + bx + c$. In addition, you now know two ways to find its equation, one using systems of equations and the other using regression.

4. If you were to fit a model to your Activity 9 data, would you need all the data you collected? What is the minimum number of points you would need? What assumption are you making?

5. Find an equation to describe the "motion segment" of the original data from Activity 9. Your equation should specify the relationship between d, the distance of the ball from the motion detector and t,

the elapsed time since the detector began recording readings (*not* translated to the release time). Describe the method that you used to get your equation. Verify your equation using residuals.

6. In your equation from Item 5, $t = 0$ has no meaning in terms of the ball's motion since it is before the release time. Use your equation from Item 5 and your value for the release height from Item 3 of Activity 9 to determine the time when the ball was released. Explain your method.

7. Your equation in Item 5 gave a relationship between d, the distance from the motion detector and t, the elapsed time since the motion detector began recording data. Let h denote the distance from the floor (not the detector) to the ball, and let t_{fall} denote the time since the ball was released.

   a) Ignoring the equations and data for a moment, explain how a graph of h versus t_{fall} would compare to a graph of d versus t if they were on the same axes.

   b) Use your description from part (a) to adjust your equation from Item 5 so that it models h versus t_{fall}.

   c) If you have not already done so, express your quadratic model from part (b) in standard form (that is, $h = at^2_{fall} + bt_{fall} + c$). Round the constants to two decimal places. Save your model for use in Activity 10.

   d) Based on your model, not the data, when did the ball reach its highest point? How can you determine this time from your equation? How high did the ball go?

You and your classmates have now found quadratic models for the data from your "toss-up" experiments. These equations differ from those in Lessons 2 and 3 in that they have three terms, not just two. Now there is a linear term, bt. To design new stunts, you need to relate the coefficients of your models directly to the motions. That way, you can select control numbers to produce the motion that you want, not just describe what has already happened.

Part I. LINKING MOTION AND MODELS

There really are two questions about the control numbers in your new models. First, do the t^2 and constant terms still mean what they did when there was no initial toss? And second, what information does the linear term provide? Answering these questions is a major goal of this mathematical exploration.

It really takes just three bits of information to describe any particular toss-up experiment. You need to know how hard the toss was, the release point for the toss, and the conditions under which the toss was performed. With that in mind, Items 1–3 ask you to describe your toss-up experiment in detail.

1. Based on your model from Item 7(c) of Individual Work 7, how high off the ground was the ball when it was released? How did you get this information from your equation?

Recall that you can use the following method to approximate the instantaneous velocity of the ball at any time t_0. Graph h, then TRACE to the point corresponding to t_0. Zoom in on this point until the graph on your screen looks like a line. Select two points on this "line" on either side of the point corresponding to t_0. Use these points to calculate the slope of this "line." This slope gives you an approximation of the instantaneous velocity at time t_0.

ACTIVITY

IN CONTROL

10

2. The velocity of the ball the instant it is released is called its **initial velocity**. Use your translated model from Individual Work 7 and the method described in the previous paragraph to approximate the initial velocity of the ball the instant it was released (in other words, when $t_{fall} = 0$).

3. Use methods you developed in Lesson 3 to determine the acceleration for the ball in your toss-up experiment.

4. Look at the coefficients in your model from Individual Work 7. Discuss the relationship between the motion of the ball and the values of the control numbers in your equation. Confirm any patterns you find in your data by checking with class-mates in other groups.

Your conclusions about the role of the control numbers in quadratic motion may be correct, at least for motion under the influence of gravity. But you have seen other motion— walks, not falls. You may wish to review your work in Individual Work 6, for example, to compare those results with Item 4, above. However, no matter how many examples you examine, you can't check them all. Is there a method of con-firming your conclusions for *all* quadratic motions, once and for all? Yes! Use the method of generalization!

5. a) For motion with constant velocity, the model (equation) for the motion is linear. Quadratic motion involves velocity that changes. Identify and describe the features of quadratic motion that permit you to know that the data are quadratic, without fitting a quadratic using quadratic regression and checking residuals.

   b) The method of generalization can help you decide whether the features you identified in part (a) really characterize every quadratic motion. Start with the general equation $d = at^2 + bt + c$. Remember that a, b, and c are all control numbers; they are constants and don't change during the motion.

ACTIVITY

IN CONTROL

10

Construct a time-location-velocity-acceleration table for this equation. A few entries in **Figure 7.32** have been done for you. For example, the entry for $t = 2$ in the Distance column was obtained by substituting 2 for t in the d equation. The entry for $t = 2$ in the Velocity column was obtained by computing the symmetric difference quotient: $((4a + 2b + c) - c)/(2 - 0) = 2a + b$.

Time	Distance	Velocity	Acceleration
0	c	xxxx	xxxx
1	$a + b + c$	$2a + b$	xxxx
2	$4a + 2b + c$		$2a$
3		$2a + b$	0
4	$16a + 4b + c$		0
5			xxxx
6		xxxx	xxxx

Figure 7.32.
Time-distance-velocity-acceleration chart for a general quadratic.

c) What does your new table tell you about the velocity and acceleration of quadratic motion? Write equations for velocity and acceleration if you can. (Note: your result is worth remembering!)

d) Compare your results to Item 4, above.

e) The vertex of the time-series graph of quadratic motion represents the point at which the motion stops going in one direction and reverses to go in the opposite direction. At that point, the velocity is 0. Use that fact, together with your equation for velocity in part (c) to come up with a formula for the "time" coodinate for the vertex of *any* quadratic motion. How does your answer compare to the one you get from the quadratic formula?

ACTIVITY

IN CONTROL

10

You now know formulas for the velocity and acceleration for any quadratic motion you ever encounter, right? Well, sort of. After all, your calculations in Item 5 are based on velocities computed as *average* velocities across intervals of one full second. They really were not instantaneous velocities.

6. Explain why the acceleration values found in Item 5 must be correct if the velocity values are correct, even if the accelerations are taken over intervals of two full seconds.

Part II. EXPLORING VELOCITY IN QUADRATIC MOTION

You have used two approaches for approximating the instantaneous velocity of an object. One method requires that you have an equation for the motion and involves "zooming in" until the curve looks like a line. You approximate the instantaneous velocity from the slope of this "line." This method seems to be most reliable; it agrees with the intuitive meaning of "instantaneous."

The second method can be used to approximate the instantaneous velocity from data even when you don't have an equation. You approximate the instantaneous velocity using symmetric difference quotients to compute a slope. How accurate is this second method compared to the first? That's what you are about to find out.

The symmetric difference quotient gets its name from the fact that the times used are chosen symmetrically, centered on the time for which the velocity is needed. In symbols, the symmetric difference quotient for the velocity at time B looks like this: $\dfrac{d(B+k) - d(B-k)}{2k}$.

Here, $d(B + k)$ is written in function notation and means "distance at time $B + k$." It does not mean multiply d by $(B + k)$. For example, to compute the velocity at time $B = 0.7$ using a symmetric difference quotient you might select $k = 0.1$. Then

ACTIVITY

IN CONTROL

10

the formula says: velocity = $(d(0.8) - d(0.6))/0.2$, or the difference between the locations at times 0.8 and 0.6, divided by the elapsed time, 0.2 seconds.

7. Within your group, make up a quadratic equation in the form $d = at^2 + bt + c$. (For example, the equation $d = -2.75t^2 + 16.5t - 5.75$ would work. However, change the constants to create a quadratic equation of your own. You may use the model from your toss-up experiment if you like.)

   a) Sketch a graph of your chosen equation. Select some "random" t-value as time B and mark the corresponding point on your graph. Each person in your group should use the same B value.

   b) Use the zoom-in method to approximate the "slope" of your parabola at $t = B$. Be as exact as you can, and agree as a group on this slope.

8. Use the same equation and value for B as in Item 7. Remember, you are trying to examine the effect of using fairly large intervals for velocity calculations. Thus, you need to vary k in the symmetric difference formula and see how things change. Split up the work on the following parts among the members of your group. Use the values for k indicated in **Figure 7.33**. Set your WINDOWs to Xmin = $B - k$ and Xmax = $B + k$ and choose appropriate values for Ymin and Ymax. Then complete parts (a) – (c) individually.

Person	"k"
1	0.01
2	0.1
3	1.0
4	2.0

Figure 7.33.
Assigned k-values for symmetric differences.

   For example, person 1 will set the WINDOW to Xmin = $B - 0.01$ and Xmax = $B + 0.01$ (using the value of B that you selected in Item 7) and choose appropriate values for Ymin and Ymax.

   a) Make a sketch of the graph in your designated window.

   b) Find the slope of the line connecting the points on your curve for times $B - k$ and $B + k$.

ACTIVITY

IN CONTROL

10

c) How does your answer compare to the value you obtained in Item 7? Are you surprised? Explain.

d) When you all have completed your individual sections of the graph, compare your graphs and computations, and summarize your conclusions.

9. Do your results depend on having a special B? Repeat Items 7 and 8 using the same equation, but one or more different points for B.

10. Do your results depend on having a special parabola? Visit with members of other groups and compare your conclusions to theirs.

11. Repeat Items 7 and 8 using an equation of your own choice that is neither linear nor quadratic. (Include a term like t^3 or t^4.)

12. Your work so far raises an interesting question about the symmetric difference quotient for parabolas. The method of generalization can be used to check your conjecture. Here's how: Instead of the equation you have been using, use the general equation $d = at^2 + bt + c$. (Remember that a, b, and c represent control numbers; they're constants.) That equation will generalize the curve. To generalize the interval, use times $B - k$ and $B + k$; that is, use letters, not numbers.

a) Find the slope. (Hint: The distance for time $B + k$ is $a(B + k)^2 + b(B + k) + c$, or $aB^2 + 2akB + bB + ak^2 + bk + c$.)

b) If you did part (a) correctly, your answer will have no ks in it. If that is true, how does changing k change the slope of the segment?

c) What does k measure in the context of the symmetric difference quotient?

ACTIVITY

IN CONTROL

10

   d) Instantaneous velocity was described as the slope of the distance-versus-time graph after you had zoomed in far enough for it to look linear. Explain what zooming in has to do with k, then interpret what your answer to part (b) means about the instantaneous slope.

13. In Item 12, you obtained a formula that allows you to compute the slope of any parabola at any B value. If that parabola represents the distance-versus-time graph of a moving object, then this slope is the instantaneous velocity of the object.

   a) Use your equation from Item 12 to check the results you got in Items 7–9. That is, substitute the values of a (the leading coefficient), b (the linear coefficient), and B (the time value) from Items 7–9 into your equation from Item 12(a). Do you get the same results using Item 12(a)'s equation as you did in Items 7–9?

   b) For each B-value that you tried in Items 7–9, list the ordered pair (B, slope). These all describe situations on the same parabola. Graph these ordered pairs and describe the graph as completely as you can. Compare your answer to your answer for Item 12(a).

   c) For the special case of $B = 0$, what does the formula in Item 12(a) tell you about the relationship between the instantaneous velocity of the object and the equation for its motion?

Well, now you have it! If an object's motion can be modeled with a quadratic equation of the form $d = at^2 + bt + c$, then its velocity is linear ($v = 2at + b$), and its acceleration is constant ($2a$). Furthermore, in this situation, symmetric difference quotients give exact values for instantaneous velocity. That pretty much characterizes quadratic motion.

INDIVIDUAL WORK 8

What Goes Up Must Come Down II

1. The calculator screens shown in **Figure 7.34** show portions of the data collected by a motion detector for a book in flight over it. Times in L1 are in seconds and are measured from the time of release. Distances in L2 are in feet.

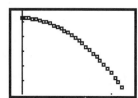

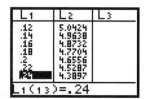

Figure 7.34.
More book-drop data.

   a) How many points do you need to get an equation to model the fall of the book?

   b) Find a d versus t equation for the book's motion.

   c) Do you believe the book was dropped or tossed? How can you tell?

   d) What do the coefficients in your model tell you?

   e) What equation would you use to model the book's velocity with respect to time?

   f) How long was the book in the air before it hit the detector's frame? How do you know?

   g) What was the book's velocity when it hit the frame?

   h) What was the book's acceleration?

2. Suppose a soccer ball on the ground is kicked with an initial upward velocity of 40 ft./sec.

   a) Write your model for the height of the soccer ball with respect to time. Use the model for this motion as if it were in a vacuum (acceleration = –32 ft./sec.^{2}): $h = -16t^2 + v_0t + h_0$. Explain what v_0 and h_0 represent in this context.

   b) How long is the ball off the ground? Write an equation that you could use to answer this question. Solve your equation using algebra. How could you solve this problem graphically?

   c) For how long is the soccer ball more than eight feet above the ground? Write an equation that is helpful in answering this question. Then solve your equation using algebra. In addition, show how you can solve this problem graphically.

d) How high does the ball go? When does it reach that height?

e) Repeat parts (b)–(d) using the value for acceleration from the book-drop data in Item 1(h). (You can use either an algebraic or graphical approach; you do not need to use both.) What model did you use? Are these results what you expected?

3. Each of the following equations could describe the motion in a vacuum of an object being acted on by gravity.

Object A: $h = -16t^2 + 90t + 100$

Object B: $h = -16t^2 + 55t + 150$

Object C: $h = -16t^2 + 20t + 200$

a) Which object reaches the greatest height?

b) Object B is thrown with an initial velocity that is more than twice that of object C. Why does Object C reach a greater height than Object B?

4. A diver stands on a platform that is 40 feet above the water. She executes her dive. Her average speed for the first 0.1 second is 18 ft./sec. upwards and she hits the water in 2.4 seconds.

a) Write an equation for her height above the water at any time. How did you determine your equation?

b) When does she attain her maximum height?

c) What is her maximum height?

d) How fast is she going the instant she reaches her maximum height? How do you know?

e) Is the situation as described above realistic? Is the modeling? Explain why or why not.

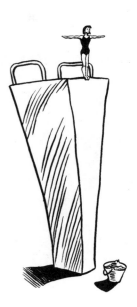

5. The equation $h = -15.2(t - 2)^2 + 40$ models the height of a rock shot straight upwards. Distance is measured in feet and time is measured in seconds. Assume that the rock started to move when $t = 0$.

a) How far was it from ground level when it was shot?

b) What was its initial velocity? How did you determine your answer?

c) What was its average velocity between 1.9 and 2.0 seconds? How about between 2.0 and 2.1 seconds?

d) Complete a velocity table like the one in **Figure 7.35**. Use symmetric difference quotients to approximate these velocities.

Figure 7.35.
Table of velocities.

Time (sec.)	1.8	1.9	2.0	2.1	2.2
Velocity (ft./sec.)					

e) Describe what you see in the table.

6. a) Return to your model from Item 7(c) of Individual Work 7, describing the motion of the basketball using times measured since its release. Determine the instantaneous velocities of the ball for the times recorded in **Figure 7.36**.

Figure 7.36.
Velocity-versus-time table

Time (sec.)	0	0.1	0.2	0.3	0.4
Velocity (ft./sec.)					

b) Sketch a graph of velocity versus time. What type of equation describes this relationship?

c) Write an equation for velocity in terms of time. Interpret the coefficients in terms of the motion.

7. Using the concepts of acceleration and velocity, discuss how you would walk so that your distance-versus-time graph would be a line. Do the same thing for a parabola.

ASSESSING THE MODEL

In Item 2 of Individual Work 8, you found that a kicked soccer ball went higher *with* air resistance than without it. That can't be correct in reality!

CONSIDER:

1. Do you think that air resistance is more of a factor with a light plastic ball or a heavier basketball? How does air resistance affect the acceleration of a ball as it travels upward? How does it affect the ball as it falls? Explain.

2. Does your answer to the previous item suggest a refinement to your models of Activity 9 and Individual Work 7, particularly if air resistance is not negligible? How might you go about using your data to find a better model?

When modeling, it is important to assess which factors are most influential. What can you ignore and what must you include in your model? For example, you know that real stunts take place in air, so there must be some air resistance. But does it matter? How much does air resistance affect a ball's motion when it is tossed upward and allowed to fall?

Depending on the kinds of balls that were used in Activity 9, you may already have the data that you need to determine the effect of air resistance. If a variety of balls were used in that experiment, move directly to Item 2. Otherwise you will need to gather more data.

1. Obtain a ball that is similar in size but significantly different in weight than the one used in Activity 9. Use the same equipment set-up and procedures as you used in Activity 9 to obtain data from the toss of this new ball.

ASSESSING THE MODEL

2. Explain why different equations might apply to the "up" and "down" portions of the data.

3. Use your new data and/or that from Activity 9 to determine four quadratic equations, one for each ball's upward motion and one for each ball's downward motion. Piece these equations together to determine piecewise-defined models, taking into account the effect of air resistance on the motions of the balls.

4. What do your piecewise models tell you about the effect of air resistance on a ball's acceleration as it moves upward and then downward?

5. Now fit a single quadratic model to the entire flight of each ball. (If you are using data from Activity 9, you did this in Individual Work 7 for at least one of the balls.) Compare the graphs of the piecewise models from Item 3 to the single-quadratic models. What do the graphs tell you about the effect of air resistance on the ball?

6. In this situation, do you think the piecewise model that accounts for the effect of air resistance is worth its added complexity? Do you think it would make a lot of difference if you used the piecewise model instead of the quadratic model in designing stunts such as Jeff Lattimore's leap from his stool? Can you think of a situation in which it would be absolutely necessary to include this feature in your model? Explain.

INDIVIDUAL WORK 9

Air Ball

1. Georgia's group just completed Activity 11. For their lightweight ball, they selected two data points collected during the ball's upward motion, the data point corresponding to the highest recorded height, and two data points collected during the ball's downward motion. These data are listed in **Figure 7.37**.

t, elapsed time since the ball's release (sec.)	h, distance from ground (ft.)
0.28754	4.13005
0.35260	4.38118
0.43857	4.50142
0.49724	4.45668
0.58221	4.23999

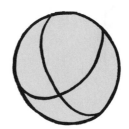

Figure 7.37.
Sample data from Georgia's group.

a) Use the data in Figure 7.37 to determine two quadratic equations, one for the ball's upward motion and the other for the ball's downward motion. Piece these two equations together to determine a model for the motion of the light ball.

b) Use the method of your choice to compute a single quadratic model for their data. Compare the graphs of your piecewise model to your single-quadratic model. Are results similar to what you found in Activity 11?

2. A stunt driver drives a car off a 40-foot-tall ramp at 60 mph. The ramp is parallel to the ground. Ignore the effect of air resistance on the falling car and assume that the car will accelerate vertically at −32 ft./sec^2.

a) Write an equation to model the height h of the car t seconds after it leaves the ramp (so $t = 0$ represents the time it leaves the ramp).

b) What equation describes the downward velocity of the car?

c) Write an equation that represents the distance d the car has traveled forward t seconds after it leaves the ramp.

INDIVIDUAL WORK 9

Time since car left the ramp (sec.), t	Height of the car above the ground (ft.), h	Horizontal distance from the ramp (ft.), d
0.1	39.84	8.8
0.2	39.36	17.6
0.3		
0.4		
0.5		
0.6		
0.7		
0.8		
0.9		
1.0		
1.1		
1.2		
1.3		
1.4		
1.5		
1.6		

Figure 7.38.
Table of times, heights, and distances.

d) Complete the entries in a copy of **Figure 7.38**.

e) Use the table to determine approximately how far the car has traveled horizontally when it hits the ground.

f) Your answers to parts (a) and (c) form a set of parametric equations that model the car's motion. Use these parametric equations directly to provide algebraic confirmation of your answer to part (e). Watch units!

Draw three graphs describing the car's motion:

g) horizontal distance-versus-time.

h) vertical height-versus-time.

i) the path of the motion (vertical location versus horizontal location).

j) If the assumption about air resistance is wrong, your answer will be off a little. Considering what you have learned so far, do you think the assumption is valid? If air resistance were not negligible, do you think the car would go farther than your predictions in parts (e) and (f)?

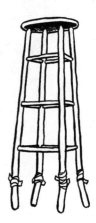

ACTIVITY

FALLING FOR YOU

12

Recall the details of Jeff Lattimore's specialty stunt, "The Leap for Life." When we left Jeff, he was stranded atop his eight-foot stool—or so he thinks. In reality, a prankster has cut it down to seven feet. Next, Jeff notices a car speeding toward him. A moment before the car crashes into his stool, Jeff jumps. His stool is snapped out from under him as the car whizzes beneath. The question is whether the stunt has a successful conclusion and Jeff lands safely on the ground, or the stunt takes a tragic turn and Jeff smashes into the top of the car.

1. Do you think he would fall more like a book or a beach ball from this height?

2. First, design Jeff's original stunt, the one using an eight-foot stool. Be sure to specify the car's dimensions, the car's velocity as it approaches the stool, Jeff's timing for his jump, and whatever else you think is important for the success of this stunt. Try to make your assumptions as reasonable as possible. (Remember that this stunt is performed in arenas at state fairs and not on interstate highways.) Demonstrate mathematically why Jeff's original stunt is successful.

3. Now, change the height of Jeff's stool to seven feet and leave everything else in Item 2 the same. Does Jeff still land safely? Justify your answer mathematically.

4. Return Jeff's stool to its original eight-foot height. What is the slowest that the car can be driven so Jeff can still make the outcome of the stunt successful if he jumps the same height as in Items 2 and 3?

5. Does Jeff really have to jump in order to land safely? Repeat Item 4 for the "no jump" possibility.

6. Design and carry out a small-scale version of the "no jump" stunt using a toy car, a plastic figure, and a suitable support stand. Compare the speed and height for your toy stunt to the same (scaled) measurements for Jeff's stunt.

When you have finished this activity, prepare to present your results to your class.

LESSON FIVE

The Grand Finale

KEY CONCEPT

Parametric equations

The Image Bank

PREPARATION READING

Stunt Time

It's time for the big event. Evel Knievel has positioned his motorcycle at the top of the ramp and is ready for his jump. He signals, then accelerates down the ramp and sails into the air. Has his landing ramp been properly positioned? Will this be another successful jump for Evel? The answer lies in the mathematics.

In this lesson you will model Evel's stunt mathematically and you'll decide where to place his landing ramp. You'll test your ideas by staging his stunt on a smaller scale using a toy car (or a racquetball or golf ball) in place of Evel. (It's not likely that you'll find a toy motorcycle suitable for this scaled-down stunt.)

In previous lessons, you have seen that forces affect moving objects, causing acceleration. You have learned how to measure the effect of gravity on a number of objects. However, in all the work you have done, the motion has been along a line (either horizontal or vertical) and the graphs of the motion have been distance-versus-time graphs. But the stunt jump takes place in two dimensions: both horizontal and vertical motions occur at the same time.

CONSIDER:

Think about the forces acting on Evel after he leaves the ramp.

1. Do any forces act on his motorcycle in the vertical ("up and down") direction?

2. How about in the horizontal ("forward and backward") direction? Describe all such forces as carefully as you can.

3. Based on your force descriptions, predict the kind of graph that will describe each component (horizontal and vertical) for the motion of the car.

ACTIVITY
13

THE STUNT MAN CAN

Back to Evel Knievel's stunt jump. For the Consider questions in the Preparation Reading, you predicted what kind of graphs would describe the horizontal and vertical components of the stunt jump. Of course, just knowing the kind of graph is not good enough for Evel. The exact equation is required in order to predict where his motorcycle will land and allow the landing ramp to be placed correctly. And that, once again, requires accurate measurements.

Part I: **THE PLAN**

In all the previous work with motion in this unit, you have always studied distance versus time. As you know, this is not the same as the path of the moving object. In fact, in Lessons 2–4, the path was along a line (objects moved vertically) and the distance-versus-time graph was a curve. In this lesson, you will also study the position of the object in the *xy*-plane (its path).

You will create a mathematical model of the position of a toy car (or a racquetball or golf ball) as it sails across your classroom. Once you have the model, you will use it to locate the best position for a ramp that will catch the car safely. **Figure 7.39** shows roughly what the set-up will look like.

The jump ramp will be set up so that the car sails off exactly

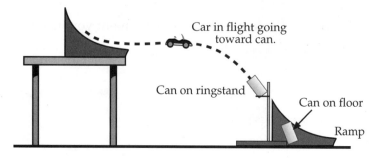

Figure 7.39.
The ramp.

ACTIVITY

THE STUNT MAN CAN

13

horizontally. Another ramp (or a can on a ring stand) will serve as the landing region for your car. Your group must decide where you will place the landing ramp. Your decision involves determining

  i. the correct direction,

 ii. the correct distance from the jump ramp, and

iii. the correct height for the landing area in order to catch the car in flight.

To develop a useful model, you will need to look at the things that affect the motion of the toy car after it leaves the jump ramp. Discuss the following questions, then plan your group's data-collection strategy.

1. What force or forces do you think affect the car's motion? In which direction(s)?

2. How will you measure the acceleration(s) that results from the force(s) you identified in Item 1?

3. a) If no force acts on the car in one particular direction, what can you say about the acceleration in that direction? What about the velocity in that direction?

   b) What measurement will you need to make in order to find a good model for motion in such a direction?

   c) Is there a direction for which you can argue that there is 0 (or almost 0) force? If so, which one; if not, why not?

   d) How will you obtain the measurement you just described in part (c)?

ACTIVITY

THE STUNT MAN CAN

13

Part II: EQUIPMENT SET-UP

Set up the equipment that you will need for your stunt. Here are some general outlines for the equipment set-up. Your teacher will provide more specific directions.

Attach the jump ramp to a table so that the jump-point of the ramp is horizontal. (Check it with a level. Use shims as needed.) The higher the jump point, the better. Secure the ramp in place with C-clamps (or masking tape).

You need to be able to release the car exactly the same way each time. One way to ensure this is to attach a small block of wood to the top of the ramp. Then to start a run, move the car up to touch this block and then let it roll down the ramp. (You may find another equally suitable method.)

Because you will need room to work around the jump ramp and the landing area, be sure both areas are unobstructed.

Using a plumb bob, find and mark the point on the floor directly under the jump point of the ramp.

Set up a photogate at the jump point of the ramp. If you do not have a photogate, make one by placing a penlight flash-light on one side of the ramp and a light probe on the other. Be sure the light source is shining directly into the probe. Adjust the height of the light so that it will hit the longest part of the car (or middle of the ball) as it rolls down the ramp.

Once you start to take measurements, be sure the jump-ramp set-up is not touched until the entire activity is completed.

THE STUNT MAN CAN

Part III: DATA COLLECTION

4. What is the acceleration due to gravity on the car (or ball) that you will be using for this activity? How did you determine your answer?

5. Make three trial runs down the ramp and find the average velocity of the car off the ramp for those trials.

   Warning: On these three test runs, use the carbon paper to find exactly where the car hits the floor. If it does not hit approximately in the same place each time, you need to refine your release methods. Once you have done so and landings are more consistent, repeat Item 5.

   When landings are consistent, find the middle of the three landing points for your test runs. Place a mark there. Mark a line from the point on the floor directly below the jump point to the landing point determined by these trial jumps. (A chalk line or line of masking tape line will work well.) This line will be the line on which you will place your landing ramp or can.

Part IV: ANALYSIS—MODEL FORMULATION

6. Using the information you found in Items 4 and 5, develop a parametric model for the flight of the car. Be sure all of your distance measurements are in the same units. Good units of measure are centimeters and seconds.

7. Based on your model from Item 6, how long will the stunt vehicle be in the air? How far from the plumb-bob mark will it land? Is your answer to this last question consistent with your test runs in Item 5?

8. What if air resistance was more of a factor than you initially thought? For example, what if the force of air resistance decreased both the horizontal and vertical acceleration on your vehicle by 0.5 cm./sec.^{2}? How would that change your models? Rework your model to see how it would change your answer to Item 7.

ACTIVITY

13

THE STUNT MAN CAN

9. Graph each component equation as a distance-versus-time graph. Then use the parametric mode of your calculator to sketch a graph of the actual path of the car, height versus distance. Discuss what the shapes of the three graphs tell you about the motion of the car. Select some particular instant during the motion, locate that instant on each graph, and interpret the graphs.

Part V: THE CONTEST

The challenge here is to place the landing area to catch your stunt jumper safely.

To make things interesting, let a student from another group specify the horizontal distance in front of the jump ramp for the landing area. (Sort of like saying, "Jump over 13 buses.") Then, together with your group, determine how high you need to set the catch ramp or ringstand for a safe landing at the specified distance.

For a little harder version, let the other group specify how high the landing area should be. In this case, you will determine how far away to place the landing area.

If you can get (or build) an additional ring, try jumping through a vertical ring before making a safe landing (**Figure 7.40**).

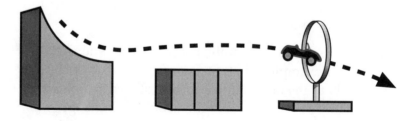

Figure 7.40.
Jumping through a vertical ring.

THE STUNT MAN CAN

ACTIVITY

13

10. Describe the challenge you are given. Then use your model to compute your solution. Explain your reasoning.

Now get ready for the big jump! But first, here are some hints before you continue:

- Be sure no one has altered the ramp, table, or other equipment since your three test runs.

- Be sure the car is released exactly the same way on the actual roll as it was on the three trials.

- Have more than one person do all the calculations, and have more than one person do the measurements for locating the landing area.

- Remember, the idea is to use your mathematical model for the flight of the car to position the landing area in this flight path so the car will land safely.

Once all the measurements have been made and the ramp or can is set, clear the area and roll the car (or ball) down the ramp. You will know if you are correct very soon!

INDIVIDUAL WORK 10

Parametric Practice

1. Willy is the kicker for your high school's football team. He is attempting the game-winning field goal. The ball is 45 yards from the crossbar. The crossbar is 10 feet above the ground. The parametric equations

$$x_t = 60.6t$$

$$y_t = 35t - 14.35t^2$$

describe the flight of his kick. Distance is measured in feet from the point of the kick, and time is in seconds.

 a) Is the kick high enough to go over the crossbar?

 b) A 6-foot tall lineman is 15 feet away when the ball is kicked. He jumps and stretches his arms to 9 feet above the ground. Could he block the kick?

 c) The lineman was offside and the kick will be taken again. First the referee moves the ball 5 yards closer to the goal because of the penalty for being offside. Does Willy make the field goal this time?

 d) Sketch a graph of the path of the ball. Approximately how long is the ball in the air?

2. Sarah is at bat in a winter league baseball game. She drives a ball to left center. The fence is 10 feet tall and 350 feet away from home plate at that point. If the parametric equations

$$x_t = 90t$$

$$y_t = 63t - 14.23t^2 + 3$$

model the flight of the ball (in feet and seconds), does she hit a home run? How long does it take the ball to get to the outfield fence?

UNIT SUMMARY

Wrapping Up Unit Seven

1. Use the distance-versus-time graphs in **Figures 7.41(a)–(c)** to answer Items (a) through (d).

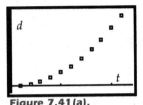

Figure 7.41(a).

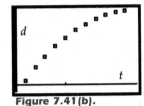

Figure 7.41(b).

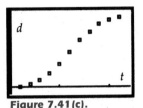

Figure 7.41(c).

a) Which of the graphs could show an object starting from a stop? Explain.

b) Which of the graphs that you selected in part (a) could show an object starting from a stop and eventually slowing down? Explain.

c) Which of the graphs could show an object slowing down? Explain.

d) Make a sketch of the graph that you selected in part (b). Draw a vertical line to show where the object first starts slowing down. Explain.

2. A parabola graph from an equation of the form $y = ax^2 + bx + c$ is known to go through the points (1, 13), (6, 23), (9, 77).

a) Find the equation of the parabola, and describe your method.

b) For this equation, algebraically find the smallest possible y-value. Explain your method, and check your answer graphically.

3. **Figure 7.42** is a distance-versus-time graph for a wind-up toy car. The data that produced this graph are shown in **Figure 7.43**.

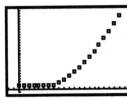

Figure 7.42.
Distance-versus-time graph for toy car

t	d
0.0	1.5
0.5	1.5
1.0	1.5
1.5	1.5
2.0	1.5
2.5	1.5
3.0	1.5
3.5	1.5
4.0	2.482
4.5	3.648
5.0	4.998
5.5	6.532
6.0	8.25
6.5	10.152
7.0	12.238
7.5	14.508
8.0	16.962
8.5	19.6
9.0	22.422
9.5	25.428
10.0	28.618

Figure 7.43.
Distance-versus-time data for toy car.

a) How many equations do you think you need to describe these data?

b) At what t-value would you switch from one equation to another?

c) Find a piecewise-defined equation for the graph. (Use the notation that was used in Unit 1, *Gridville*.)

d) Assume that the motion started at the time that you stated in part (b). Which piece of your equation in part (c) is relevant to the actual motion? Translate this piece by determining the relationship between d and t_{new} where t_{new} is the elapsed time since the motion began. Explain your method.

4. You and your friends are watching drivers at the stoplight closest to your school. With stopwatches and tape measures, your group makes the following measurements for two different drivers.

a) For Driver 1:

Time (sec.)	Distance from start (ft.)	Velocity (Units?)	Acceleration (Units?)
0	0.0		
1	3.4		
2	13.9		
3	33.0		
4	56.1		
5	90.0		
6	127.0		
7	175.0		

Figure 7.44.
Data from Driver 1.

i) Copy and complete the table in **Figure 7.44**. (Add units to the last two columns.) Use symmetric difference quotients to approximate the velocities and accelerations.

ii) Draw graphs of distance versus time, velocity versus time, and acceleration versus time.

iii) Find an equation for distance in terms of time.

b) For Driver 2:

Time (sec.)	Distance from start (ft.)	Velocity (Units?)	Acceleration (Units?)
0	0		
1	3.5		
2	10.0		
3	21.5		
4	40.0		
5	67.5		
6	106.0		
7	157.5		

Figure 7.45.
Data from Driver 2.

   i) Copy and complete the table in **Figure 7.45**. (Add units to the last two columns.) Use symmetric difference quotients to approximate the velocities and accelerations.

   ii) Draw graphs of distance versus time, velocity versus time, and acceleration versus time.

   iii) Find an equation for distance in terms of time.

c) Plot the residuals for each equation you found in parts (a)(iii) and (b)(iii). Which equation is the best fit? How do the residuals show this? What is a likely explanation for these results?

d) Approximately how fast (in miles per hour) was each car going after six seconds? Remember the units in the tables are in feet and seconds.

e) Which car traveled the greatest distance in seven seconds? Which car reached the greatest velocity?

5. Suppose that Eric tosses a water balloon out a window. He hurls it up and away from the building. Suppose that he decides that the parametric equations $x = 3t$, $y = 10.5 + 5.2t - 4.9t^2$, where x and y are horizontal and vertical distances (meters) and t is the elapsed time (seconds) since the balloon was released, do a good job in describing the motion of the balloon. Assuming that he is correct, answer the following questions.

   a) How long was the balloon in the air?

   b) How far from the building did it land?

   c) What was the initial upward velocity of the balloon? What about the initial horizontal velocity?

   d) When does it reach its highest point? How high is it? Give an algebraic justification for your answer. Then draw a graph of the path of the balloon and confirm your answer based on your graph.

   e) Suppose Jason is five meters away from the building standing directly under the flight line of the balloon. Jason jumps to try and catch the balloon. Is there a chance he can catch it? Explain your reasoning.

   f) What assumptions did Eric make when he decided on his model? Do you think these are reasonable assumptions?

Mathematical Summary

*I*n this unit, you have applied principles of mathematical modeling to determine equations describing various motions. You found that time-distance data from motion detectors proved useful. For example, linear distance-versus-time graphs are produced by walking at a constant rate, and curved graphs are produced by walking at a non-constant rate.

The average velocity from time 1 to time 2 may be calculated using the ratio (distance 2 – distance 1)/(time 2 – time 1). Similarly, after determining velocities at time 1 and time 2, you can determine the average acceleration by calculating the ratio (velocity 2 – velocity 1)/(time 2 – time 1).

The simplest motion studied was that of an object moving along a line at a constant rate. This type of motion can be modeled by distance-versus-time equations of the form $d = vt + d_0$, where d is the distance and t is the elapsed time. The slope, v, of the graph tells you the object's velocity, and the constant term d_0 gives the object's initial distance (its distance when $t = 0$). Since velocity for this kind of motion is constant, both instantaneous and average velocity are easily interpreted as slopes.

The motion of an object moving under the influence of gravity is somewhat more complex to analyze because the object's velocity is not constant. Hence, a linear model is not an appropriate model for this type of motion. Instead, such motion is modeled by quadratic equations; therefore they have parabolic distance-versus-time graphs. In general, motion described by quadratic equations is characterized by constant acceleration, linear velocity-versus-time relationships, and parabolic distance-versus-time graphs.

The instantaneous velocity at any time may still be thought of as the "slope" of the distance graph; such slope can be approximated by "zooming in" on the point of interest until the graph looks linear. Since the distance-time relationship is quadratic, the velocity-time relationship is linear. The slope of this linear relationship (i.e., the rate of change of the velocity with respect to time) is the acceleration. For falling objects, acceleration is constant.

The coefficients of the quadratic model of the form $h = at^2 + v_0t + h_0$, where h is the height and t is the elapsed time since the object began moving, tell the story of the motion. The value of a is half the acceleration, v_0 is the initial velocity (the velocity when $t = 0$), and h_0 is the initial height. Thus, the linear term is the key to identifying whether an object began from a stationary "drop" or had an initial "push" up or down.

Parametric equations allow you to model the horizontal and vertical components of motion separately and then put them together to describe the original motion.

Glossary

ACCELERATION:
The rate of change of velocity with respect to time.

AVERAGE ACCELERATION:
Average acceleration is the rate of change of velocity from time 1 to time 2:
(velocity 2 – velocity 1)/(time 2 – time 1), or (change in velocity)/(elapsed time). Note that force is the cause of acceleration. Typical forces are due to gravity, friction, or the push of a hand.

AVERAGE VELOCITY:
Average velocity is the rate of change of distance from time 1 to time 2:
(distance 2 – distance 1)/(time 2 – time 1), or (distance traveled)/(elapsed time).

INITIAL VELOCITY:
The velocity of an object the instant it is released.

INSTANTANEOUS VELOCITY AT TIME T_0:
The rate of change of location with respect to time at the instant $t = t_0$. You can approximate the instantaneous velocity by tracing to the point on the distance-versus-time graph corresponding to t_0, zooming in on this point until the graph resembles a line, selecting two points on this "line," and using these points to compute the slope. This "slope" is your approximation of the instantaneous velocity at t_0.

SYMMETRIC DIFFERENCE QUOTIENT:
A symmetric difference quotient is an average rate of change calculated using endpoints equidistant (left and right) from a specified point. It is a good approximation for instantaneous rates of change, for example, instantaneous velocity or acceleration, and is exact for quadratic data.

TRANSLATION:
A translation is a transformation of an equation or graph that leaves its shape, orientation, and size unchanged, but alters its location.

Index

Glossary terms and the pages on which they are defined in the text appear in boldface.

Acknowledgements

PROJECT LEADERSHIP:

Solomon Garfunkel
COMAP, INC., LEXINGTON, MA

Landy Godbold,
THE WESTMINSTER SCHOOLS, ATLANTA, GA

Henry Pollak
TEACHERS COLLEGE, COLUMBIA UNIVERSITY, NY

EDITOR:

Landy Godbold

AUTHORS:

Allan Bellman
WATKINS MILL HIGH SCHOOL, GAITHERSBURG, MD

John Burnette
KINKAID SCHOOL, HOUSTON, TX

Horace Butler
GREENVILLE HIGH SCHOOL, GREENVILLE, SC

Claudia Carter
MISSISSIPPI SCHOOL FOR MATH AND SCIENCE,
COLUMBUS, MS

Nancy Crisler
PATTONVILLE SCHOOL DISTRICT, ST. ANN, MO

Marsha Davis
EASTERN CONNECTICUT STATE UNIVERSITY,
WILLIMANTIC, CT

Gary Froelich
SECONDARY SCHOOL PROJECTS MANAGER, COMAP, INC.,
LEXINGTON, MA

Landy Godbold
THE WESTMINSTER SCHOOLS, ATLANTA, GA

Bruce Grip
ETIWANDA HIGH SCHOOL, ETIWANDA, CA

Rick Jennings
EISENHOWER HIGH SCHOOL, YAKIMA, WA

Paul Kehle
INDIANA UNIVERSITY, BLOOMINGTON, IN

Darien Lauten
OYSTER RIVER HIGH SCHOOL, DURHAM, NH

Sheila McGrail
CHARLOTTE COUNTRY DAY SCHOOL, CHARLOTTE, NC

Geraldine Oliveto
THOMAS JEFFERSON HIGH SCHOOL FOR
SCIENCE AND TECHNOLOGY, ALEXANDRIA, VA

Henry Pollak
TEACHERS COLLEGE, COLUMBIA UNIVERSITY, NY

J.J. Price
PURDUE UNIVERSITY, WEST LAFAYETTE, IN

Joan Reinthaler
SIDWELL FRIENDS SCHOOL, WASHINGTON, D.C.

James Swift
ALBERNI SCHOOL DISTRICT, BRITISH COLUMBIA, CANADA

Brandon Thacker
BOUNTIFUL HIGH SCHOOL, BOUNTIFUL, UT

Paul Thomas
MINDQ, FORMERLY OF THOMAS JEFFERSON HIGH SCHOOL
FOR SCIENCE AND TECHNOLOGY, ALEXANDRIA, VA

REVIEWERS:

Dédé de Haan, Jan de Lange,
Henk van der Kooij
FREUDENTHAL INSTITUTE, THE NETHERLANDS

David Moore
PURDUE UNIVERSITY, WEST LAFAYETTE, IN

Henry Pollak
TEACHERS COLLEGE, COLUMBIA UNIVERSITY, NY

ASSESSMENT:

Dédé de Haan, Jan de Lange,
Kees Lagerwaard, Anton Roodhardt,
Henk van der Kooij
THE FREUDENTHAL INSTITUTE, THE NETHERLANDS

REVISION TEAM

Marsha Davis, Gary Froelich,
Landy Godbold, Bruce Grip

EVALUATION:

Barbara Flagg
MULTIMEDIA RESEARCH, BELLPORT, NY

TEACHER TRAINING:

Allan Bellman, Claudia Carter,
Nancy Crisler, Beatriz D'Ambrosio,
Rick Jennings, Paul Kehle,
Geraldine Oliveto, Paul Thomas

FIELD TEST SCHOOLS AND TEACHERS:

Clear Brook High School,
Friendswood, TX
JEAN FRANKIE, TOM HYLE, LEE YEAGER

Dr. James Hogan Senior High School,
Vallejo, CA
GEORGIA APPLEGATE, PAM HUTCHISON, JERRY LEGE,
TOM LEWIS

Foxborough High School,
Foxborough, MA
BERT ANDERSON, SUE CARLE, MAUREEN DOLAN,
JOHN MARINO, MARY PARKER, DAVE WALKINS,
LEN YUTKINS

Frontier Regional High School,
South Deerfield, MA
LINDA DIDGE, DON GORDEN, PATRICIA TAYLOR

Gresham Union High School,
Gresham, OR
DAVE DUBOIS, KAY FRANCIS, ERIN HALL,
THERESA HUBBARD, RICK JIMISON, GAYLE MEIER,
CRAIG OLSEN

Jefferson High School, Portland, OR
STEVE BECK, DAVE DAMCKE, LYNN INGRAHAM,
MARTHA LANSDOWNE, JOHN OPPEDISANO, LISA WILSON

Lincoln School, Providence, RI
JOAN COUNTRYMAN

Mills E. Godwin High School,
Richmond, VA
KEVIN O'BRYANT, ANN W. SEBRELL

New School of Northern Virginia,
Fairfax, VA
JOHN BUZZARD, VICKIE HAVELAND,
BARBARA HERR, LISA TEDORA

Northside High School, Fort Wayne, IN
ROBERT LOVELL, EUGENE MERKLE

Ossining High School, Ossining, NY
JOSEPH DICARLUCCI

Pattonville High School,
Maryland Heights, MO
SUZANNE GITTEMEIER, ANN PERRY

Price Laboratory School, Cedar Falls, IA
DENNIS KETTNER, JIM MALTAS

Rex Putnam High School,
Milwaukie, OR
JEREMY SHIBLEY, KATHY WALSH

Sam Barlow High School, Gresham, OR
BRAD GARRETT, KATHY GRAVES, COY ZIMMERMAN

Simon Gratz High School,
Philadelphia, PA
LINDA ANDERSON, ANNE BOURGEOIS, WILLIAM ELLERBEE

Ursuline Academy, Dallas, TX
SUSAN BAUER, FRANCINE FLAUTT, DEBBIE JOHNSTON,
MARGARET KIDD, ELAINE MEYER, MARGARET NOULLET,
MARY PAWLOWICZ, SHARON PIGHETTI, PATTY WALLACE,
KATHY WARD

COMAP STAFF

Solomon Garfunkel, Laurie Aragon,
Sheila Sconiers, Gary Froelich,
Roland Cheyney, Roger Slade,
George Ward, Frank Giordano,
Linda Vahey, Susan Judge, Emily Sacca,
Daiva Kiliulis, David Barber, Gail Wessell,
Gary Feldman, Clarice Callahan,
Brenda McDonald, George Jones,
Rafael Aragon, Peter Bousquet

PHOTOGRAPHY/ART

The Image Bank
BOSTON, MA

Corbis
BELLEVUE, WA

Paul Klaver Collection
ARLINGTON, MA

Evel Knievel Fan Club
LARGO, FL

Frank & Ernest Comics
SOLANA BEACH, CA

Randy Fredner
EARLYSVILLE, VA

David Barber Illustrations
MARBLEHEAD, MA

COVER ART

The Image Bank
BOSTON, MA

INDEX EDITOR

Seth Maislin
FOCUS PUBLISHING SERVICES, WATERTOWN, MA

References

UNIT 1:

Dean, Anabel. 1978. *Fire, How Do They Fight It?* Philadelphia, PA: The Westminster Press.

UNIT 6:

Manning, Anita, "Ebola experts count epidemic's days, lessons," USA Today (Tuesday, August 8, 1995): 1D.

UNIT 7:

Young, Mark C., Ed. 1996. *The Guinness Book of World Records 1957–1997.* Stamford, CT: Guinness Media, Inc.